Intercellular Communication in Plants

Annual Plant Reviews

A series for researchers and postgraduates in the plant sciences. Each volume in this series focuses on a theme of topical importance, and emphasis is placed on rapid publication.

Editorial Board:

Professor Jeremy A. Roberts (Editor-in-Chief), Plant Science Division, School of Biosciences, University of Nottingham, Sutton Bonington Campus, Loughborough, Leicestershire LE12 5RD, UK; **Professor Hidemasa Imaseki**, Obata-Minami 2 4 19, Moriyama-ku, Nagoya 463, Japan; **Dr Michael T. McManus**, Institute of Molecular BioSciences, Massey University, Palmerston North, New Zealand; **Dr Jocelyn K. C. Rose**, Department of Plant Biology, Cornell University, Ithaca, New York 14853, USA.

Titles in the series:

1. Arabidopsis
Edited by M. Anderson and J. A. Roberts

2. Biochemistry of Plant Secondary Metabolism
Edited by M. Wink

3. Functions of Plant Secondary Metabolites and their Exploitation in Biotechnology
Edited by M. Wink

4. Molecular Plant Pathology
Edited by M. Dickinson and J. Beynon

5. Vacuolar Compartments
Edited by D. G. Robinson and J. C. Rogers

6. Plant Reproduction
Edited by S. D. O'Neill and J. A. Roberts

7. Protein–Protein Interactions in Plant Biology
Edited by M. T. McManus, W. A. Laing and A. C. Allan

8. The Plant Cell Wall
Edited by J. K. C. Rose

9. The Golgi Apparatus and the Plant Secretory Pathway
Edited by D. G. Robinson

10. The Plant Cytoskeleton in Cell Differentiation and Development
Edited by P. J. Hussey

11. Plant–Pathogen Interactions
Edited by N. J. Talbot

12. Polarity in Plants
Edited by K. Lindsey

13. Plastids
Edited by S. G. Møller

14. Plant Pigments and their Manipulation
Edited by K. M. Davies

15. Membrane Transport in Plants
Edited by M. R. Blatt

16. Intercellular Communication in Plants
Edited by A. J. Fleming

Intercellular Communication in Plants

Edited by

ANDREW J. FLEMING
Professor of Plant Sciences
University of Sheffield, UK

Blackwell
Publishing

CRC Press

© 2005 by Blackwell Publishing Ltd

Editorial offices:
Blackwell Publishing Ltd, 9600 Garsington Road, Oxford OX4 2DQ, UK
  Tel: +44 (0)1865 776868
Blackwell Publishing Asia Pty Ltd, 550 Swanston Street, Carlton, Victoria 3053, Australia
  Tel: +61 (0)3 8359 1011

ISBN 1-4051-2068-1
ISSN 1460-1494

Published in the USA and Canada (only) by CRC Press LLC, 2000 Corporate Blvd., N.W., Boca Raton, FL 33431, USA
Orders from the USA and Canada (only) to CRC Press LLC

USA and Canada only:
ISBN 0-8493-2363-0
ISSN 1097-7570

The right of the Author to be identified as the Author of this Work has been asserted in accordance with the Copyright, Designs and Patents Act 1988.

All rights reserved. No part of this publication may be reproduced, stored in a retrieval system, or transmitted, in any form or by any means, electronic, mechanical, photocopying, recording or otherwise, except as permitted by the UK Copyright, Designs and Patents Act 1988, without the prior permission of the publisher.

This book contains information obtained from authentic and highly regarded sources. Reprinted material is quoted with permission, and sources are indicated. Reasonable efforts have been made to publish reliable data and information, but the author and the publisher cannot assume responsibility for the validity of all materials or for the consequences of their use.

Trademark notice: Product or corporate names may be trademarks or registered trademarks, and are used only for identification and explanation, without intent to infringe.

First published 2005

Library of Congress Cataloging-in-Publication Data:
A catalog record for this title is available from the Library of Congress

British Library Cataloguing-in-Publication Data:
A catalogue record for this title is available from the British Library

Set in 10/12 pt Times
by Techbooks
Printed and bound in Great Britain
by MPG Books Ltd, Bodmin, Cornwall

The publisher's policy is to use permanent paper from mills that operate a sustainable forestry policy, and which has been manufactured from pulp processed using acid-free and elementary chlorine-free practices. Furthermore, the publisher ensures that the text paper and cover board used have met acceptable environmental accreditation standards.

For further information on Blackwell Publishing, visit our website:
www.blackwellpublishing.com

Contents

Contributors	**xi**
Preface	**xiii**

1 Auxin as an intercellular signal   **1**
JIŘÍ FRIML AND JUSTYNA WIŚNIEWSKA

1.1 Introduction	1
1.2 Auxin transport – known pathways	2
1.2.1 Polar auxin transport pathway	2
1.2.2 Chemiosmotic model	4
1.2.3 Multicomponent auxin efflux carrier system	5
1.3 Molecular components	6
1.3.1 Auxin influx – AUX1 proteins	7
1.3.2 Auxin efflux – PIN proteins	7
1.3.3 ABC transporters	9
1.4 Subcellular dynamics of auxin carriers	9
1.4.1 Constitutive recycling of PIN proteins	9
1.4.2 AEIs and vesicle trafficking	11
1.4.3 GNOM and PIN dynamics	12
1.5 The role of auxin gradients in plant development	12
1.5.1 Monitoring of auxin distribution in planta	13
1.5.2 Embryonic axis formation	13
1.5.3 Postembryonic organ formation	15
1.5.4 Root meristem maintenance	16
1.5.5 Tropisms	17
1.5.6 Downstream of auxin gradients	18
1.5.7 Auxin as morphogen	21
1.6 Conclusions	21
Acknowledgements	22
References	22

2 Peptides as signals   **27**
YIJI XIA

2.1 Introduction	27
2.2 Peptide signals in plants and their biological functions	28
2.2.1 Systemins mediate systemic and local wound responses	28
2.2.2 RALF regulates plant growth and development	31
2.2.3 ENOD40 regulates nodulation and cell proliferation	31

	2.2.4 PSK (phytosulfokine) is a mitogenic factor	34
	2.2.5 CLAVATA 3 (CLV3) regulates stem cell homeostasis	35
	2.2.6 S-locus cysteine-rich proteins determine specificity of self-incompatibility in the Brassicaceae	38
2.3	Proteolytic processing of prohormones	40
2.4	Technologies for discovering new peptide signals	41
2.5	Concluding remarks	43
	References	44

3 RNA as a signalling molecule — 49
PATRICE DUNOYER AND OLIVIER VOINNET

3.1	Intercellular movement of plant mRNAs	49
	3.1.1 Cell-to-cell movement of plant mRNAs	49
	3.1.1.1 Plant plasmodesmata	49
	3.1.1.2 Cell-to-cell movement of a transcription factor with its mRNA	50
	3.1.2 Long-distance transport of plant mRNAs	51
3.2	Intercellular movement of viroids	53
	3.2.1 What are viroids?	53
	3.2.2 Intercellular movement of viroids	54
	3.2.2.1 Cell-to-cell movement of viroids	54
	3.2.2.2 Long-distance movement of viroids	54
	3.2.3 Cellular factors involved in viroid movement	55
	3.2.3.1 Phloem Lectin 2	55
	3.2.3.2 VirP1	55
3.3	Intercellular movement of RNA silencing	56
	3.3.1 Mechanism of RNA silencing	56
	3.3.1.1 Co-suppression in petunia	57
	3.3.1.2 Double-stranded RNA: trigger molecule of RNA silencing	57
	3.3.1.3 Short interfering (si)RNAs are the specificity determinants of RNA silencing	58
	3.3.1.4 RNA-induced silencing complex RISC	59
	3.3.1.5 Transitive RNA silencing	60
	3.3.1.6 Biological functions of RNA silencing in plants	61
	3.3.2 The discovery of systemic RNA silencing	61
	3.3.3 Initiation of systemic RNA silencing	63
	3.3.3.1 Spontaneous activation of systemic RNA silencing	63
	3.3.3.2 Exogenously induced systemic silencing	64
	3.3.4 Propagation of systemic RNA silencing	65
	3.3.4.1 Long-distance movement of RNA silencing	65
	3.3.4.2 Cell-to-cell movement of RNA silencing	67
	3.3.5 Maintenance of systemic silencing	71
	3.3.6 What is the nucleic acid component of the silencing signal?	72
	3.3.6.1 Cell-to-cell movement and phloem transport of silencing involve separate mechanisms and, most likely, separate signals	73
	3.3.6.2 Possible nature of the RNA species involved in cell-to-cell movement of silencing	73

	3.3.6.3 No specific RNA species has been correlated with long-distance transport of silencing in plants	74
	3.3.7 Plant factors required for movement of RNA silencing	75
	3.3.8 Biological functions of non-cell autonomous RNA silencing in plants	75
	3.3.8.1 Antiviral defence	75
	3.3.8.2 A role in non-cell autonomous regulation of gene expression?	76
	References	77

4 The plant extracellular matrix and signalling 85
ANDREW J. FLEMING

4.1	Introduction	85
4.2	The cell wall and signalling	86
4.3	The cell wall as a potential source of chemical signals	87
	4.3.1 Polysaccharide signals	88
	4.3.2 Arabinogalactan proteins as signals	90
	4.3.3 Cutin and signalling	94
	4.3.4 Uncharacterised cell wall determinants involved in signalling	96
4.4	The cell wall and biophysical signalling	98
	4.4.1 Connections between the cell wall and the cytosol as a conduit for intercellular signalling	101
4.5	Conclusions	102
	Acknowledgements	103
	References	103

5 Plasmodesmata – gateways for intercellular communication in plants 109
TRUDI GILLESPIE AND KARL J. OPARKA

5.1	Introduction	109
	5.1.1 Plasmodesmata – key components of the symplast	109
	5.1.2 Plasmodesmata: simple description, complex function	110
	5.1.3 Discovery of plasmodesmata	110
5.2	Structure	110
	5.2.1 The general ultrastructure of plasmodesmata	110
	5.2.2 Primary and secondary; simple or branched	113
	5.2.3 Plasmodesmal frequency and distribution: gain and loss	116
	5.2.4 Plasmodesmal components	117
	5.2.5 Passage through the cytoplasmic sleeve	120
5.3	Macromolecular trafficking	120
	5.3.1 Passive transport and the basal SEL	122
	5.3.2 Selective transport and gating: modulation of the SEL	124
	5.3.3 Physiological modulation of SEL	124
	5.3.4 Fine regulation of plasmodesmal SEL – role of the cytoskeleton	129
	5.3.5 'Coarse' regulation by callose	130
	5.3.6 Phosphorylation, protein unfolding and chaperones	131
	5.3.7 The emerging picture of plasmodesmata	134
	Acknowledgements	134
	References	135

CONTENTS

6 Lessons from the vegetative shoot apex — 147
JOHN F. GOLZ

- 6.1 Introduction — 147
- 6.2 Structure of the angiosperm shoot apical meristem — 147
  - 6.2.1 Zones of the meristem — 148
  - 6.2.2 Layers of the meristem — 149
  - 6.2.3 Symplastic fields within the meristem — 149
- 6.3 Periclinal chimaeras reveal a role for signalling in plant development — 149
- 6.4 Signalling involved in meristem maintenance — 151
  - 6.4.1 The CLAVATA mutants — 151
  - 6.4.2 The CLAVATA signalling pathway — 154
  - 6.4.3 The wuschel mutant — 156
  - 6.4.4 The CLAVATA–WUSCHEL regulatory loop — 156
- 6.5 Maintaining indeterminate cells in the meristem requires homeobox genes — 158
- 6.6 Interactions between *KNOX* genes and hormones regulate meristem activity — 161
- 6.7 Signals involved in organ formation — 162
  - 6.7.1 Models of phyllotaxis — 162
  - 6.7.2 The role of auxin in phyllotaxis — 163
  - 6.7.3 Organ outgrowth involves physical forces — 166
- 6.8 Signalling between organ primordia and the meristem — 166
- 6.9 Conclusion — 169
- Acknowledgements — 170
- References — 170

7 Intercellular communication during floral initiation and development — 178
GEORGE COUPLAND

- 7.1 Introduction — 178
- 7.2 Long-distance signaling during the induction to flowering — 179
  - 7.2.1 Discovery of a role for long-distance signaling in the induction of flowering — 179
  - 7.2.2 Mutations that impair long-distance signaling in pea and maize — 180
  - 7.2.3 Molecular genetic analysis of flowering-time control in *Arabidopsis* places the long-distance signal within a regulatory hierarchy — 181
    - 7.2.3.1 A network of pathways controls flowering of *Arabidopsis* — 181
    - 7.2.3.2 Spatial regulation of flowering-time control — 184
    - 7.2.3.3 Identifying the floral stimulus: a perspective from *Arabidopsis* molecular genetics — 187
- 7.3 Intercellular communication during floral development — 188
  - 7.3.1 Some of the transcription factors that control floral meristem or organ identity act non-cell autonomously in the developing flower — 189
  - 7.3.2 Movement of transcription factors between cells defines one mechanism for short-distance signaling in the developing flower — 190
- 7.4 Perspectives — 193
- References — 194

8	**Lessons from the root apex** MARTIN BONKE, SARI TÄHTIHARJU AND YKÄ HELARIUTTA	**199**
	8.1 Introduction	199
	8.2 Organization of the root	199
	8.2.1 Anatomy of the root meristem and procambium in the apex of a growing root	199
	8.2.2 Cellular organization of the root is established during embryonic development	200
	8.2.3 Development of secondary roots	203
	8.3 Cell fate studies of the growing root	203
	8.4 Molecular genetics of root development	205
	8.4.1 Distal patterning	205
	8.4.2 Genetic control of initiation of secondary roots	210
	8.4.3 Molecular genetics of epidermal patterning	211
	8.4.4 Patterning of ground tissue	215
	8.4.5 Vascular patterning	218
	8.5 Future prospects	220
	Acknowledgements	220
	References	220
9	**Lessons from leaf epidermal patterning in plants** BHYLAHALLI PURUSHOTTAM SRINIVAS AND MARTIN HÜLSKAMP	**225**
	9.1 Overview	225
	9.2 Introduction	225
	9.3 Mechanisms of trichome patterning	225
	9.3.1 Trichome differentiation	226
	9.3.2 Why is a mechanism postulated to explain the trichome spacing pattern and what kind of underlying principles are operating?	226
	9.3.3 Analysis of trichome initiation mutants	227
	9.3.3.1 Positive regulators of trichome initiation	228
	9.3.3.2 Negative regulators of trichome initiation	229
	9.3.4 Interactions between the trichome initiation genes	229
	9.3.5 Local cell–cell interactions leading to cell fate decisions: a model	231
	9.3.6 Long-range control of trichome initiation by hormones	233
	9.4 Stomatal development and patterning	233
	9.4.1 Cell division pattern during stomata patterning	235
	9.4.2 Cell signalling and the control of asymmetric cell divisions during stomata development	236
	9.5 Perspective	237
	References	237
10	**Lessons on signalling in plant self-incompatibility systems** ANDREW G. MCCUBBIN	**240**
	10.1 Introduction	240

10.2	S-RNase-based single-locus gametophytic SI	241
	10.2.1 S-RNases encode S-specificity in the pistil	242
	10.2.2 S-RNase structure/function	242
	10.2.3 The pollen S-gene	244
	10.2.4 Non-S-linked components of S-RNase-based SI	246
	10.2.5 Model for the operation of S-RNase-based SI	247
10.3	Gametophytic self-incompatability in the Papaveraceae	249
	10.3.1 The *S*-gene controlling stigma function in *P. rhoeas*	250
	10.3.2 Structure/function of S-proteins	251
	10.3.3 Biochemical responses in pollen following self-recognition	251
	10.3.3.1 Ca^{2+} signalling in the SI response	252
	10.3.3.2 Protein kinase activity and the SI response	253
	10.3.4 S-protein-binding proteins in pollen	254
	10.3.5 Changes in the actin cytoskeleton	255
	10.3.6 PCD in the SI response	255
	10.3.7 Model for the mechanism of self-incompatibility in *P. rhoeas*	257
10.4	Sporophytic self-incompatability	259
	10.4.1 *Brassica S*-locus glycoproteins	260
	10.4.2 SRK encodes *S*-haplotype specificity in the stigma	260
	10.4.3 SCR/SP11 encodes pollen *S*-haplotype specificity	261
	10.4.4 Regulation of SRK	263
	10.4.5 SRK substrates	264
	10.4.6 Model for the action of SSI in Brassica	265
10.5	Summary	267
References		268

Index   277

The colour plate section follows page 146

Contributors

Dr Martin Bonke	Plant Molecular Biology Laboratory, Institute of Biotechnology, POB 56, University of Helsinki, FIN-00014, Finland
Professor George Coupland	Max Planck Institute for Plant Breeding, Carl von Linne Weg, 10, D-50829 Cologne, Germany
Dr Patrice Dunoyer	Institut de Biologie Moléculaire des Plantes du CNRS, 12, rue du Général Zimmer, 67084 Strasbourg Cedex, France
Professor Andrew J. Fleming	Department of Animal and Plant Sciences, University of Sheffield, Western Bank, Sheffield S10 2TN, UK
Dr Jiří Friml	Department of Developmental Genetics, University of Tübingen, Auf der Morgenstelle 3, D-72076 Tübingen, Germany
Dr Trudi Gillespie	Scottish Crop Research Institute, Invergowrie, Dundee DD2 5DA, UK
Dr John Golz	School of Biological Sciences, University of Victoria, PO Box 18, Victoria 3800, Australia
Dr Ykä Helariutta	Plant Molecular Biology Laboratory, Institute of Biotechnology, POB 56, University of Helsinki, FIN-00014, Finland
Professor Martin Hülskamp	Botanical Institute, University of Cologne, Gyrhofstrasse 15, D-50931 Cologne, Germany
Dr Andrew McCubbin	School of Biological Sciences, Washington State University, Pullman, WA 99164-4236, USA
Professor Karl J. Oparka	Scottish Crop Research Institute, Invergowrie, Dundee DD2 5DA, UK
Dr Bhylahalli Purushottam Srinivas	Botanical Institute, University of Cologne, Gyrhofstrasse 15, D-50931 Cologne, Germany

Dr Sari Tähtiharju	Plant Molecular Biology Laboratory, Institute of Biotechnology, POB 56, University of Helsinki, FIN-00014, Finland
Dr Olivier Voinnet	Institut de Biologie Moléculaire des Plantes du CNRS, 12, rue du Général Zimmer, 67084 Strasbourg Cedex, France
Dr Justyna Wiśniewska	Institute of General and Molecular Biology, Nicholaus Copernicus University, 87-100 Torun, Poland
Dr Yiji Xia	Donald Danforth Plant Science Center, 975 North Warson Road, St Louis, Missouri, MO 63132, USA

Preface

Intercellular communication in plants plays a vital role in the coordination of processes leading to the formation of a functional organism. The signalling systems must function at a local level to coordinate events of cellular differentiation over long distances to coordinate developmental and physiological responses in different parts of the plant, and they must even operate between separate individuals – for example, to control fertilisation as part of the evolutionary strategy of a particular species. To cope with the diverse requirements for intercellular signalling, plants have evolved a spectrum of molecular mechanisms, and significant progress has been made over the last few years in our understanding of these processes.

One intriguing area has been the identification and characterisation of novel signalling compounds, most notably the finding that polypeptides and RNA can move between plant cells and act in an instructive manner. Chapters 2 and 3 of this volume focus on the exciting progress being made in these new areas of signalling research. This does not mean that the classical growth factors previously implicated in intercellular signalling should be ignored. Indeed, fantastic new insights into the coordination of plant growth processes and organogenesis have arisen from research on auxin, which has been recognised as an important signalling molecule for almost 80 years. Chapter 1 describes the significant advances that have been made in this area. It has also become apparent that the plant cell wall plays an important role in intercellular communication, as a potential source and modulator of signals (Chapter 4) and because organelles embedded in the cell wall (the plasmodesmata) can act as regulators of signal movement (as described in Chapter 5).

The different signalling systems involved in intercellular communication must be integrated into particular biological systems, and particular systems will rely to different extents on different signalling systems. This cross-talk and integration of signalling systems is a major topic for future research, and the final chapters of this volume provide insights into different aspects of the plant where significant progress has been made. This ranges from local regulation of cell proliferation and differentiation in the shoot (Chapter 6), root (Chapter 8) and leaf (Chapter 9) to control the floral transition (Chapter 7) and pollen/stigma interactions (Chapter 10).

Intercellular communication in plants is a massive topic and, inevitably, a volume of this size can only focus on a few aspects. Topics have been selected to demonstrate research areas showing significant recent progress and promise. They have also been selected to focus on the actual process of intercellular signalling rather than on subsequent intracellular signal transduction. The signalling interactions of plants with other organisms (symbiotic and parasitic) are not covered because the focus here is on the plant as an independent organism in which a developmental program is unfurled, which requires coordination (thus signalling) at several levels of organisation.

Intercellular communication is a major topic in plant biology and one in which fantastic progress is being made, providing novel and often unexpected insights into basic biological processes. The aim of this book is to provide a taste of this excitement, an idea of how much is there still to discover, and to act as an encouragement to researchers, old and new, to tackle the fascinating problem of how plant cells communicate with each other.

Andrew J. Fleming

1 Auxin as an intercellular signal
Jiří Friml and Justyna Wiśniewska

1.1 Introduction

Auxin has a prominent position among the classical plant hormones since it mediates multiple aspects of plant growth and development. At the cellular level, auxin is required for cell division as well as for elongation. Moreover, it plays a role in cell fate acquisition and hence in multiple patterning processes. At the whole plant level, auxin acts as a correlative signal between cells, tissues and organs. Thus, macroscopically, auxin mediates a surprising variety of processes such as cell divisions in the cambium and in tissue culture, vascular tissue differentiation, embryo development, organ initiation and growth, tropic responses as well as apical dominance (Davies, 1995).

The discovery of auxin dates back to Charles and Francis Darwin's experiments on the phototropism of canary grass coleoptiles, which indicated the existence of a transported signal (Darwin & Darwin, 1881). They demonstrated the transmission of some 'influence' from the place of light perception at the tip to the other tissues where differential growth (i.e. elongation) was induced. Later, Boysen-Jensen (1913) was able to pass this 'influence' through gelatine, and thus demonstrated the chemical nature of this growth-promoting substance, which was therefore termed auxin (Gk. *auxein* – to increase, to grow). Despite the multitude of auxin effects, its chemical nature is rather simple. The most abundant naturally occurring auxin is indole-3-acetic acid (IAA); other auxins are indole-3-butyric acid (IBA) (Fig. 1.1) and 4-chloroindole-3-acetic acid (4-Cl-IAA). In agricultural applications, synthetic auxins with higher metabolic stability, such as 1-naphthylacetic acid (1-NAA) and 2,4-dichlorophenoxyacetic acid (2,4-D), are commonly used (Fig. 1.1). Besides the physiologically active, free forms, conjugates to sugars, amino acids or to small peptides have been isolated (Cohen & Bandurski, 1982). Several lines of evidence support the hypothesis of regulated degradation of free IAA (Tam *et al.*, 1995). Consequently, the momentary concentration of auxin in target cells depends on the net outcome of the metabolic processes such as biosynthesis, degradation, conjugation and, most importantly, auxin transport. Auxin is unique among plant hormones in being directionally transported from the source organs in young, apical regions (Ljung *et al.*, 2001) to target cells in most plant tissues (Davies, 1995). Almost a century of physiological as well as genetic studies revealed that regulated auxin distribution between cells underlies most of the known auxin effects and thus highlighted the central role of intercellular auxin transport in plant biology.

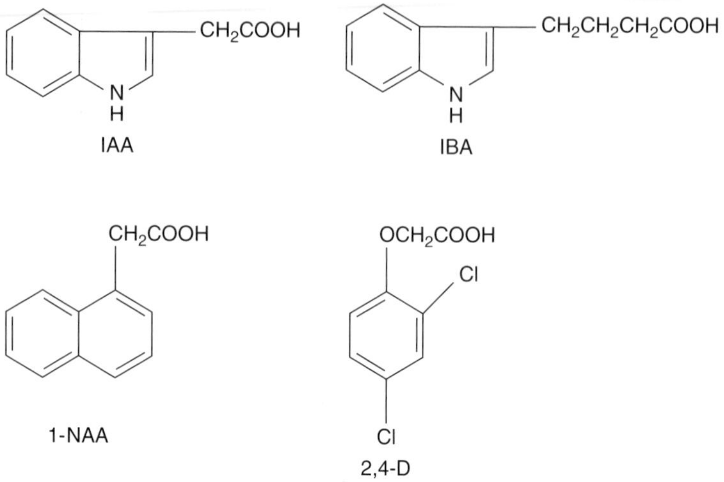

Figure 1.1 Chemical structures of important auxins. IAA: indole-3-acetic acid; IBA: indole-3-butyric acid; 1-NAA: 1-naphthylacetic acid; 2,4-D: 2,4-dichlorophenoxyacetic acid.

1.2 Auxin transport – known pathways

Two physiologically distinct and spatially separated pathways act together to transport auxin over long distances through plants (Fig. 1.2). Firstly, auxin is translocated rapidly by mass flow with other metabolites in the mature phloem. Secondly, auxin is transported downwards toward the root tips from immature tissues close to the shoot apex by a much slower, carrier-dependent, cell-to-cell polar transport (Goldsmith, 1977).

1.2.1 Polar auxin transport pathway

Uniquely amongst plant signaling molecules, auxins can be transported in a strictly regulated, polar fashion in plant tissues. Unlike auxin translocation in phloem, the polar auxin transport (PAT) is slower, specific for active free auxins and has a strictly unidirectional character. The main (basipetal) PAT stream runs from the apex with a velocity of 5–20 mm/h toward the base of the plant (Lomax et al., 1995). Using radioactively labeled auxin, this kind of transport was mainly detected in the cambium and adjacent, partially differentiated xylem vessels and xylem parenchyma. In the shoot, in addition to the basipetal stream, the movement of auxin was detected also in the lateral direction (Morris & Thomas, 1978). In the root, the auxin stream continues toward the root tip (acropetally), where a part of the auxin is redirected backwards from the root cap and transported basipetally through the root epidermis to the elongation zone (Fig. 1.2) (Rashotte et al., 2000). Auxin transport assays revealed that PAT requires energy, is saturable and sensitive to protein synthesis inhibitors. These results together with a cell-to-cell character of PAT suggested the

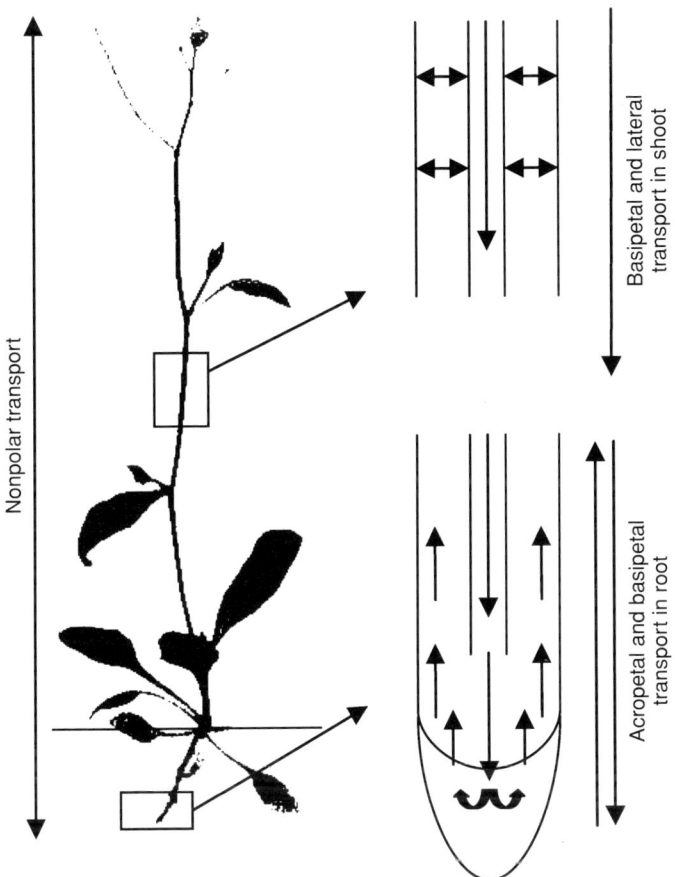

Figure 1.2 Nonpolar and polar auxin transport pathways. Auxin is distributed throughout plant by a nonpolar pathway in phloem and by a slow polar movement from the apex to the base (basipetal) and further to the root tip (acropetal) through vasculature and from the tip backwards (basipetal) through outer layers.

existence of specific auxin uptake and efflux proteins (Rubery & Sheldrake, 1974; Raven, 1975). Indeed, both saturable cellular auxin efflux and influx were demonstrated in single plant cells and tissue segments (Benning, 1986; Lomax et al., 1995). Cellular efflux readily transports, besides naturally occurring IAA, also the synthetic analogue NAA. On the other hand, 2,4-D is a preferred substrate for auxin uptake carriers but not, in most species, for auxin efflux carriers (Delbarre et al., 1996). Auxin influx and efflux pathways can also be physiologically distinguished using auxin efflux inhibitors (AEIs). These pharmacological tools arose from correlative exploration of structure–activity profiles of chemicals with auxin-like activity (Katekar & Geissler, 1977) and were demonstrated to inhibit efflux of auxin from cells and shoots segments. The most widely recognized AEIs are 1-N-naphthylphthalamic acid (NPA) and 2,3,5-triiodobenzoic acid (TIBA). Also, some naturally occurring flavonoid compounds show similar effects as AEIs and have been considered to be natural regulators of PAT (Jacobs & Rubery, 1988). Despite very limited data on the molecular mechanism of AEI function, they facilitated establishment of the central role of auxin efflux in plant development. When applied to plants, AEIs interfere with a number of processes such as axis establishment during embryogenesis (Hadfi et al., 1998), lateral root and aerial organ initiation (Reinhardt et al., 2000; Casimiro et al., 2001), root meristem patterning (Kerk & Feldman, 1994; Ruegger et al., 1997), vascular patterning (Mattson et al., 1999), hypocotyl and root elongation in light (Jensen et al., 1998), apical hook formation (Lehman et al., 1996) and most prominently tropic responses (Marchant et al., 1999). Owing to the lack of similar inhibitors of auxin influx, the characterization of the role of auxin influx in plant development was slowed down. Nevertheless, recently, compounds such as 1-naphthoxyacetic acid (1-NOA) and 3-chloro-4-hydroxyphenylacetic acid (CHPAA) were shown to specifically inhibit auxin uptake in tobacco culture cells (Imhoff et al., 2000) and were used to demonstrate a role of auxin influx in root gravitropism (Parry et al., 2001). The manipulation of PAT using inhibitors established PAT as the major, developmentally important route for auxin translocation.

1.2.2 Chemiosmotic model

The wealth of physiological findings on PAT such as its cell-to-cell character, saturability, energy and protein synthesis dependence, together with the known chemical nature of auxin, led to the formulation of the chemiosmotic hypothesis in the middle of the 1970s (Rubery & Sheldrake, 1974; Raven, 1975). The chemiosmotic hypothesis provides a coherent model for the polar transport of auxin through cell files and postulates the existence of auxin-specific carrier proteins (Fig. 1.3). In the relatively acidic environment of the cell wall (pH around 5.5), part of IAA exists in its protonated form (IAAH). This noncharged, lipophilic molecule passes easily through the plasma membrane by diffusion. In the more basic cytoplasm (pH around 7), the majority of IAAH dissociates and hence the resulting polar IAA^- anion is 'trapped' in the cell because of its poor membrane permeability and can leave the cell only by the activity of specific efflux carriers. The unidirectionality of auxin flow was explained by postulating asymmetrically distributed auxin efflux carriers within the cells. Thus the polar auxin export at the single cell level would multiply within a file of cells and result in the directed translocation of auxin throughout plant tissues. In addition, auxin influx was proposed to participate in PAT (Goldsmith, 1977) and later physiologically demonstrated (Benning, 1986). This classical model was reinforced after identification and characterization of candidate proteins for auxin influx (AUX1/LAX family) and efflux (PIN family) (Bennett et al., 1996; Gälweiler et al., 1998; Luschnig et al., 1998). Numerous circumstantial evidence demonstrate their role in auxin transport (Friml & Palme, 2002) and most strikingly both AUX1 and PIN proteins have been shown to be asymmetrically localized in auxin transport competent cells in accordance with the known directions of auxin flux (Gälweiler

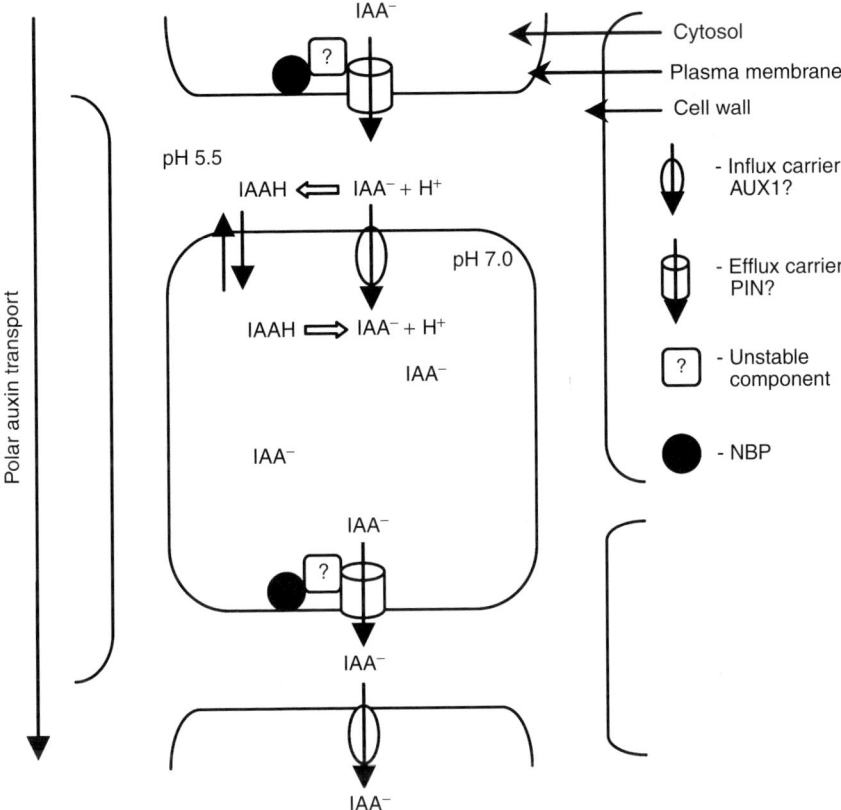

Figure 1.3 Model for cell-to-cell polar auxin transport. According to the chemiosmotic model, a membrane pH gradient (maintained by plasma membrane H^+-ATPases) drives accumulation of IAA within cell. At the higher pH of the cytoplasm, some of the protonated auxin molecules dissociate and are 'trapped' inside the cell and can only exit by efflux carrier systems. Asymmetry in the distribution of the efflux carrier results in a directional transport of auxin through the cell file. Efflux carrier is a protein complex containing auxin efflux catalyst and NPA-binding protein (NBP) linked with an unstable component.

et al., 1998; Müller *et al.*, 1998; Swarup *et al.*, 2001; Friml *et al.*, 2002a,b). These findings provide a molecular confirmation that at least in some tissues, influx and efflux in concert facilitate the polar translocation of auxin (Friml, 2003).

1.2.3 Multicomponent auxin efflux carrier system

Subsequent physiological studies, especially those focused on the mechanism of the effect of AEIs such as TIBA or NPA on PAT, provided important insights into the composition of the auxin efflux carrier complex. Finding that plant protein extracts specifically bind NPA, taken together with the NPA effect on auxin efflux, suggested the existence of an NPA-binding protein (NPB), forming a part of the auxin efflux carrier complex (Fig. 1.3). Despite extensive studies on the NPB, little is known about its nature. Studies on NPB revealed different localizations for NPB, either at the periphery of the plasma membrane associated with the cytoskeleton (Cox & Muday, 1994) or as an integral membrane protein (Bernasconi *et al.*, 1996). On the other hand, several proteins, which bind NPA with different affinities, have been isolated from *Arabidopsis* protein extracts by NPA affinity chromatography. These include multidrug resistance (MDR) proteins of the ATP-binding cassette (ABC) transporter family (AtMDR1 and AtPGP1), aminopeptidase (AMP1) and others (Murphy *et al.*, 2002). In addition, an *Arabidopsis* mutant named *tir3* was isolated in a genetic screen for NPA-insensitive roots. This mutant was shown to have reduced NPA-binding activity and auxin transport capacity, suggesting the possibility that *TIR3* encodes the NPB or a functionally related protein (Ruegger *et al.*, 1997). Nonetheless, the characterization of the *TIR3* gene product as a homolog of *Drosophila* CALOSSIN/PUSHOVER (CAL/O) protein, termed BIG for its extraordinary size (Gil *et al.*, 2001), did not clarify its functional connection to auxin efflux or NPA binding. In summary, both biochemical and genetic approaches provided candidates for the NPB, but the identity of the NPB and the mechanism by which NPA inhibits polar auxin transport still remains a matter of debate. Based on additional experiments using protein synthesis inhibitors, such as cycloheximide, the existence of a third unstable component of the auxin efflux carrier complex has been proposed (Fig. 1.3). Short-term cycloheximide treatment does not affect auxin efflux or the saturable binding of NPA to microsomal membranes, but it interferes with the inhibitory effect of NPA on efflux (Morris *et al.*, 1991). These observations suggest that the NBP and the efflux carrier itself are separate proteins with low turnover, interacting through a third, unstable protein (Delbarre *et al.*, 1998; discussed by Morris, 2000). Surprisingly, inhibitors of vesicle traffic to the plasma membrane, monensin and Brefeldin A (BFA), interfere with cellular auxin efflux (Delbarre *et al.*, 1998; Morris, 2000) and with polar (basipetal) auxin transport in stem segments (Robinson *et al.*, 1999). Significantly, BFA does not affect saturable NPA binding to microsomal preparations, providing additional evidence that the NBP and the auxin efflux catalyst are different proteins (Robinson *et al.*, 1999). Taken together, these observations indirectly suggest that an essential component of the auxin efflux carrier system, but not the NPB, is targeted to the plasma membrane through the BFA-sensitive secretory system, and that this component turns over very rapidly at the plasma membrane without the need for concurrent protein synthesis. This implies that part of auxin efflux carrier component rapidly cycles between the plasma membrane and internal pools (Delbarre *et al.*, 1998; Morris & Robinson, 1998; Robinson *et al.*, 1999).

1.3 Molecular components

Most instrumental in identifying molecular components of PAT were genetic approaches, especially in *Arabidopsis thaliana*. Various screening strategies have been successfully applied to identify mutants affected in PAT. Some mutants have been selected on the basis of abnormal responses to auxin transport inhibitors or were identified fortuitously in screens for developmental alterations and only later was the connection to PAT discovered (Friml & Palme, 2002).

1.3.1 Auxin influx – AUX1 proteins

A mutant called *auxin1* (*aux1*), which confers a root agravitropic and auxin-resistant phenotype, was instructive for identification of a gene possibly encoding an auxin influx carrier. The *AUX1* gene encodes a 485 amino acid long protein sharing significant similarity with plant amino acid permeases consistent with the role for AUX1 in the uptake of the tryptophan-like IAA (Bennett *et al.*, 1996). So far the definitive biochemical proof of AUX1 function as an auxin uptake carrier is lacking, but several lines of evidence (mainly based on detailed analysis of the *aux1* phenotype) strongly support that AUX1 is required for auxin influx. Strikingly, the *aux1* root agravitropic phenotype can be restored by treatment with a membrane permeable auxin NAA in contrast to less permeable 2,4-D. Moreover, this rescue coincides with restoration of basipetal auxin transport, which is defective in *aux1* (Yamamoto & Yamamoto, 1998; Marchant *et al.*, 1999). In addition, the main features of the *aux1* phenotype can be mimicked by growing seedlings on inhibitors of auxin influx (Parry *et al.*, 2001). Other evidence that AUX1 participates in auxin influx came from auxin uptake assays in *aux1* and wild-type roots. They revealed that *aux1* roots accumulated significantly less radioactively labeled 2,4-D than did wild-type, and that this difference was not found when the membrane-permeable 1-NAA or the IAA-like amino acid tryptophan were assayed (Marchant *et al.*, 1999). Recently, the AUX1 protein was localized within *Arabidopsis* root tissue (Swarup *et al.*, 2001). The AUX1 protein was detected in a subset of stele, columella, lateral root cap and epidermal cells exclusively in root tips. Considering the localized expression of AUX1 only in root tips, it is surprising that *aux1* mutant root tips contain lower auxin levels, which rather suggests defects in long-distance supply to the root tip (Swarup *et al.*, 2001). This paradox, taken together with localization of AUX1 at the upper side of protophloem cells (see Plate 1.1A, following page 146), suggests a role of the AUX1 protein in unloading of the phloem flow via the protophloem to the root apical meristem (Swarup *et al.*, 2001). Thus, AUX1 would appear to provide a molecular connection between nonpolar and polar auxin transport routes. *AUX1* is a member of the small gene family in *Arabidopsis*. However, the characterization of the three other *LIKE AUX1* (*LAX*) genes has not yet been reported.

1.3.2 Auxin efflux – PIN proteins

Another *Arabidopsis* mutant, *pin-formed* (*pin1*), with its characteristic needle-like stem had already been functionally associated with auxin efflux on the basis of its dramatic morphological aberrations, which can be phenocopied by inhibition of auxin efflux. In addition, *pin1* inflorescences show a drastic reduction in basipetal auxin transport (Okada *et al.*, 1991). The *PIN1* gene was cloned by transposon tagging and found to encode a 622 amino acid protein with up to 12 putative transmembrane segments with similarity to a group of transporters from bacteria (Gälweiler et al., 1998). Simultaneously, a homologous gene was identified independently by several groups – the *PIN2/EIR1/AGR1* gene (Chen et al., 1998; Luschnig et al., 1998; Müller et al., 1998; Utsuno et al., 1998) – and analysis of additional homologs (*PIN3*, *PIN4* and *PIN7*) followed (Friml et al., 2002a,b, 2003). In total, the *Arabidopis PIN* gene family consists of eight members and homologous genes were found in other plant species, e.g. maize, rice, soybean and others. The proposed function for PIN proteins as efflux carriers has not ultimately been proven; nonetheless, several lines of evidence strongly support their role in PAT:

1. Topology and localization of PIN proteins: The PIN proteins share more than 70% similarity and have almost identical topology – a large hydrophilic loop is symmetrically flanked by two conserved, highly hydrophobic domains with five to six transmembrane segments. Transporters of the major membrane facilitator class display similar topology (Chen et al., 1998; Luschnig et al., 1998; Müller et al., 1998; Utsuno et al., 1998). When localized in planta, most PIN proteins show asymmetric cellular localization (Plate 1.1B–D), impressively correlating with the known direction of PAT in these tissues. This polar localization was predicted by chemiosmotic hypothesis for auxin efflux proteins (Rubery & Sheldrake, 1974; Raven, 1975).

2. Heterologous expression of PIN proteins: To date, the only experimental system used to address PIN transport activity is yeast assay (Luschnig et al., 1998). Yeast carrying a mutation in the *GEF1* gene (resulting in an altered ion homeostasis) shows enhanced resistance to the yeast toxin fluoroindole, when overexpressing PIN2/EIR1/AGR1. Fluoroindole shows some (albeit limited) structural similarity to auxin (Luschnig et al., 1998). The yeast also retains less radioactively labeled auxin than does control yeast (Chen et al., 1998). Nonetheless, measurements of auxin efflux instead of auxin retention have not been demonstrated so far.

3. PAT is defected in *pin* mutants: All defects observed so far in *pin* mutants occur in processes known to be regulated by PAT and they can be phenocopied by treatment of wild-type plants with AEIs. One of the strongest arguments for the involvement of PIN proteins in auxin transport is a reduction of PAT in *pin* mutants, which directly correlates with loss of PIN expression in corresponding tissue, as was demonstrated for basipetal auxin transport in stem of *pin1* mutant or in root of *pin2* mutant (Okada et al., 1991; Rashotte et al., 2000). In addition, local distribution and accumulation of auxin monitored both by the activity of an auxin responsive construct (e.g. *DR5::GUS*; Sabatini et al., 1999) and by direct measurements of auxin content (Friml et al., 2002b) or using radioactive IAA preloaded root tips (Chen et al., 1998) have been shown to be affected in *pin* mutants.

The data accumulated so far provide an extensive body of evidence to argue that PIN proteins are involved in some important aspects of auxin transport, probably in auxin efflux. Nevertheless, the central question of whether PIN proteins represent transport or regulatory component of auxin efflux machinery still remains a topic for future investigations.

1.3.3 ABC transporters

Recently, another protein family has been implicated in auxin transport – MDR proteins, a subfamily of the ABC transporters (Noh *et al.*, 2001). Members of this family are known to enhance the export of chemotherapeutic substances in mammalian systems. Two of them, AtMDR1 and AtPGP1, were isolated by NPA-affinity chromatography and were also able to bind NPA *in vitro* or when expressed in yeast cells. Moreover, the corresponding mutants and double mutants show lower rate of PAT and other phenotypic aberrations somewhat resembling defects in PAT. Nevertheless, *Atmdr1* mutants still exhibited up to 60% of the NPA binding found in wild type and NPA is effective in reducing PAT to background levels in the mutant (Noh *et al.*, 2001). So the question whether MDR/PGPs represent the elusive NPB, which regulates PAT, remains open. A mechanism how the MDR/PGP proteins modulate PAT has not been identified so far, but a direct auxin transport function or a positive regulation of a PIN-type auxin efflux carriers has been suggested. Another possibility is that MDR proteins are involved in correct localization of auxin carriers. Indeed, it has been reported that in all cell types of *Atmdr1* and *Atpgp1* hypocotyls the usual basal localization of PIN1 is replaced with more punctuate pattern (Noh *et al.*, 2003). However, previous reports identified PIN expression associated only with vascular tissue (Friml *et al.*, 2002a), and so the biological meaning of this observation remains unclear.

1.4 Subcellular dynamics of auxin carriers

The widely accepted chemiosmotic model on PAT conveys the idea that auxin influx and efflux carriers reside in the plasma membrane where they exhibit transport functions. However, physiological studies already suggested that a fraction of the auxin efflux complexes have a short half-life in the plasma membrane and cycle rapidly through an unidentified intracellular compartment (Morris, 2000). These findings were unexpected and difficult to reconcile with classical models. After isolation of PAT components and with the availability of tools to visualize them, the subcellular dynamics of these proteins was addressed directly using cell biological approaches.

1.4.1 Constitutive recycling of PIN proteins

The fungal toxin BFA is in both animal and plant systems a well-characterized compound that interferes with subcellular vesicle trafficking. The molecular targets of

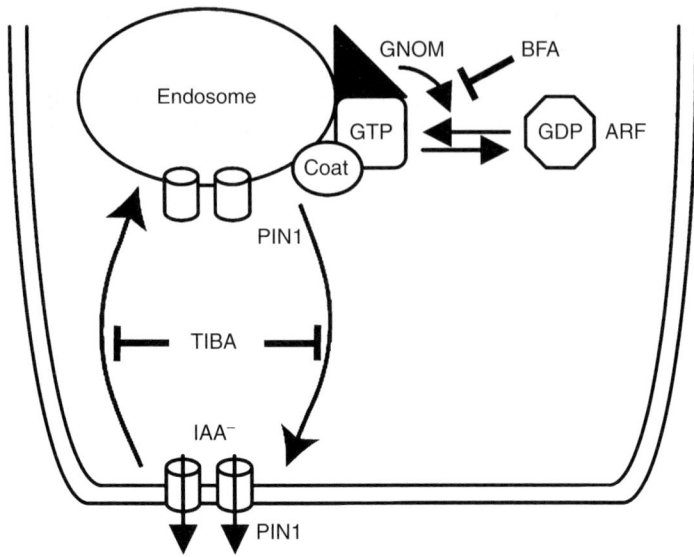

Figure 1.4 Subcellular movement of PIN proteins. Schematic model to explain internalization of PIN proteins upon BFA treatment. BFA blocks GNOM ARF-GEF responsible for activation of endosomal ARF GTPases, which mediate recycling of PIN1 to the plasma membrane. Ongoing endocytosis is BFA insensitive. AEIs such as TIBA interfere with both steps of PIN cycling.

BFA are guanine nucleotide exchange factors (GEFs), which activate small GTPases of the ARF family – important regulators of vesicle budding. In *Arabidopsis* roots, PIN1 protein is rapidly and reversibly internalized from the plasma membrane in response to BFA (Fig. 1.4) (Geldner *et al.*, 2001). This also occurs in the presence of a protein synthesis inhibitor, thereby demonstrating that internalized PIN1 originated from the plasma membrane and that PIN1 is rapidly cycling between the plasma membrane and a 'BFA' compartment, recently characterized as an accumulation of endosomes (Plate 1.1E) (Geldner *et al.*, 2003). The action of drugs that disrupt the structure of the cytoskeleton indicated that PIN1-containing vesicles are transported predominantly along the actin cytoskeleton. However, in dividing cells, tubulin is also required for correct PIN1 traffic (Geldner *et al.*, 2001). The actin-dependent recycling was also demonstrated for PIN3 (Friml *et al.*, 2002a) and PIN2 (Grebe *et al.*, 2003) proteins. The PIN recycling phenomenon has been further corroborated by electron microscopy studies, which detected PIN3 not only at the plasma membrane but frequently also in intracellular vesicles (Friml *et al.*, 2002a). BFA is also known to rapidly interfere with auxin efflux and could phenocopy the effect of AEIs (Delbarre *et al.*, 1998; Morris & Robinson, 1998; Geldner *et al.*, 2001). These surprising findings can be hardly incorporated in the old static models, with influx and efflux carrier complexes residing and functioning at the plasma membrane. The crucial question remains of what the biological relevance of the PIN cycling is. Here different scenarios can be conceived:

1. A high turnover of PAT components would provide the flexibility to allow rapid changes in polarity of carrier distribution and provide a mechanism for the rapid redirection of auxin fluxes in response to environmental or developmental cues (Friml & Palme, 2002). Indeed, PIN3 was shown to rapidly relocate in response to gravity stimulation (Plate 1.1M) (Friml *et al.*, 2002a).
2. Components of polar auxin transport may have a dual receptor/transporter function (Hertel, 1983). In this case, cycling may be part of a mechanism for signal transduction and receptor regeneration, as is known for some other kinds of receptors. Dual sensor and transport functions have been proposed for sugar transporters in yeast and plants (Lalonde *et al.*, 1999).
3. The most exciting possibility is that vesicle trafficking itself is a part of the PAT mechanism and that, in a manner analogous to the mechanism of neurotransmitter release in animals, auxin is a vesicle cargo, released from cells by polar exocytosis (Friml & Palme, 2002). In this model, instead of being 'auxin channels', PIN proteins would mediate the accumulation or retention of auxin in the vesicles in which auxin would be translocated. Some support of this scenario comes from the BIG protein, which is involved in PAT and PIN1 subcellular trafficking, since its homolog in *Drosophila* mediates vesicle recycling during synaptic transmission.

Regardless of how well any of the scenarios described above eventually turns out to fit the true picture, an understanding of the cellular mechanisms controlling the subcellular dynamics of the auxin carriers will be crucial for our understanding of PAT process.

1.4.2 AEIs and vesicle trafficking

Another surprising outcome of the cell-biological studies on PIN cycling concerns effects of AEIs such as TIBA on vesicle trafficking (Fig. 1.4). Despite the fact that AEIs were major tools for physiological studies on PAT, the mechanism of their action remains elusive. The finding that TIBA inhibits PIN1 recycling (Geldner *et al.*, 2001) raised an attractive possibility that TIBA inhibits auxin efflux by interfering with the recycling of auxin efflux components. However, AEIs also interfere with vesicle-mediated traffic of PAT unrelated proteins and much higher concentrations are needed for trafficking inhibition than for PAT inhibition. In addition, observations on BY-2 tobacco cells have revealed that the inhibition of auxin efflux by NPA is much more efficient than the inhibition caused by the well-established inhibitor of protein traffic BFA (Petrášek *et al.*, 2003). These findings argue against a causal link between a general role of NPA and other phytotropins in vesicle trafficking and auxin efflux inhibition. Thus, it still remains open whether inhibition of vesicle traffic and PAT by AEIs are functionally related.

1.4.3 GNOM and PIN dynamics

Significant complementary evidence for the importance of subcellular vesicle trafficking in the process of PAT, and especially for its importance in plant development, came from the analysis of the *Arabidopsis* mutant *gnom*. The *gnom* (*gn*) mutant (Plate 1.1G) was isolated from a screen for defects in early apical–basal patterning in *Arabidopsis* seedlings (Mayer *et al.*, 1991). Embryos of *gn* mutants display a variety of aberrations, including missing roots and fused or improperly placed cotyledons. Most of these defects are reminiscent to defects observed when embryos are cultivated in the presence of polar auxin transport inhibitors (Hadfi *et al.*, 1998; Friml *et al.*, 2003). The *GN* gene was identified as a GEF for ARF (ADP-ribosylation factor) GTPases – essential regulators of vesicle trafficking in many organisms. ARFs participate in the formation of transport vesicles from donor compartments and the selection of their protein cargo. ARF proteins are present in two forms: an active GTP-bound and an inactive GDP-bound form. Conversion of GDP-bound to the GTP-bound form is mediated by specific GEFs (Donaldson & Jackson, 2000). The function of ARF-GEFs such as GNOM is inhibited by BFA, which binds to an ARF-GDP/ARF-GEF complex (Peyroche *et al.*, 1999). At first, it was difficult to reconcile the biochemical function of GNOM in vesicle trafficking with the auxin-transport-related phenotype of the *gnom* mutant. A detailed analysis of *gn* revealed that at the subcellular level, the coordinated polar localization of PIN1 was defective in *gnom* mutant embryos (Steinmann *et al.*, 1999). This finding, taken together with a BFA-sensitive subcellular cycling of PIN proteins, suggested a role for GNOM ARF-GEF in the regulation of subcellular trafficking of PAT components such as PIN proteins (Fig. 1.4). Consistently with this, GNOM localizes to and maintains the integrity of endosomes through which PIN proteins recycle (Geldner *et al.*, 2003). To determine whether GNOM specifically controls PIN1 traffic, a single amino acid substitution (696M to L) was introduced into the originally BFA-sensitive GNOM, to generate a fully functional, but BFA-insensitive variant of the GNOM protein. This allowed specifically dissecting the function of GNOM from other BFA-sensitive trafficking steps. When plants expressing the BFA-resistant GNOM are treated with BFA, PIN1 remains correctly localized to the cell surface (Fig. 1.4), demonstrating direct involvement of GNOM in PIN1 recycling. In addition, BFA-resistant GNOM renders both auxin efflux and auxin-mediated growth insensitive to BFA inhibition (Geldner *et al.*, 2003). Thus, these findings directly linked a component of membrane traffic – GNOM ARF-GEF to the PIN recycling and PAT process, highlighting the importance of polarized trafficking in fundamental processes of plant development.

1.5 The role of auxin gradients in plant development

It remained a mystery for years how a simple molecule like auxin can influence the variety of seemingly unrelated developmental processes such as embryonic axis formation, organogenesis, meristem maintenance, tropisms, root and shoot elongation, apical hook formation and others. The availability of molecular tools for visualization and manipulation of PAT components as well as for visualization of auxin and its activity enabled researchers to address more specifically the role of PAT-dependent auxin distribution in each of these developmental processes. These studies convincingly demonstrated that regulated, local auxin acummulation underlies most of auxin-mediated development.

1.5.1 Monitoring of auxin distribution in planta

One of the major obstacles in studies addressing the role of auxin in plant development was the inability to visualize auxin distribution *in planta*. An important breakthrough was brought through the discovery of genes that are rapidly upregulated by auxin (reviewed in Hagen & Guilfoyle, 2002). The consensus sequence TGTCTC (auxin response element, AuxRE) was identified within promoters of these genes, which confers the auxin responsiveness (Ulmasov *et al.*, 1995). Multiple repeats of the AuxREs yielded synthetic promoters *DR5* or *DR5rev* (another variant with inverse repeats), which are highly responsive to auxin and, therefore, can be used for indirect monitoring of auxin levels. Indeed, multiple strategies were utilized to show the correlation between *DR5* activity and auxin accumulation. Firstly, direct measurements of auxin content within *Arabidopsis* roots demonstrated elevated levels at the tip, where the highest *DR5* activity was also detected (Casimiro *et al.*, 2001). Secondly, exogenously supplied auxin was able to induce *DR5* activity in all cells and inhibition of PAT changed the *DR5* expression pattern, suggesting that the spatially restricted signals in untreated plants visualize differences in auxin levels between cells. Finally, the accumulation of auxin itself (monitored using an anti-IAA antibody) mirrors the *DR5* pattern (Plates 1.1J and 1.1K). In summary, despite the theoretical limitations for the use of *DR5* and related tools, so far it seems that *DR5* activity can be used as a reasonable approximation for auxin levels at least in embryonic and meristematic tissues. Indeed, the *DR5::GUS* and/or *DR5rev::GFP* constructs have been instrumental in detecting the spatial pattern of auxin accumulation in many tissues including *Arabidopsis* embryos (Plate 1.1I), roots (Plate 1.1J) and organ primordia (Plate 1.1H), and have demonstrated a universal role for auxin gradients in plant development (Sabatini *et al.*, 1999; Friml *et al.*, 2002a,b, 2003; Benková *et al.*, 2003).

1.5.2 Embryonic axis formation

Embryogenesis is a process that transforms a single-celled zygote into the embryo containing all basic pattern elements of the future plant. The mature embryo displays an axis of polarity, with the shoot meristem at the apical end and root meristem at the basal end. This remarkably uniform apical–basal pattern has been traced in *Arabidopsis* back to the earliest stages of embryogenesis. The first manifestation of the apical–basal axis is the asymmetric division of the zygote, which produces a small apical cell and a larger basal cell. The apical cell divides vertically and generates the 'proembryo', which later gives rise to the most regions of the seedling. The basal cell continues to divide horizontally and produces the suspensor – a file of cells that attaches the proembryo to maternal tissue. At the early globular stage, the uppermost cell is specified to become the hypophysis – the founder of the root meristem. At the triangular stage, the early patterning is finished with the initiation of two symmetrically positioned embryonic leaves – cotyledons (Jürgens, 2001). A role for auxin in embryo patterning has been suspected for some time, not the least because embryo defects can be induced by blocking PAT (Hadfi et al., 1998). In addition, genetic disruption of auxin response in *Arabidopsis* mutants such as *monopteros* (*mp*) and *bodenlos* (*bdl*) led to defects in embryonic axis formation. Molecular analysis of these mutants revealed that *MP* encodes a transcriptional activator – the auxin response factor 5 (ARF5), and *BDL* encodes the corresponding transcriptional repressor – IAA12, both components of auxin signaling (Hardtke & Berleth, 1998; Hamann et al., 2002). To really pinpoint the role of auxin in embryogenesis, the distribution of auxin and its response was monitored using anti-IAA antibodies and *DR5rev::GFP*. Immediately after the division of the zygote, when the apical cell is specified, auxin accumulates in this cell and during subsequent development persists in the proembryo. At around the 32-cell stage, when the basal embryo pole is being specified, the gradient of auxin accumulation suddenly reverses and forms a new maximum in the uppermost suspensor cells, including the hypophysis (Plate 1.1I). At later stages of embryogenesis, additional *DR5* reporter gene signals appear in the tips of the developing cotyledons (Friml et al., 2003). Both chemical (AEIs) and genetic (*gn* or multiple *pin* mutants) inhibition of PAT interfere with this dynamic distribution of auxin during embryogenesis. Furthermore, they cause identical developmental defects, ranging from cup-shaped embryos with misspecified apical structures and a nonfunctional root pole (Plates 1.1F and 1.1G), to ball-shaped embryos without any discernible apical–basal axis. These findings, together with analysis of PIN expression and localization, completed the picture, indicating a role for PIN-dependent auxin distribution in embryo patterning. At early stages (Fig. 1.5A), auxin is actively provided to the apical cell from the adjacent basal cell by the action of apically localized PIN7. This apical–basal auxin gradient is required for the specification of the apical cell. At subsequent stages, the cells of the suspensor continue to localize PIN7 at their apical side, while in the proembryo another protein, PIN1, is expressed without apparent polarity. But after the 32-cell stage (Fig. 1.5B), PIN1 becomes localized to the basal membranes of the provascular cells, suggesting downward transport toward the region of the future root pole. Simultaneously, the asymmetric localization of PIN7 is reversed within the basal cells, mediating auxin transport out of the embryo. Subsequently, PIN4 expression starts at the basal pole of the embryo, supporting the action of both PIN1 and PIN7. As a result of these changes in auxin flow, the auxin gradient reverses, displaying its new maximum in the uppermost suspensor cell, which in response to auxin is specified to become the hypophysis – the founder of the future root meristem. Thus, developmentally regulated changes in polarity of PIN proteins result in the redirection of auxin fluxes for local auxin accumulation, which is then required first for specification of the apical and later for the basal pole of the embryonic apical–basal axis (Friml et al., 2003).

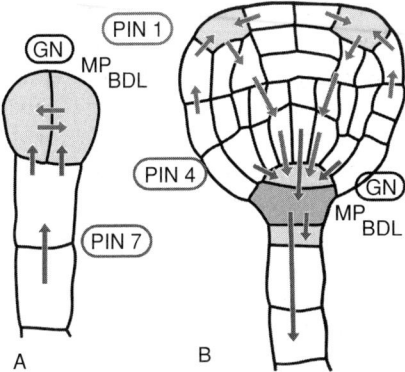

Figure 1.5 Auxin transport and distribution during embryogenesis. Sites of auxin accumulation are shadowed. Arrows indicate routes of auxin efflux mediated by PIN1, PIN4 and PIN7. Also depicted are proteins involved in embryo patterning and related to auxin transport (encircled) or auxin response. (A) Two cell stage embryo – apical cell specification. Auxin accumulates in the proembryo through PIN7-dependent transport via the suspensor. Auxin response (*mp*, *bdl*) and transport (*gn*) mutants show defects in the establishment of the apical cell. (B) Triangular stage – from early globular stage on auxin accumulates, in a PIN1- and PIN4-dependent manner, in the hypophysis, which is specified and is further transported through the suspensor via a PIN7-dependent route. New sites of auxin accumulation emerge at the tips of forming cotyledons. *mp*, *bdl* and *gn* show defects in root pole as well as cotyledon establishment. Adapted from Friml *et al.* (2003), with permission.

1.5.3 Postembryonic organ formation

Embryo development establishes the basic body plan of both animals and plants. However, the adult form of a plant also depends largely on postembryonic development. Plants, unlike animals, can postembryonically initiate new organs such as leaves, flowers, flower organs, ovules and lateral roots. The regular initiation pattern of leaves and flowers (called *phyllotaxis*) is the major determinant of adult plant architecture. During organ formation, first a site of primordium initiation is selected and then a new growth axis of the organ primordium is established. It seems that PIN-dependent redirection of auxin flow and local accumulation of auxin play a fundamental role in both of these processes. Exogenous auxin application is sufficient to trigger leaf or flower formation in the shoot apex (Reinhardt *et al.*, 2000) or lateral root initiation (Laskowski *et al.*, 1995), and endogenous accumulation of auxin and its response was detected at the initiation site of incipient organs in shoots and roots (Benková *et al.*, 2003). On the other hand, interference with PAT (AEIs, *pin* mutants) or auxin response (*mp*, *solitary* root) blocks organ formation (Okada *et al.*, 1991; Przemeck *et al.*, 1996; Fukaki *et al.*, 2002). In the shoot, PIN1 localization in the outermost layer (L1) of the meristem undergoes dynamic rearrangement toward these loci of auxin accumulation (Reinhardt *et al.*, 2003). The pattern of auxin accumulation and PIN localization suggests that auxin is transported toward the meristem through the L1 cell layer. There, auxin becomes absorbed by

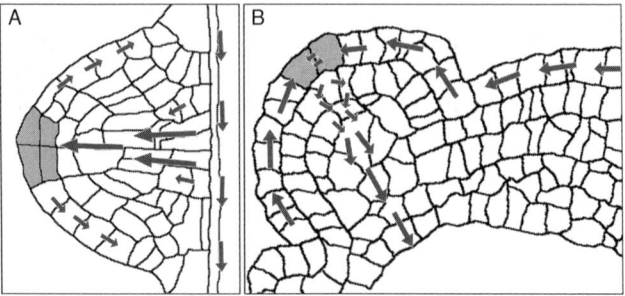

Figure 1.6 Auxin transport and distribution during organogenesis. Sites of auxin accumulation are shadowed. Presumptive routes of auxin transport are depicted by arrows. (A) Lateral root primordium: Auxin is provided by PIN1-dependent auxin transport through the primordium interior toward the tip, where it accumulates. From here, part of the auxin is retrieved through the outer layers. (B) Auxin is provided to the primordium tip through the outer layers. From the tip, auxin is drained through a gradually established transport route toward the vasculature. Reproduced from Benková et al. (2003), with permission.

already initiated primordia, which transport auxin into their interior and further toward already differentiated vasculature (Benková et al., 2003). Therefore, auxin is depleted from the surroundings of the primordia (sink function) and its highest concentration remains at the most distant position, where a new primordium is initiated. Thus, a positive feedback represented by auxin accumulation in combination with lateral inhibition provides a mechanism for reiterativity and stability of the spiral or other types of phylotactic pattern (Reinhardt et al., 2003). In the root, auxin also accumulates by a PAT-dependent mechanism at sites of organ initiation (Benková et al., 2003), but how it is precisely regulated remains so far unclear.

Once sites of organ initiation are selected, cell division is activated and the organ primordium develops along a new growth axis. In all types of primordia, PIN polar localization reorganizes and the new direction of PIN-mediated auxin transport determines the growth axis of the developing organ, establishing an auxin gradient with its maximum at the tip (Benková et al., 2003). Interestingly, the main auxin transport routes appear to be reversed, when lateral root primordia are compared with other types of organs. In aerial organs, auxin is supplied through the outer layer and accumulates at the primordium tip (Fig. 1.6B). From the tip, auxin is transported into the interior of the primordium. In the case of lateral roots, auxin is provided to the tip through the inner cells and distributed away through the primordium surface (Fig. 1.6A). Regardless of these differences, a common auxin-gradient-dependent mechanism seems to underlie organ initiation as well as primordium development in all plant organs regardless of their mature morphology or developmental origin.

1.5.4 Root meristem maintenance

Plants, in contrast to animals, not only can initiate new organs postembryonically, but also possess specialized, permanently dividing and differentiating tissues called meristems, which enable a perpetuation of postembryonic growth. Meristems contain populations of 'stem' cells, which, according to their position, differentiate into different cell types. The *Arabidopsis* root meristem displays a highly regular and predictable pattern of cell divisions and differentiation and is, therefore, an ideal system for studies on mechanisms of meristem patterning. Manipulation of auxin distribution as well as auxin signaling have demonstrated a role for auxin in regulating the root meristem activity (Kerk & Feldman, 1994; Ruegger *et al.*, 1997; Sabatini *et al.*, 1999). Accumulation of auxin and auxin response was detected in the columella initial and first columella layer cells (Sabatini *et al.*, 1999; Benková *et al.*, 2003). Interestingly, the PIN4 auxin efflux regulator displays a polar localization pointing toward the same area, suggesting a role of PIN4 in maintenance of this auxin accumulation. In support of this, *pin4* mutation or chemical inhibition of PAT disrupts this auxin accumulation, which results in changes in cell fate specification (Friml *et al.*, 2002b). These observations suggest that PIN4 mediates PAT through the central root meristem tissues, thus actively maintaining an auxin gradient with its maximum in the distal root tip (Plate 1.1M). The PIN4 function appears to be also necessary for local auxin turnover, since *pin4* mutant root tips display elevated auxin levels and fail to canalize exogenously applied auxin properly (Friml *et al.*, 2002b). Following the most plausible scenario, auxin from upper tissues is actively concentrated by PIN-dependent transport in the distal root tip, which serves as an 'auxin sink'. There, part is probably rendered inactive by an as yet unknown mechanism and other part is redistributed back by PIN2 action through the outer layers (Plate 1.1M).

1.5.5 Tropisms

Tropisms are growth responses to external stimuli such as light (phototropism) or gravity (gravitropism), resulting from differential elongation rates on either side of a plant organ. The Cholodny–Went hypothesis proposed that differential growth rates result from the asymmetrical distribution of auxin, which subsequently promotes or inhibits cell growth and elongation (Went, 1974). Indeed, differential auxin or auxin response distributions were visualized in various plant organs including gravity stimulated tobacco shoots (Li *et al.*, 1991), light and gravity stimulated *Arabidopsis* hypocotyls (Friml *et al.*, 2002a) or developing peanut gynophores (Moctezuma, 1999). Because AEIs interfere with the asymmetric distribution of auxin as well as with tropisms, PAT has been implicated as the process underlying asymmetric auxin distribution (Lehman *et al.*, 1996; Friml *et al.*, 2002a) and the existence of a lateral auxin transport in shoots has been proposed. This would facilitate the exchange of auxin between the basipetal stream in vasculature and peripheral regions, where control of elongation occurs (Fig. 1.2). The analysis of localization and function of the auxin efflux regulator PIN3 provided molecular support for this concept (Friml *et al.*, 2002a). The *pin3* mutants are defective in hypocotyl phototropism and gravitropism as well as root gravitropism, although these defects are rather subtle, suggesting functional redundancy. In addition, PIN3 is predominantly localized at the lateral side of shoot endodermis cells, where it is perfectly positioned to regulate lateral auxin flow (Friml *et al.*, 2002a).

In roots, the situation is more complex, since gravity is perceived in the root cap but the growth response occurs in the elongation zone where elevated auxin levels on the lower side inhibit growth, resulting in downward bending (Chen et al., 2002). It seems that following gravity stimulation, auxin is redistributed laterally toward the lower side of the root cap, from where it is transported to the elongation zone (Sabatini et al., 1999; Rashotte et al., 2000). Localization and mutant analyses suggest that the auxin influx component AUX1 facilitates auxin uptake in the lateral root cap and epidermis region and that PIN2 efflux regulator mediates directional auxin translocation toward the elongation zone (Müller et al., 1998; Swarup et al., 2001). The notion that both auxin influx and efflux are required for root gravitropism is further supported by the experiments with inhibitors of both processes (Parry et al., 2001). But how is gravity perception linked to the initial lateral auxin redistribution in the root cap? Gravity is perceived by sedimentation of starch-containing organelles (statoliths) in the columella root cap cells (Chen et al., 2002). PIN3 is localized in these cells under normal growth conditions without any apparent asymmetry at the cell boundaries. Intriguingly, when roots are gravistimulated, already within 2 min PIN3 changes its position and relocates, presumably to the new lower cell boundary (Plate 1.1M) (Friml et al., 2002a). It is conceivable that the rapid recycling of PIN3 between endosomes and the plasma membrane along the actin cytoskeleton provides a mean for its rapid retargeting in response to the environmental stimulus. However, how the statolith sedimentation and PIN3 relocation are linked remains unclear. It is possible that actin reorganization following the statoliths sedimentation would redirect intracellular traffic of PIN3 along the sedimentation routes and PIN3 would preferentially accumulate at the lower side of the cell. It remains to be demonstrated whether PIN3 relocation also mediates the auxin redistribution in the shoot. However, it is likely, since both statoliths and PIN3 are present in the shoot endodermis (Friml et al., 2002a), which is essential for shoot tropism responses (Fukaki et al., 1998). Also the link between light perception and lateral auxin redistribution during phototropism remains a topic for future investigations.

1.5.6 Downstream of auxin gradients

Recent studies on a role of PAT in plant development revealed that local gradients in auxin accumulation underlie the developmental responses to auxin. So far, this mechanism could be demonstrated for embryonic axis and postembryonic organ formation, meristem pattern maintenance and tropisms. However, how can the accumulation of a structurally rather simple molecule as IAA lead to such a wide variety of different responses? Since it is clear that different cells respond to auxin by activation of different developmental programs, part of the answer lies downstream of auxin gradients. The ability of auxin to bring about diverse responses appears to result from the existence of several independent mechanisms for auxin perception and from a complex transcriptional network at the lower end of the auxin signaling pathway.

So far, we know little about auxin perception, since, despite the fact that several auxin-binding proteins (most prominent among them ABP1) have been isolated from various plant species, their role as auxin receptors have not been demonstrated unequivocally (for overview, see Timpte, 2001). The elucidation of auxin signal transduction pathway has been more successful for very downstream events. Auxin can induce the expression of a variety of primary response genes including members of *AUX/IAA*, *GH3* and *SAUR* gene families. The TGTCTC auxin response element (AuxRE) that confers auxin inducibility was identified in the promoters of these genes. Transcription factors of the ARF family bind specifically to AuxREs (Hagen & Guilfoyle, 2002). There are at least 22 *ARF* genes in *Arabidopsis*, which contain an N-terminal DNA-binding domain (Guilfoyle & Hagen, 2001). Results obtained in assays with transfected protoplasts demonstrated that some ARFs function as activators, while others are repressors (Tiwari *et al.*, 2003). ARFs contain a C-terminal dimerization domain, which is related in amino acid sequence to motifs III and IV that have been found in short-lived nuclear proteins of the AUX/IAA family (Guilfoyle & Hagen, 2001; Liscum & Reed, 2002). There are 24 genes in *Arabidopsis* that are predicted to encode Aux/IAA proteins, and these contain four conserved motifs (which are referred to as domains I through IV). Domain I is an active repression domain that is transferable and dominant over activation domains (Tiwari *et al.*, 2004). Domain II and domains III and IV play roles in protein stability and dimerization, respectively. Yeast two-hybrid interaction assays suggest that ARF C-terminal dimerization domains and AUX/IAA domains III and IV can homo- and heterodimerize (Kim *et al.*, 1997; Ouellet *et al.*, 2001; Tiwari *et al.*, 2004). Thus, the multitude of mutually competing interactions between transcription factors and their repressors represent a network for activating different sets of genes in response to auxin. But how does an auxin-derived signal enter and activate this transcriptional network? Genetic studies in *Arabidopsis* implied that ubiquitin-mediated protein degradation is required for auxin response. Mutations in AXR1 and TIR1, both components of the ubiquitination pathway, confer reduced auxin response. TIR1 encodes an F-box protein that interacts with the cullin AtCUL1 and a SKP1-like protein (ASK1 or ASK2) to form an SCF ubiquitin protein ligase (E3) (Ruegger *et al.*, 1998). AXR1, on the other hand, encodes a part of the enzyme that regulates activity of SCF ubiquitin ligases by RUB1 conjugation of the AtCUL1. SCF^{TIR1} physically interacts with domain II of AUX/IAA proteins and mutations affecting the SCF^{TIR1} complex increase stability of AUX/IAA proteins (Leyser *et al.*, 1993). Auxin treatment stimulates the interaction between SCF^{TIR1} and AUX/IAA proteins and promotes their degradation. In support of this, mutations within domain II abolish the interaction between AUX/IAAs and SCF^{TIR1} and stabilize AUX/IAA repressors in presence of auxin (Gray *et al.*, 2001). Stabilized AUX/IAA protein variants such as *bodenlos* (*bdl*, stabilized IAA12) or *short hypocotyl 2* (*shy2*, stabilized IAA3) confer dominant effects on various aspects of auxin-dependent development (Tian & Reed, 1999; Hamann *et al.*, 2002). By integrating all these findings, a current model on how auxin regulates gene expression emerged (Fig. 1.7). When auxin concentrations are below a certain threshold, early auxin response genes are actively repressed, because AUX/IAA repressors are dimerized to ARF transcriptional activators, preventing gene transcription

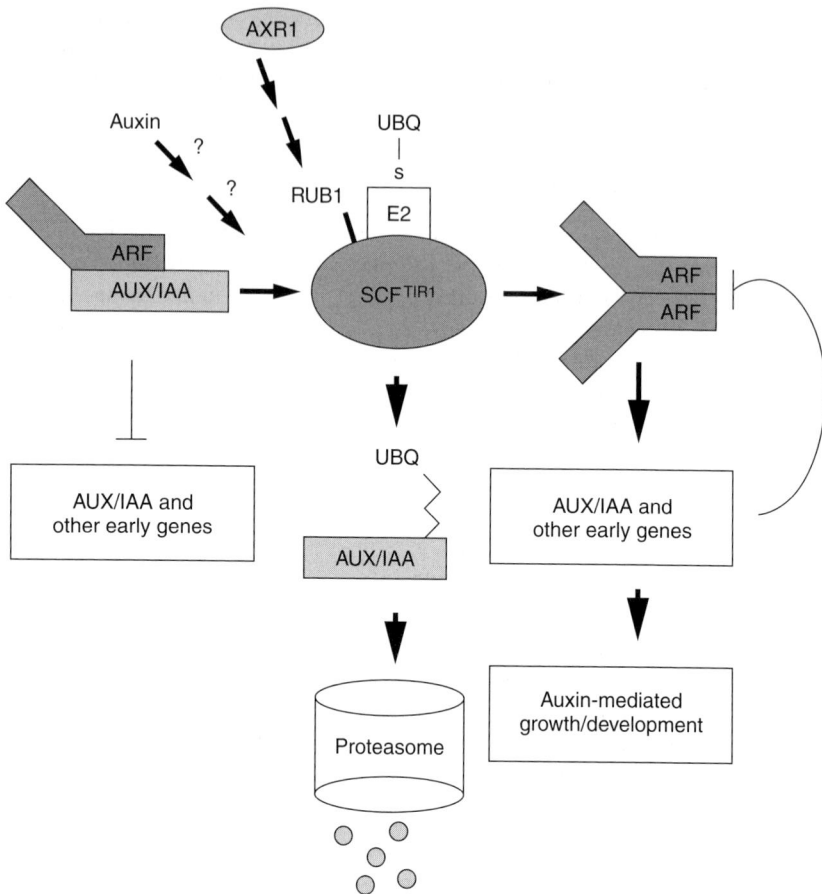

Figure 1.7 Model for auxin response. AUX/IAA proteins repress the auxin-response pathway by inhibiting ARF transcription factors. Auxin promotes the ubiquitination of AUX/IAAs by promoting their interaction with the SCF^{TIR1} ubiquitin ligase. The subsequent degradation of AUX/IAA proteins results in activation of ARFs and derepression of the auxin-response pathway. Because AUX/IAA genes themselves are rapidly induced by auxin, a negative-feedback loop exists with the newly synthesized AUX/IAA proteins restoring repression upon the pathway. Reproduced from Gray *et al.* (2001), with permission.

(Guilfoyle & Hagen, 2001). When auxin levels increase, the AUX/IAA repressors are degraded via SCF^{TIR1}-dependent ubiquitination (reviewed by Kepinski & Leyser, 2002). This releases ARF transcriptional regulators from their AUX/IAA-based inhibition and results in the derepression of the auxin response genes (Gray *et al.*, 2001; Zenser *et al.*, 2001; Tiwari *et al.*, 2004). It is unclear if or how the other part of ARF family (the transcriptional repressors) functions in auxin signaling. They may act on genes that are downregulated in response to auxin or have functions that are not yet identified. Despite the considerable gaps in our knowledge on auxin perception and signaling, the elucidated complex transcriptional network of activating and repressing ARFs modulated by interacting AUX/IAA inhibitors provide sufficient diversity to elicit the various developmental programs, which are known to be mediated by auxin.

1.5.7 Auxin as morphogen

The term *morphogen* was introduced as a theoretical term in mathematical models of self-organizing systems. The concept of positional information proposes concentration gradients of morphogens, which instruct cells within a homogeneous field about their position, thus leading to their commitment to different cell types (Wolpert, 1998). In contrast to plants, in animals, several molecules with morphogen properties have been identified, such as Decapentaplegic (DPP) or Wingless (WG) in *Drosophila* imaginal discs (reviewed in Teleman *et al.*, 2001). With increasing knowledge on these 'classical morphogens', the meaning of the term morphogen has shifted. The general description defines a morphogen as a substance which forms a concentration gradient and which is involved in patterning (Wolpert, 1998). More stringent definitions provide conditions, which 'real' morphogens must meet; i.e., they must show a stable concentration gradient and show direct and concentration-dependent instruction to responding cells (Teleman *et al.*, 2001; Entchev & González-Gaitán, 2002). Auxin comes close to meeting some of these criteria. Auxin is certainly a compound linked to patterning in multiple developmental contexts. In addition, by several independent approaches, differences in auxin concentration between neighboring cells have been detected, and these can be described as concentration gradients (Uggla *et al.*, 1998; Sabatini *et al.*, 1999; Casimiro *et al.*, 2001; Benková *et al.*, 2003; Friml *et al.*, 2003). This gradient is actively maintained and controlled by a PIN-dependent auxin transport. Interestingly, also in the animal field, the classical morphogens such as WG or DDP have been shown to move actively through a field of cells (Entchev & González-Gaitán, 2002). The question remains whether such an auxin gradient is instructive for patterning. There is clear correlation: genetic or chemical interference with the gradient also results in patterning defects – cells are not properly specified, often display mixed fates and are in wrong positions (Sabatini *et al.*, 1999; Friml *et al.*, 2002a, 2003). Nonetheless, a direct and concentration-dependent auxin effect on the responding cells has not been demonstrated based on our current limited knowledge on auxin-gradient perception and downstream signaling. Whatever the outcome will be of rather academic discussions on whether auxin is a morphogen or not, the crucial role of auxin and its graded distribution in many plant patterning processes is now firmly established.

1.6 Conclusions

More than a century of physiological work, as well as recent contributions from molecular genetics, has clearly shown that auxin is the most prominent intercellular signal in plants, and that it regulates a wide variety of developmental processes.

Auxin distributed over long distances largely contributes to the coordination and integration of growth and development at the level of the whole plant (it coordinates, for example, initiation of side roots with growth of aerial tissues). Auxin distributed over short distances by the polar transport system forms local gradients and thus mediates various patterning events. Furthermore, growth responses to external stimuli such as light or gravity also utilize auxin as a signal and appear to use the same mechanism involving auxin gradients. These auxin gradients are established, maintained and modulated by an active transport system, requiring regulators from the PIN family. The activity of these auxin transport regulators can be modulated at the single cell level by changes in their vesicle-trafficking-dependent polar targeting. Thus, the directional throughput of this complex auxin distribution network can be modulated by both endogenous and exogenous signals and provides, by means of mediating auxin fluxes and creating local gradients, a common mechanism for the plasticity and adaptability of plant development. Another level of control in the whole system occurs downstream of the gradients where a large variety of mutually interacting transcription activators and repressors provide the molecular basis for the multitude of developmental programs, which can be initiated and controlled by auxin.

Acknowledgements

The authors acknowledge the support for their work from VolkswagenStiftung (J.F.) and the Foundation for Polish Science (J.W.).

References

Benková, E., Michniewicz, M., Sauer, M., Teichmann, T., Seifertova, D., Jürgens, G. & Friml, J. (2003) Local, efflux-dependent auxin gradients as a common module for plant organ formation. *Cell*, **26**, 591–602.

Bennett, M.J., Marchant, A., Green, H.G., May, S.T., Ward, S.P., Millner, P.A., Walker, A.R., Schulz, B. & Feldmann, K.A. (1996) *Arabidopsis AUX1* gene: a permease-like regulator of root gravitropism. *Science*, **273**, 948–950.

Benning, C. (1986) Evidence supporting a model of voltage-dependent uptake of auxin into *Cucurbita pepo* vesicles. *Planta*, **169**, 228–237.

Bernasconi, P., Patel, B.C., Reagan, J.D. & Subramanian, M.V. (1996) The *N*-1-naphthylphthalamic acid-binding protein is an integral membrane protein. *Plant Physiol.*, **111**, 427–432.

Boysen-Jensen, P. (1913) Uber die Leitung des phototropischen Reizes in der Avenakoleoptile. *Ber. Deut. Bot. Ges.*, **31**, 559–566.

Casimiro, I., Marchant, A., Bhalerao, R.P., Beeckman, T., Dhooge, S., Swarup, R., Graham, N., Inze, D., Sandberg, G., Casero, P.J. & Bennett, M. (2001) Auxin transport promotes *Arabidopsis* lateral root initiation. *Plant Cell*, **13**, 843–852.

Chen, R., Guan, C., Boonsirichai, K. & Masson, P.H. (2002) Complex physiological and molecular processes underlying root gravitropism. *Plant Mol. Biol.*, **49**, 305–317.

Chen, R., Hilson, P., Sedbrook, J., Rosen, E., Caspar, T. & Masson, P.H. (1998) The *Arabidopsis thaliana AGRAVITROPIC 1* gene encodes a component of the polar-auxin-transport efflux carrier. *Proc. Natl. Acad. Sci. U.S.A.*, **95**, 15112–15117.

Cohen, J.D. & Bandurski, R.S. (1982) Chemistry and physiology of the bound auxins. *Annu. Rev. Plant Physiol.*, **33**, 403–430.

Cox, D.N. & Muday, G.K. (1994) NPA binding-activity is peripheral to the plasma membrane and is associated with the cytoskeleton. *Plant Cell*, **6**, 1941–1953.

Darwin, C. & Darwin, F. (1881) The power of movement in plants (German translation: Das Bewegungsvermögen der Planze). In: *Darwins gesammelte Werke*, Vol. 13. Schweizer-bart'sche Verlagsbuchhandlung, Stuttgart.

Davies, P.J. (ed) (1995) *Plant Hormones, Physiology, Biochemistry and Molecular Biology*. Martinus Nijhoff, Dordrecht.

Delbarre, A., Muller, P. & Guern, J. (1998) Short-lived and phosphorylated proteins contribute to carrier-mediated efflux, but not to influx, of auxin in suspension-cultured tobacco cells. *Plant Physiol.*, **116**, 833–844.

Delbarre, A., Muller, P., Imhoff, V. & Guern, J. (1996) Comparison of mechanisms controlling uptake and accumulation of 2,4-dichlorophenoxy acetic acid, naphthalene-1-acetic acid, and indole-3-acetic acid in suspension-cultured tobacco cells. *Planta*, **198**, 532–541.

Donaldson, J.G. & Jackson, C.L. (2000) Regulators and effectors of the ARF GTPases. *Curr. Opin. Cell Biol.*, **12**, 475–482.

Entchev, E.V. & González-Gaitán, M.A. (2002) Morphogen gradient formation and vesicular trafficking. *Traffic*, **3**, 98–109.

Friml, J. (2003) Auxin transport – shaping the plant. *Curr. Opin. Plant Biol.*, **6**, 7–12.

Friml, J. & Palme, K. (2002) Polar auxin transport – old questions and new concepts? *Plant Mol. Biol.*, **49**, 273–284.

Friml, J., Benková, E., Blilou, I., Wisniewska, J., Hamann, T., Ljung, K., Woody, S., Sandberg, G., Scheres, B., Jürgens, G. & Palme, K. (2002a) AtPIN4 mediates sink-driven auxin gradients and root patterning in *Arabidopsis*. *Cell*, **108**, 661–673.

Friml, J., Vieten, A., Sauer, M., Weijers, D., Schwarz, H., Hamann, T., Offringa, R. & Jürgens, G. (2003) Efflux-dependent auxin gradients establish the apical–basal axis of *Arabidopsis*. *Nature*, **13**, 147–513.

Friml, J., Wisniewska, J., Benková, E., Mendgen, K. & Palme, K. (2002b) Lateral relocation of auxin efflux regulator PIN3 mediates tropism in *Arabidopsis*. *Nature*, **415**, 806–809.

Fukaki, H., Tameda, S., Masuda, H. & Tasaka, M. (2002) Lateral root formation is blocked by a gain-of-function mutation in the SOLITARY-ROOT/IAA14 gene of *Arabidopsis*. *Plant J.*, **29**, 153–168.

Fukaki, H., Wysocka-Diller, J., Kato, T., Fujisawa, H., Benfey, P.N. & Tasaka, M. (1998) Genetic evidence that the endodermis is essential for shoot gravitropism in *Arabidopsis thaliana*. *Plant J.*, **14**, 425–430.

Gälweiler, L., Guan, C., Müller, A., Wisman, E., Mendgen, K., Yephremov, A. & Palme, K. (1998) Regulation of polar auxin transport by AtPIN1 in *Arabidopsis* vascular tissue. *Science*, **282**, 2226–2230.

Geldner, N., Anders, N., Wolters, H., Keicher, J., Kornberger, W., Muller, P., Delbarre, A., Ueda, T., Nakano, A. & Jürgens, G. (2003) The *Arabidopsis* GNOM ARF-GEF mediates endosomal recycling, auxin transport, and auxin-dependent plant growth. *Cell*, **112**, 219–230.

Geldner, N., Friml, J., Stierhof, Y.D., Jürgens, G. & Palme, K. (2001) Auxin transport inhibitors block PIN1 cycling and vesicle trafficking. *Nature*, **413**, 425–428.

Gil, P., Dewey, E., Friml, J., Zhao, Y., Snowden, K.C., Putterill, J., Palme, K., Estelle, M. & Chory, J. (2001) BIG: a calossin-like protein required for polar auxin transport in *Arabidopsis*. *Genes Dev.*, **15**, 1985–1997.

Goldsmith, M.H.M. (1977) The polar transport of auxin. *Annu. Rev. Plant Physiol.*, **28**, 439–478.

Gray, W.M., Kepinski, S., Rouse, D., Leyser, O. & Estelle, M., (2001) Auxin regulates SCF TIR1-dependent degradation of Aux/IAA proteins. *Nature*, **414**, 271–276.

Grebe, M., Xu, J., Möbius, W., Ueda, T., Nakano, A., Geuze H.J., Rook, M.B. & Scheres, B. (2003) *Arabidopsis* sterol endocytosis involves actin-mediated trafficking via ARA6-positive early endosomes. *Curr. Biol.*, **13**, 1378–1387.

Guilfoyle, T.J. & Hagen, G. (2001) Auxin response factors. *J. Plant Growth Regul.*, **10**, 281–291.

Hadfi, K., Speth, V. & Neuhaus, G. (1998) Auxin-induced developmental patterns in *Brassica juncea* embryos. *Development*, **125**, 879–887.

Hagen, G. & Guilfoyle, T. (2002) Auxin-responsive gene expression: genes, promoters and regulatory factors. *Plant Mol. Biol.*, **49**, 373–385.

Hamann, T., Benková, E., Baurle, I., Kientz, M. & Jürgens, G. (2002) The *Arabidopsis* BODENLOS gene encodes an auxin response protein inhibiting MONOPTEROS-mediated embryo patterning. *Genes Dev.*, **16**, 610–1615.

Hardtke, C.S. & Berleth, T. (1998) The *Arabidopsis* gene *MONOPTEROS* encodes a transription factor mediating embryo axis formation and vascular development. *EMBO J.*, **17**, 1405–1411.

Hertel, R. (1983) The mechanism of auxin transport as a model for auxin action. *Z. Pflanzenphysiol.*, **112**, 53–67.

Imhoff, V., Muller, P., Guern, J. & Delbarre, A. (2000) Inhibitors of the carrier-mediated influx of auxin in suspension-cultured tobacco cells. *Planta*, **210**, 580–588.

Jacobs, M. & Rubery, P.H. (1988) Naturally occurring auxin transport regulators. *Science*, **241**, 346–349.

Jensen, P.J., Hangarter, R.P. & Estelle, M. (1998) Auxin transport is required for hypocotyl elongation in light-grown *Arabidopsis. Plant. Physiol.*, **16**, 455–462.

Jürgens, G. (2001) Apical–basal pattern formation in *Arabidopsis* embryogenesis. *EMBO J.*, **20**, 3609–3616.

Katekar, G.F. & Geissler, A.E. (1977) Auxin transport inhibitors. *Plant Physiol.*, **60**, 826–829.

Kepinski, S. & Leyser, O. (2002) Ubiquitination and auxin signaling: a degrading story. *Plant Cell*, **14**, 81–95.

Kerk, N. & Feldman, L. (1994) The quiescent centre in roots of maize – initiation, maintenance, and role in organization of the root apical meristem. *Protoplasma*, **183**, 100–106.

Kim, J., Harter, K. & Theologis, A. (1997). Protein–protein interactions among the Aux/IAA proteins. *Proc. Natl. Acad. Sci. U.S.A.*, **94**, 11786–11791.

Lalonde, S., Boles, E., Hellmann, H., Barker, L., Patrick, J.W., Frommer, W.B. & Ward, J.M. (1999) The dual function of sugar carriers: transport and sugar sensing. *Plant Cell*, **11**, 707–726.

Laskowski, M., Williams, M., Nusbaum, H. & Sussex, I. (1995) Formation of lateral root meristems is a two-stage process. *Development*, **121**, 3303–3310.

Lehman, A., Black, R. & Ecker, J.R. (1996) *HOOKLESS1*, an ethylene response gene, is required for differential cell elongation in the *Arabidopsis* hypocotyls. *Cell*, **85**, 183–194.

Leyser, H.M., Lincoln, C.A., Timpte, C., Lammer, D., Turner, J. & Estelle, M. (1993) *Arabidopsis* auxin-resistance gene AXR1 encodes a protein related to ubiquitin-activating enzyme E1. *Nature*, **8**, 161–164.

Li, Y., Hagen, G. & Guilfoyle, T.J. (1991) An auxin-responsive promoter is differentially induced by auxin gradients during tropisms. *Plant Cell*, **3**, 1167–1176.

Liscum, E. & Reed, J.W. (2002) Genetics of Aux/IAA and ARF action in plant growth and development. *Plant Mol. Biol.*, **49**, 387–400.

Ljung, K., Bhalerao, R.P. & Sandberg, G. (2001) Sites and homeostatic control of auxin biosynthesis in *Arabidopsis* during vegetative growth. *Plant J.*, **28**, 465–474.

Lomax, T.L., Muday, G.K. & Rubery, P.H. (1995) Auxin transport. In: *Plant Hormones, Physiology, Biochemistry and Molecular Biology* (ed. P.J. Davies), pp. 509–530. Martinus Nijhoff, Dordrecht.

Luschnig, C., Gaxiola, R., Grisafi, P. & Fink, G. (1998) EIR1, a root specific protein involved in auxin transport, is required for gravitropism in *Arabidopsis thaliana. Genes Dev.*, **12**, 2175–2187.

Marchant, A., Kargul, J., May, S.T., Muller, P., Delbarre, A., Perrot-Rechenmann, C. & Bennett, M.J. (1999) AUX1 regulates root gravitropism in *Arabidopsis* by facilitating auxin uptake within root apical tissues. *EMBO J.*, **18**, 2066–2073.

Mattson, J., Sung, Z.R. & Berleth, T. (1999) Responses of plants vascular system to auxin transport inhibition. *Development*, **126**, 2979–2991.

Mayer, U., Ruiz, R.A.T., Berleth, T., Misera, S. & Jürgens. G. (1991) Mutations affecting body organization in the *Arabidopsis* embryo. *Nature*, **353**, 402–407.

Moctezuma, E. (1999) Changes in auxin patterns in developing gynophores of the peanut plant (*Arachis hypogaea* L.). *Ann. Bot.*, **83**, 235–242.

Morris, D.A. (2000) Transmembrane auxin carrier systems – dynamic regulators of polar auxin transport. *Plant Growth Regul.*, **32**, 161–172.

Morris, D.A. & Robinson, J.S. (1998) Targeting of auxin carriers to the plasma membrane: differential effects of brefeldin A on the traffic of auxin uptake and efflux carriers. *Planta*, **205**, 606–612.

Morris, D.A. & Thomas, A.G. (1978) A microautoradiographic study of auxin transport in the stem of intact pea seedlings (*Pisum sativum* L.). *J. Exp. Bot.*, **29**, 147–157.

Morris, D.A., Rubery, P.H., Jarman, J. & Sabater, M. (1991) Effects of inhibitors of protein synthesis on transmembrane auxin transport in *Cucurbita pepo* L. hypocotyl segments. *J. Exp. Bot.*, **42**, 773–783.

Müller, A., Guan, C., Gälweiler, L., Tänzler, P., Huijser, P., Marchant, A., Parry, G., Bennett, M., Wisman, E. & Palme, K. (1998) AtPIN2 defines a locus of *Arabidopsis* for root gravitropism control. *EMBO J.*, **17**, 6903–6911.

Murphy, A.S., Hoogner, K.R., Peer, W.A. & Taiz, L. (2002) Identification, purification and molecular cloning of *N*-1-naphthylphthalamic acid-binding plasma membrane-associated aminopeptidases from *Arabidopsis. Plant Physiol.*, **128**, 935–950.

Noh, B., Bandyopadhyay, A., Peer, W.A., Spalding, E.P. & Murphy, A.S. (2003) Enhanced gravi- and phototropism in plant *mdr* mutants mislocalizing the auxin efflux protein PIN1. *Nature*, **423**, 999–1002.

Noh, B., Murphy, A.S. & Spalding, E.P. (2001) *Multidrug resistance*-like genes of *Arabidopsis* required for auxin transport and auxin-mediated development. *Plant Cell*, **13**, 2441–2454.

Okada, K., Ueda, J., Komaki, M.K., Bell, C.J. & Shimura, Y. (1991) Requirement of the auxin polar transport system in the early stages of *Arabidopsis* floral bud formation. *Plant Cell*, **3**, 677–684.

Ouellet, F., Overvoorde, P.J. & Theologis, A. (2001) IAA17/AXR3: biochemical insight into an auxin mutant phenotype. *Plant Cell*, **13**, 829–842.

Parry, G., Delbarre, A., Marchant, A., Swarup, R., Napier, R., Perrot-Rechenmann, C. & Bennett, M.J. (2001) Novel auxin transport inhibitors phenocopy the auxin influx carrier mutation *aux1*. *Plant J.*, **25**, 399–406.

Petrášek, J., Černá, A., Schwarzerová, K., Elčkner, M., Morris, D.A. & Zažímalová, E. (2003) Do phytotropins inhibit auxin efflux by impairing vesicle traffic? *Plant Physiol.*, **131**, 254–263.

Peyroche, A., Antonny, B., Robineau, S., Acker, J., Cherfils, J. & Jackson, C.L. (1999) Brefeldin A acts to stabilize an abortive ARFGDP-Sec7 domain protein complex: involvement of specific residues of the Sec7 domain. *Mol. Cell*, **3**, 275–285.

Przemeck, G.K.H., Mattsson, J., Hardtke, C.S., Sung, Z.R. & Berleth, T. (1996) Studies on the role of the *Arabidopsis* gene MONOPTEROS in vascular development and plant cell axialization. *Planta*, **200**, 229–237.

Rashotte, A.M., Brady, S.R., Reed, R.C., Ante, S.J. & Muday, G.K. (2000) Basipetal auxin transport is required for gravitropism in roots of *Arabidopsis. Plant Physiol.*, **122**, 481–490.

Raven, J.A. (1975) Transport of indoleacetic acid in plant cells in relation to pH and electrical potential gradients, and its significance for polar IAA transport. *New Phytol.*, **74**, 163–172.

Reinhardt, D., Mandel, T. & Kuhlemeier, C. (2000) Auxin regulates the initiation and radial position of plant lateral organs. *Plant Cell*, **12**, 507–518.

Reinhardt, D., Pesce1, E.R., Stieger, P., Mandel, T., Baltensperger, K., Bennett, M., Traas, J., Friml, J. & Kuhlemeier, C. (2003) Regulation of phyllotaxis by polar auxin transport. *Nature*, **426**, 255–260.

Robinson, J.S., Albert, A.C. & Morris, D.A. (1999) Differential effects of brefeldin A and cycloheximide on the activity of auxin efflux carriers in *Cucurbita pepo* L. *J. Plant Physiol.*, **155**, 678–684.

Rubery, P.H. & Sheldrake, A.R. (1974) Carrier-mediated auxin transport. *Planta*, **188**, 101–121.

Ruegger, M., Dewey, E., Gray, W.M., Hobbie, L., Turner, J. & Estelle, M.. (1998) The TIR1 protein of *Arabidopsis* functions in auxin response and is related to human SKP2 and yeast grr1p. *Genes Dev.*, **12**, 198–207.

Ruegger, M., Dewey, E., Hobbie, L., Brown, D., Bernasconi, P., Turner, J., Muday, G. & Estelle, M. (1997) Reduced naphthylphthalamic acid binding in the *tir3* mutant of *Arabidopsis* is associated with a reduction in polar auxin transport and diverse morphological defects. *Plant Cell*, **9**, 745–757.

Sabatini, S., Beis, D., Wolkenfelt, H., Murfett, J., Guilfoyle, T., Malamy, J., Benfey, P., Leyser, O., Bechtold, N., Weisbeek, P. & Scheres, B. (1999) An auxin-dependent distal organizer of pattern and polarity in the *Arabidopsis* root. *Cell*, **99**, 463–472.

Steinmann, T., Geldner, N., Grebe, M., Mangold, S., Jackson, C.L., Paris, S., Gälweiler, L., Palme, K. & Jürgens, G. (1999) Coordinated polar localization of auxin efflux carrier PIN1 by GNOM ARF GEF. *Science*, **286**, 316–318.

Swarup, R., Friml, J., Marchant, A., Ljung, K., Sandberg, G., Palme, K. & Bennett, M. (2001) Localization of the auxin permease AUX1 suggests two functionally distinct hormone transport pathways operate in the *Arabidopsis* root apex. *Genes Dev.*, **15**, 2648–2653.

Tam, Y.Y., Slovin, J.P. & Cohen, J.D. (1995) Selection and characterization of α-methyltryptophan resistant lines of *Lemma gibba* showing a rapid rate of indole-3-acetic acid turnover. *Plant Physiol.*, **107**, 77–85.

Teleman, A.A., Strigini, M. & Cohen, S.M. (2001) Shaping morphogen gradients. *Cell*, **105**, 559–562.

Tian, Q. & Reed, J.W. (1999) Control of auxin-regulated root development by the *Arabidopsis thaliana* SHY2/IAA3 gene. *Development*, **126**, 711–721.

Timpte, C. (2001) Auxin binding protein: curiouser and curiouser. *Trends Plant Sci.*, **6**, 586–590.

Tiwari, S.B., Hagen, G. & Guilfoyle, T. (2003) The roles of auxin response factor domains in auxin-responsive transcription. *Plant Cell*, **15**, 533–543.

Tiwari, S.B., Hagen, G. & Guilfoyle, T.J. (2004) Aux/IAA proteins contain a potent transcriptional repression domain. *Plant Cell*, **16**, 533–543.

Uggla, C., Mellerowicz, E.J. & Sundberg, B. (1998) Indole-3-acetic acid controls cambial growth in scots pine by positional signalling. *Plant Physiol.*, **117**, 113–121.

Ulmasov, T., Liu, Z.B., Hagen, G. & Guilfoyle, T.J. (1995) Composite structure of auxin response elements. *Plant Cell*, **7**, 1611–1623.

Utsuno, K., Shikanai, T., Yamada, Y. & Hashimoto, T. (1998) *AGR*, an *Agravitropic* locus of *Arabidopsis thaliana*, encodes a novel membrane-protein family member. *Plant Cell Physiol.*, **39**, 1111–1118.

Went, F.W. (1974) Reflections and speculations. *Annu. Rev. Plant Physiol.*, **25**, 1–26.

Wolpert, L. (1998) *Principles of Development*. Elsevier Science, Oxford.

Yamamoto, M. & Yamamoto, K.T. (1998) Differential effects of 1-naphthaleneacetic acid, indole-3-acetic acid and 2,4-dichlorophenoxyacetic acid on the gravitropic response of roots in an auxin-resistant mutant of *Arabidopsis*, *aux1*. *Plant Cell Physiol.*, **39**, 660–664.

Zenser, N., Ellsmore, A., Leasure, C. & Callis, J. (2001) Auxin modulates the degradation rate of Aux/IAA proteins. *Proc. Natl. Acad. Sci. U.S.A.*, **98**, 11795–11800.

2 Peptides as signals

Yiji Xia

2.1 Introduction

Peptides are a major class of signal molecules in animals that regulate a wide variety of physiological and developmental processes. The first peptide hormone, insulin, was discovered in 1922 (Banting & Best, 1922). Like other hormones, insulin is produced in specific organs (endocrine glands), is released into the bloodstream, and acts on distant target cells to regulate glucose metabolism. The subsequent discovery of a large number of other peptide hormones in animals was aided by the relative ease in isolating the specific hormone-producing organs and tissues such as the pituitary gland, the pancreas, and the thyroid, as well as the availability of biological assays for their activities. Although most peptide hormones are small, some peptide hormones are relatively large, with a size of up to 300 amino acids (Habener, 1987).

Peptide hormones are generally synthesized as inactive precursors that undergo proteolytic processing in the endoplasmic reticulum (ER) and Golgi apparatus to form mature hormones. Mature hormones are then released by exocytosis and enter the bloodstream in secretory granules. Peptide hormones are polar molecules that cannot diffuse easily through the plasma membrane and, therefore, usually do not enter target cells. Peptide signals are perceived on the cell surface by cognate receptors and the signals are then relayed through second messengers to convert the extracellular signals into intracellular physiological responses. The largest family of cell surface receptors in animals is the G-protein-coupled receptor (GPCR) family that are characterized by seven-transmembrane-segment (STMS) topography (Neves et al., 2002). Binding of a hormone to GPCR activates a heterotrimeric guanyl nucleotide-binding protein (G-protein). G-protein then activates a variety of effectors, including adenylate cyclase that catalyzes the formation of cyclic AMP (cAMP) and phospholipases that catalyze the phosphoinositide signaling cascade.

Many peptide hormones exert their effects by binding to receptors that have tyrosine kinase activity in their intracellular domains. The ligand binding results in dimerization of such a receptor tyrosine kinase (RTK) and leads to autophosphorylation and then phosphorylation of cytoplasmic target proteins (Schlessinger, 2000). The RTK signaling pathways are often mediated by a monomeric G-protein named Ras. Binding of the cognate ligand to RTK activates Ras, which then relays the signal through a kinase cascade to activate mitogen-activated protein kinases (MAPKs). Activated MAPKs translocate from the cytosol to the nucleus to activate transcription factors.

The advantages of using peptides as signals include their enormous structural diversity and the existence of various cellular mechanisms for controlling their activities (Alberts *et al.*, 1994; Bisseling, 1999). However, it was once thought that plants have not evolved signaling systems that use a peptide as a signal. Instead, plant hormone signaling was believed to be mediated by small molecules including the five classical phytohormones: auxin, cytokinin, gibberellin, ethylene, and abscisic acid. The question of how a small number of chemically simple molecules could account for the observed diversity of cellular responses remains a paradox (Miklashevichs *et al.*, 1996; Lindsey *et al.*, 2002). Ryan and colleagues at Washington State University set a milestone by discovering the first plant peptide hormone, called systemin, that induces the systemic wounding response in tomato (Pearce *et al.*, 1991). Since the discovery of systemin, several additional peptide signals have been found to function as signals in cell-to-cell communication in plant developmental and physiological processes. Growing evidence indicates that many more peptide signals are likely involved in a diverse range of biological pathways.

2.2 Peptide signals in plants and their biological functions

2.2.1 Systemins mediate systemic and local wound responses

Plants, being sessile organisms, cannot move to escape attack by pathogens and predators. Thus, plants have evolved complex defense mechanisms to protect themselves. In addition to preformed mechanisms such as physical barriers, plant cells develop different inducible mechanisms in response to an attack. Wounding of plant tissues by herbivory or mechanical damage activates a battery of defense responses including production of protease inhibitors (pin) I and II which block protein degradation in insects' digestive systems and affect larval growth (Johnson *et al.*, 1989). The defense response is activated not only at the wound site but also in distal tissues, suggesting that the local wound response leads to generation of a mobile signal that activates the systemic response.

Application of an extract from wounded tomato leaves to an excised tomato plant through a cut stem induces pin production in the leaves, indicating the existence of a mobile wound signal in the extract. This biological assay was used to purify the systemic signal, systemin (Pearce *et al.*, 1991). Systemin was found to be a proline-rich 18-amino acid peptide derived from its 200-amino acid precursor through limited proteolysis (Fig. 2.1). When isotope-labeled systemin was applied to leaves through fresh wounds, it was loaded into the phloem and transported into distal leaves of the plant within 1–2 h. Systemin is biologically active at a femtomole concentration. This strong inducing activity, together with its high mobility, makes it a powerful systemic wound signal.

Transgenic tomato in which the systemin gene is suppressed through the antisense-mediated gene suppression technology is defective in the systemic wound response, which demonstrates its key role in distal signaling (McGurl *et al.*, 1992).

Systemin   AVQSKPPSKRDPPKMQTD

TobHypSys I   RGANLPOOSOASSOOSKE
TobHypSys II KNKPLSOOSOKPADGQRP

TomHypSys I   RTOYKTOOOOTSSSOTHQ
TomHypSys II   GRHDYVASOOOOKPQDEQRQ
TomHypSys III GRHDSVLPOOSOKTD

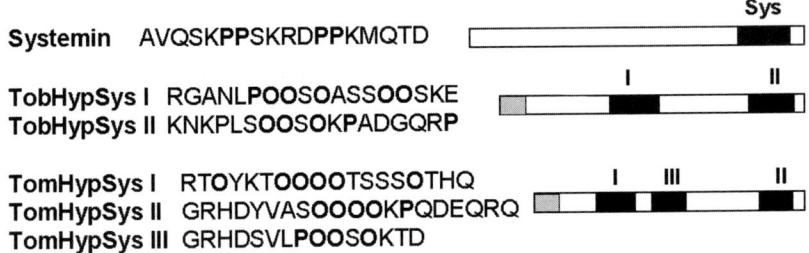

Figure 2.1 The sequences of systemins and the structures of their precursors. Prolines (P) and hydroxyprolines (O) are in boldface. In the precursor diagram, the systemin peptides are shown as black boxes. The gray boxes represent the leader sequences. After Ryan and Pearce (2003).

As a result, the transgenic plant is compromised in defense against *Manduca sexta* larvae.

Systemin homologues have been identified in many other species of the Solaneae subtribe in the Solanaceae family but not in other subtribes of this family, such as tobacco. In addition, the tomato systemin does not induce the wound response in tobacco, which suggests that it is highly diverged from the tobacco wound signal and not perceived by tobacco cells. Tobacco appears to be missing a strong leaf-to-leaf long-distance wound signaling system, although strong leaf-to-root and local wound signaling systems are present (Pearce *et al.*, 1993; Zhang & Baldwin, 1997; Constabel *et al.*, 1998).

Exogenous application of systemin to tomato suspension cultures causes a rapid increase in the pH of the medium. This alkalinization is a convenient assay for the biological activity of systemin. Although the tomato systemin does not generate an alkalinization response when applied to tobacco suspension cells, extracts from tobacco leaves can do so, suggesting that a systemin-like activity exists in tobacco. Facilitated by the alkalinization assay, the Ryan group isolated two peptides from tobacco, Tobacco Systemin I and Tobacco Systemin II (which were later renamed TobHypSys I and II), that function to induce the localized wound response (Pearce *et al.*, 2001a). Both tobacco systemins are hydroxyproline-rich 18-amino acid peptides generated from proteolytic processing of a single 165-amino acid precursor (Fig. 2.1). TobHypSys I and II do not share significant sequence similarity to each other or to the tomato systemin. In addition, unlike the tomato systemins, the tobacco systemins are glycosylated. However, they are all proline-rich (or hydroxyproline-rich in the case of tobacco systemins) and have a –PPS– (or –OOS–) motif. Prolines and hydroxyprolines may allow some limited conformations in their secondary structures, which could play important roles in interaction with their receptors (Pearce & Ryan, 2003; Narvaez-Vasquez & Ryan, 2004).

Although the suppression of the tomato systemin gene blocks systemic wounding signaling, it does not have a significant affect on local wound signaling (McGurl *et al.*, 1992), which suggests that local wound signaling in tomato is independent of systemin. The search for additional wound signaling molecules in tomato has led to the identification of three new peptide signals, tomato hydroxyproline-rich systemin (TomHypSys) I, II, and III (Pearce & Ryan, 2003). These three peptides are 20, 18, and 15 amino acids in length respectively (Fig. 2.1). Like the TobHypSys systemins, they are hydroxyproline-rich and glycosylated. All three TomHypSys systemins are derived from a single wound-inducible 146-residues precursor that shares sequence similarity to the precursor of the tobacco systemins, which suggests they evolved from a common ancestor gene.

The systemin precursor, prosystemin, is synthesized in phloem tissues and localized in the cytosol (Narvaez-Vasquez & Ryan, 2004). In contrast, the precursors of the TomHypSys and TobHypSys signals are likely synthesized on the rough ER, processed and modified in ER and the Golgi apparatus, and then enter the secretory pathway. Therefore, it is plausible that the TomHypSys and TobHypSys systemins move from cell to cell through intercellular fluid and act as short-range signals to induce the local wound response, whereas the localization of systemin provides the advantage of rapid transportation through phloem to activate the systemic wound response.

Jasmonate is another mobile signal molecule that interplays with systemin to amplify wound signaling (Ryan & Pearce, 2003). Wounding causes the generation and release of systemin from damaged cells. It moves through the apoplast and is perceived by neighboring cells, resulting in the induction of jasmonate synthesis. As jasmonate moves through the plant, it induces the production of more prosystemin. This systemin–jasmonate cycle amplifies the wound signaling process to achieve a strong systemic wound response and effectively defend the plant from its predator. However, the question as to how prosystemin is released from the cytosol of unwounded distal cells to the apoplast remains unanswered. Since polypeptides usually do not diffuse across the plasma membrane, a specific transporter would be required to transport prosystemin to the intercellular space where it is activated by proteolytic processing. In species such as tobacco that lacks systemin, jasmonate is mainly responsible for long-range wound signaling. The peptide signals such as TobHypSys I and II help to amplify wound signaling and activate the local wound response.

The receptor of systemin was recently identified by its binding to systemin. The receptor is called SR160, a 160-kDa transmembrane protein with an extracellular leucine-rich repeat (LRR) domain and an intracellular serine/threonine kinase domain (Scheer & Ryan, 2002). It turns out that SR160 shares a very high sequence similarity to BRI1 in *Arabidopsis*. BRI1 was previously identified as the cell surface receptor of brassinolide, a steroid hormone that regulates plant growth and development (Li & Chory, 1997). Further evidence that SR160 is indeed the tomato brassinolide receptor came from another independent study aimed to isolate the tomato gene *CU-3* that functions like *BRI1* (Montoya et al., 2002). The study found that the cloned *CU-3* gene, mutation of which results in insensitivity to brassinolide, encodes SR160. It was further revealed that the *cu-3* mutant is not only defective in brassinolide signaling but also compromised in systemin signaling (Scheer et al., 2003). The above studies demonstrate that BRI1/SR160 has a dual function.

It is possible that SR160/BRI1 initially functions in the defense response but was co-opted as a component in brassinolide signaling during the course of evolution, although the reverse may be the case (Ryan & Pearce, 2003). Nevertheless, it raises an interesting question as to how perception of systemin and brassinolide by the same receptor activates two distinctive intracellular biochemical pathways.

It is yet to be determined if SB160 is also the receptor for the other systemins. However, even if different receptors are involved, it is clear that the signaling pathways mediated by different systemins converge at an early step and, therefore, they activate a similar physiological response. As in many signaling pathways mediated by peptide hormones in animals, a kinase cascade may be involved in activation of the plant wound response. Both the tomato systemin and the TobHypSys systemins activate a 48-kDa MAPK (Schaller & Ryan, 1996; Stratmann et al., 2000; Ryan & Pearce, 2003). Other intracellular events involved in the systemin signaling pathway include the release of Ca^{2+} from vacuoles to the cytosol as well as activation of phospholipase A2. These signaling events lead to generation of jasmonate from linolenic acid, which, together with ethylene, activates defense genes.

2.2.2 RALF regulates plant growth and development

During the purification of the tobacco systemins, another peptide from tobacco leaves was found to cause rapid alkalinization of the medium of the tobacco suspension cells. The peptide was named rapid alkalinization factor (RALF) (Pearce et al., 2001b). RALF is a 49-amino acid peptide derived from a 115-amino acid preproprotein. RALF induces a stronger and more rapid alkalinization response than the two tobacco systemins. However, unlike the systemins, it does not induce the tobacco PIN genes and, therefore, is not a signal of the wound response.

To determine its possible biological function, RALF was applied to excised plants and germinating seeds. It was found that RALF caused an arrest of seedlings' root growth by affecting the elongation zone and the root meristem (Pearce et al., 2001b). It also blocked root hair formation. RALF appears to be expressed in a variety of tissues and its homologues are present in plants species across the plant kingdom. The mechanism by which RALF regulates root growth remains elusive. Neither is it clear whether RALF affects other growth and developmental process.

2.2.3 ENOD40 regulates nodulation and cell proliferation

Some nitrogen fixation bacteria (rhizobia) in soil establish an endosymbiotic association with legumes. Biological nitrogen fixation is carried out by rhizobia in a specific type of organs called root nodules, which form following rhizobia infection. The nodule formation involves communication between the bacteria and the host as well as between different types of root cells. Flavonoids secreted by legume roots activate the rhizobial genes involved in production of Nod factors. Nod factors share the similar basic structure comprising lipo-chitin oligomers with a four or five b-1,4-linked *N*-acetyl glucosamine backbone in which the *N*-acetyl group of the

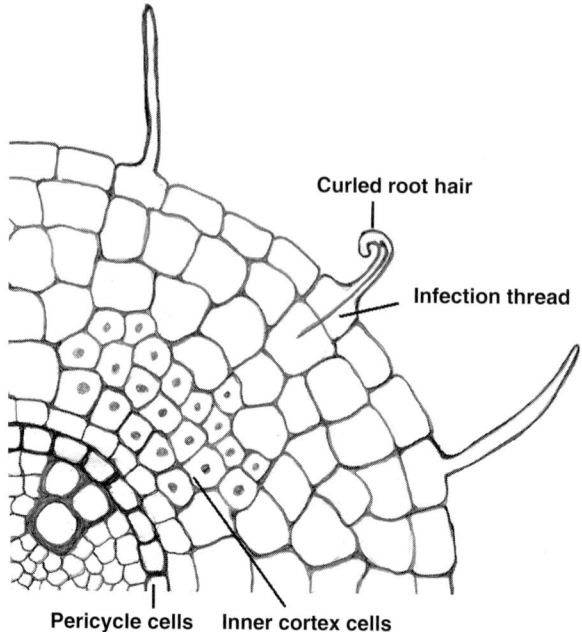

Figure 2.2 Schematic drawing of a part of a root cross section undergoing nodulation. Bacterial infection induces root hair curling and formation of an infection thread that carries the bacteria into inner cell layers. ENOD40 expression is first detected in the pericycle cells, which undergo limited cell division following the infection. The inner cortex cells dedifferentiate to enter the cell cycle, leading to formation of a nodule primordium. After Geurts and Bisseling (2002).

terminal sugar is replaced by an acyl chain (Geurts & Bisseling, 2002). Nod factors made by different rhizobial species are modified in a variety of ways and are a major determinant of host specificity.

Nod factors are the signaling molecules secreted by rhizobia to the root surface where they are recognized by specific receptors of host cells. They induce many host genes termed nodulin or nodule-specific genes and initiate a series of coordinated events, leading to reprogramming of differentiated root cells and formation of a nodule primordium (Fig. 2.2) (Geurts & Bisseling, 2002).

Rhizobia enter plants through root hairs. Rhizobia infection causes root hair to deform by reorienting tip growth (root hair curling). An infection thread containing the bacteria is then formed within the curled root hair. The infection thread grows toward the base of the root hair cell and eventually to the nodule primordium. The response of the epidermal cells to rhizobia is induced within minutes after Nod factor application. The response of inner cell layers to Nod factors can be detected within an hour. Following the rhizobial infection, pericycle cells undergo a limited number of cell divisions. The inner cortex cells then dedifferentiate to enter the cell division cycle, leading to formation of the nodule primordium. The nodule primordium differentiates to form a nodule after bacteria are released from the infection thread.

ENOD40 is one of the early nodulin genes (*ENOD*). The first *ENOD40* gene, *GmENOD40*, was identified from soybean (Yang et al., 1993). *GmENOD40* is strongly induced during nodule development prior to the start of N_2 fixation. This suggests that the gene is involved in the nodulation process but not N_2-fixation process. During early stages of nodule development, the *GmENOD40* gene is expressed first in the pericycle of root vascular bundle and then in dividing root cortical cells and the nodule primordium.

ENOD40 produces a 0.7-kb transcript that was initially thought to function as a nontranslatable RNA because it lacked a significant open reading frame (ORF) (Crespi et al., 1994). Further insights into the possible function of ENOD40 come from the study on the *ENOD40* gene from *Medicago truncatula* (*MtENOD40*) (Sousa et al., 2001). When *MtENOD40* was delivered into epidermal cells and the outermost cortical layer by ballistic microtargeting, it induced division of inner cortical cells to initiate nodule formation. This phenomenon was used as an assay for the MtENOD40 activity. The alfalfa *ENOD40* contains two small ORFs (sORF I and sORF II) that encode 13 and 27 amino acids respectively. The sORF I region is highly conserved among the *ENOD40* genes from different species. It was found that sORF I or sORF II alone is sufficient for ENOD40 activity. In addition, their activity requires translation of the sORFs. The other regions of the transcript may play a role in controlling stability or regulating the translation process. Therefore, the study indicates that the short peptides encoded by the sORFs are responsible for ENOD40 activity.

A study using *in vitro* translation of soybean *ENOD40* in wheat germ extracts revealed that two peptides (Peptides A and B) of 12 and 24 residues, respectively, were translated from the single mRNA (Rohrig et al., 2002). Both peptides are translated from the region conserved among the *ENOD40* genes. The studies on MtENOD40 and Soybean ENOD40 demonstrate that multiple peptides are produced from the polycistronic transcripts by translation of multiple small ORFs. Interestingly, both Peptides A and B bind to nodulin 100, which is a subunit of sucrose synthase (Rohrig et al., 2002). The discovery suggests that ENOD40 peptides may control photosynthate use, but it does no provide an obvious answer as to how the regulation of sucrose synthase activity by ENOD40 leads to cortical cell dedifferentiation and proliferation.

The *ENOD40* gene is expressed not only during nodulation but also in phloem cells of stems. In addition, *ENOD40* genes have been isolated from non-legume dicot species and monocot species such as tomato, rice, and maize that do not form nodules (van de Sande et al., 1996; Kouchi et al., 1999). Therefore, ENOD40 may have a general role as a regulator of cell proliferation in plants.

The peptides derived from *ENOD40* are apparently localized in the cytosol. However, expression of ENOD40 gene in outer cell layers of alfalfa roots induces cell division of inner layers, suggesting that the ENOD40 peptide signals produced in the epidermal cells may move to the inner cell layers to induce cell proliferation (Sousa et al., 2001). Such a short-range movement can be achieved by diffusing through plasmodesmata. Alternatively, the ENOD40 action may generate a secondary signal that transmits the information to other cells to induce cell division. More work is needed to understand the mode of action underlying the ENOD40-mediated nodulation and cell proliferation processes.

2.2.4 PSK (phytosulfokine) is a mitogenic factor

Synthetic hormones such as auxin and cytokinin are usually added to cultured cells to stimulate cell proliferation. In addition, dispersed culture cells must reach a required initial cell density in order to proliferate. Low-density suspension cell cultures display strikingly low mitotic activity that could not be improved by supplementation with known plant hormones. However, adding conditioned medium derived from rapidly growing cell culture to a low-density culture can stimulate proliferation, suggesting that a mitogenic factor is secreted into the medium.

A suspension culture system of mesophyll cells prepared from *Asparagus officinalis* was used to purify the mitogenic factor (Matsubayashi & Sakagami, 1996). Two mitogenic factors were identified to be a sulfated pentapeptide (H-Tyr(SO3H)-Ile-Tyr(SO3H)-Thr-Gln-OH) and a sulfated tetrapeptide (H-Tyr(SO3H)-Ile-Tyr(SO3H)-Thr-OH). The two peptides are derived from a common precursor. The peptides were named phytosulfokine-a and -b (PSK-a and -b). Truncated PSK which is missing the two C-terminal residues still retained up to 20% activity, whereas truncation of any of the three N-terminal residues abolished its activity, indicating that the N-terminal tripeptide fragment is the active core (Matsubayashi et al., 1996). In addition, the sulfate group was found to be essential for its activity.

Incubation of low-density cells with either of the PSKs at nanomolar concentrations in combination with other hormones stimulates cell proliferation. The failure in proliferation of low-density cell culture is likely caused by the dilution of PSK below its active concentration in initial cultures. It has been proposed that PSKs are synthesized in culture cells in response to auxin and cytokinin and secreted into medium to regulate cell dedifferentiation and activate the cell cycle (Matsubayashi et al., 1999).

PSK-a was also identified in the conditioned medium of rice culture cells (Matsubayashi et al., 1997). The cloned rice PSK gene encodes a 89-amino acid product, preprophytosulfokine, which contains a 22-amino acid cleavable leader sequence (Yang et al., 1999). The PSK sequence is located near the C-terminus of the precursor. The PSK gene is expressed at a relatively high level in rice culture cells. It is also expressed at a low level in rice seedlings, indicating that the hormone play an important role in cell proliferation in culture *and* intact plants. Rice culture cells overexpressing the rice PSK gene were found to divide approximately twofold faster than wild-type cells. In contrast, antisense-mediated suppression of the rice gene caused a decrease in cell division of the rice cultures.

A homology search for *Arabidopsis* genes encoding the PSK sequence led to the identification of at least four putative *AtPSKs*, named *AtPSK1*, *AtPSK2*, *AtPSK3*, and *AtPSK5*, whose designated numbers reflect their chromosomal location (Yang *et al.*, 2001). These four genes encode 67–87-amino acid polypeptides. The deduced *Arabidopsis* PSK precursors and the rice PSK precursor share little overall sequence similarity except that they are identical in an 8-amino acid region (termed the PSK domain), which includes the PSK sequence. In *Arabidopsis* cell cultures, both PSK-a and PSK-b are secreted to the medium. The *Arabidopsis* PSK genes are expressed in apical meristem, leaves, hypocotyls, and roots, again indicating that the PSK signals function not only in culture cells but also in intact plants.

Specific high-affinity binding activities for PSK-a have been detected in the outer surface of rice's plasma membrane. The specific binding was altered by the pH condition and ionic interaction, suggesting that the ligand-receptor binding is controlled by ionic interaction (Matsubayashi & Sakagami, 2000). The photoaffinity cross-linking study revealed the existence of two putative PSK receptors in rice. They are glycosylated proteins with molecular weights of 120 and 160 kDa respectively. The carrot cell line NC contains a relatively high concentration of high-affinity PSK-binding protein and therefore was used to purify the PSK receptor. Photoaffinity labeling of membrane proteins from the carrot cell line with a photoactivatable PSK analog revealed that a 120-kDa protein and a minor 150-kDa protein, both of which are glycosylated proteins, bind to PSK. The two binding proteins were purified and identified as an LRR-containing receptor kinase that shares sequence similarity to those of the known hormone receptors such as BRI1 and CLV1 (Matsubayashi *et al.*, 2002). The 150-kDa protein is likely a modified form of the 120-kDa receptor.

The PSK receptor gene encodes a 1021-amino acid polypeptide with a predicted cleavable signal peptide. It contains an extracellular N-terminus with 21 LRR repeats, a transmembrane domain, and a cytoplasmic serine/threonine kinase domain. Suppression of the receptor gene by an antisense construct caused inhibition of PSK-mediated cell proliferation, further demonstrating its function as the PSK receptor.

Apparently, multiple hormones including auxin, cytokinin, and PSK interact to regulate differentiation and proliferation of cultured cells. It is intriguing to speculate that many other auxin- and/or cytokinin-mediated biological pathways in plants may involve peptide signals. Identification of downstream components in the PSK signaling pathway should provide novel insights into molecular mechanisms that control cell differentiation and proliferation.

2.2.5 *CLAVATA 3 (CLV3) regulates stem cell homeostasis*

Embryogenesis in animals generates a miniature version of an adult organism. In contrast, a plant embryo has a much simpler structure that contains two stem-cell populations: the shoot apical meristem and the root apical meristem (Clark, 2001). The body patterns of an adult plant are largely established by postembryonic pattern formation at these meristems. The shoot apical meristem (SAM) acts as a

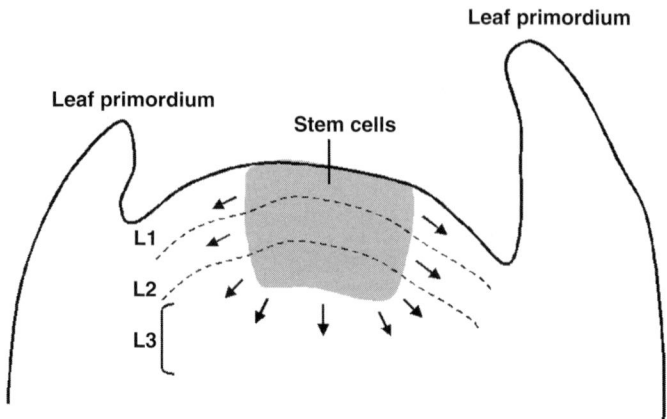

Figure 2.3 Schematic drawing of organization of an apical meristem. The meristem is organized into three layers: L1, L2, and L3. The stem cells are indicated as the gray area. The arrows indicate the flow of cells as a result of stem cell proliferation. Two leaf primordia are formed on the flanks of the meristem. After Clark (2001).

self-renewing source of undifferentiated stem cells whose descendents become incorporated into organ and tissue primordia and acquire different fates, leading to formation of the above-ground organs such as leaves, flowers, vasculature and other tissues of the stem, whereas the root meristem is responsible for development of the root system (Sharma *et al.*, 2003).

In order for the shoot meristem to form organs continuously, different regions of the shoot meristem have to establish constant communication so that a balance between stem cell proliferation through cell division and cell departure from the meristem to form lateral organs is maintained (see also Chapter 6). Cells of SAM consist of three clonally distinct cell layers (Fig. 2.3) (Clark, 2001). Cells of the L1 layer (the epidermal layer) and the L2 layer (the subepidermal layer) divide in an anticlinal fashion, and therefore they are single cell thick. Cells of the underlying L3 layer divide in various orientations. Since all the differentiated cell types of the adult plant are derived from a small number of stem cells, cell lineage patterns do not play a critical role in regulating cell fate. Instead, cell fate is determined by a positional effect that requires communication within and between the different layers of cells.

Genetic screens for *Arabidopsis* mutants with an altered SAM morphology have identified three *CLAVATA* genes (*CLV1*, *CLV 2*, and *CLV3*) whose loss-of-function mutations result in an enlarged meristem (Clark *et al.*, 1993, 1995; Kayes & Clark, 1998). Some of the *clv* mutants accumulate over 1000-fold more undifferentiated cells in SAM than do wild-type plants. Genetic analysis revealed that these three genes function in the same pathway by preventing unrestricted stem cell proliferation. For instance, the phenotypes of *clv1* and *clv3* null mutants and *clv1 clv3* double mutants are nearly identical.

The three *CLV* genes have been cloned. The *CLV3* gene encodes a 96-amino acid protein that shares no significant sequence similarity to other proteins with known functions (Fletcher *et al.*, 1999). It contains a predicted 18-amino acid N-terminal signal peptide that directs the protein into the secretory pathway. *CLV1* encodes a receptor kinase and, like BRI1 and the PSK receptor, it has an extracellular LRR domain and an intracellular kinase domain (Clark *et al.*, 1997). CLV2 is similar to CLV1 but lacks a cytoplasmic kinase domain. The identification of the CLV proteins suggests that CLV1 and CLV2 function as the receptors, and CLV3 acts as the extracellular ligand. CLV1 and CLV2 likely form a heterodimer (Jeong *et al.*, 1999; Trotochaud *et al.*, 1999). Binding of CLV3 with CLV1/CLV2 activates the receptor, which then recruits additional signaling components to form a 450-kDa signaling complex, leading to activation of downstream effectors (Trotochaud *et al.*, 2000; Clark, 2001).

Studies on expression patterns of the *CLV* genes have generated important insights into their functions. *CLV3* is expressed predominantly in the L1 and L2 layers at the apex of the shoot apical meristem. However, *CLV1* is predominantly expressed in the L3 cells, largely beneath the *CLV3* expression domain (Fletcher *et al.*, 1999). The results indicate that CLV3 acts as a short-distance signal that is secreted from the L1 and L2 layers and diffused to the inner cell layers where it is perceived by CLV1/CLV2 to restrict stem cell proliferation.

If the CLV pathway functions to restrict stem cell proliferation, then how is stem cell homeostasis maintained? Sharma *et al.* (2003) has proposed a feedback loop model in which the CLV pathway and a positive, stem-cell-promoting pathway interact to maintain the balance between cell loss and cell division. The homeodomain transcription factor WUSCHEL (WUS) is a key component of the cell-promoting pathway. WUS is required for maintaining the stem cell reservoir, and its loss-of-function mutant is unable to maintain a pool of pluripotent stem cells, resulting in meristem termination (Laux *et al.*, 1996; Mayer *et al.*, 1998). The WUS-mediated positive pathway acts to promote stem cell proliferation. The resulting enlargement of the stem cell population leads to increased production of the CLV3 ligand. As a consequence, it activates the CLV-mediated negative pathway. The enhanced CLV-signaling then causes reduction in the level of WUS transcription, which in turn reduces production of CLV3. The resulting reduction in the CLV-mediated negative pathway then activates the positive pathway. Through this mutual regulation equilibrium is attained.

The identification of the components in the CLV1 450-kDa complex would lead to the identification of additional players involved in the CLV pathway. At least two proteins have been identified from the complex (Trotochaud *et al.*, 1999). The first one is the kinase-associated protein phosphatase (KAPP). In contrast to protein kinases, phosphatases often switch off a response through dephosphorylation. KAPP likely negatively regulates CLV1 action by dephosphorylating CLV1. Another component of the CLV1 complex is a Rho/Rac-GTPase-related protein. Members of this family such as Ras are important components in relaying signals through a series of intracellular biochemical reactions including the kinase cascade.

2.2.6 S-locus cysteine-rich proteins determine specificity of self-incompatibility in the Brassicaceae

During a compatible pollen–pistil interaction, a desiccated pollen grain adheres tightly to the surface of stigma, where it hydrates and germinates. The pollen tube then elongates into the female tissues. The pollen tube elongation is guided toward individual ovules where sperm cells are released to fertilize the egg cell and central cells, leading to development of embryo and endosperm (Pruitt, 1999).

Self-incompatibility (SI) is a phenomenon in which self-pollen is recognized and rejected by pistil whereas non-self pollen is accepted (see also Chapter 10). SI is used by many plant species to prevent inbreeding and maintain genetic variability. The SI system is genetically controlled by a single locus called the *S* locus, which contains multiple genes that could be multi-allelic. Different mechanisms govern pollen recognition and rejection. The SI system of the Solanaceae species is called gametophytic self-incompatibility. In this system, if the single *S* allele carried by the haploid pollen matches either of the two *S* alleles present in the diploid tissues of the pistil, the pollen will be rejected (Matton *et al.*, 1994). Usually self-pollen can germinate but pollen tube elongation is inhibited after they enter the style.

In contrast, in the sporophytic self-incompatibility system used by the Brassicaceae species, the genotype of the male parent, not that of the pollen itself, determines the outcome. If either of the two *S* alleles present in the male parent matches either of the two *S* alleles carried by the pistil, pollen hydration and germination is arrested. The self-incompatible response in this SI system occurs on the stigma surface by blocking the very early events of pollen–pistil interactions, suggesting that cell surface molecules are involved in pollen recognition. The molecular mechanism that determines sporophytic self-incompatibility in the Brassicaceae is described here.

Genetic and molecular studies carried out a decade ago identified two multi-allelic genes associated with the *S* locus in *Brassica* (Nasrallah *et al.*, 1987; Stein *et al.*, 1991). One of them encodes *S*-locus glycoprotein (SLG), and the other encodes *S*-locus receptor kinase (SRK). SLG is a glycoprotein that is localized to the cell wall of papillae, the epidermal cells of the pistil where pollen–pistil interaction occurs (Kandasamy *et al.*, 1989). SRK is a transmembrane protein which contains an extracellular domain (S-domain) with high sequence similarity to SLG and an intracellular serine/threonine kinase domain. SLG and SRK are tightly linked and highly polymorphic. Like SLG, SLK is predominantly expressed in the papilla cells. Suppression of LSG or LRK was found to convert an otherwise incompatible interaction into self-compatible (Takayama & Isogai, 2003), demonstrating that both proteins function in pollen recognition. It is believed that SRK is the female determinant of SI specificity in the stigma, and that SLG enhances the strength of the SI response by facilitating SRK maturation and stability (Dixit *et al.*, 2000; Dixit & Nasrallah, 2001).

After the identification of the female determinant, many efforts were directed toward identifying the male determinant of self-incompatibility. The following criteria are applied to define a male determinant (Takayama & Isogai, 2003): (1) it is encoded by a gene located at the *S* locus; (2) it exhibits allelic diversity reflecting different variants of the *S* locus (designated *S* haplotypes); (3) it is expressed in sporophytic cells because the SI is determined by sporophytic genotypes; (4) it is localized on the surface of pollen grains; and (5) it physically interacts with SRK.

In a search for the male determinant, a gene located between *SLG8* and *SRK8* of the *Brassica S*8 haplotype was identified as a good candidate (Schopfer *et al.*, 1999). This gene exhibits polymorphism associated with different *S* haplotypes and is specifically expressed in anthers. It encodes a small cysteine-rich protein with a molecular weight of approximately 9 kDa. The protein was named *S*-locus cysteine-rich protein (SCR). A definitive proof that SCR functions as the male determinant of SI came from transgenic studies. Pollen of an *S*2 haplotype plant is normally compatible with stigma of an *S*6 plant. However, when *SCR6* was transformed into an *S*2 haplotype, pollen from the transgenic plants was rejected by *S*6 stigma.

The SCR genes are expressed in the microspores and tapetal cells, the diploid maternal tissues surrounding the developing male gametophytes (pollen) (Takayama *et al.*, 2000). The tapetal cells undergo programmed cell death during microgametogenesis and provide their constituents to pollen coating. The SCR expression pattern explains why the *Brassica* SI system is sporophytically determined. Further studies revealed that SCR is localized mainly in the pollen coat (Takayama & Isogai, 2003).

SCR has a cleavable leader sequence and is secreted. SCRs encoded by different alleles generally share approximately 40% sequence identity. This indicates an extensive divergence of this protein, consistent with their role as the SI specificity determinants. SCRs generally contain eight cysteine residues that form four disulphide bonds (Takayama *et al.*, 2001). It is speculated that specificity of SI might be determined by other residues. Following pollination, SCR is diffused from pollen coats through the papilla wall and binds to the cognate SRK (Kachroo *et al.*, 2001; Takayama *et al.*, 2001). The ligand binding induces a conformational change of SRK, which activates the kinase through autophosphorylation. Activated SRK phosphorylates downstream targets to initiate a signaling cascade and activate specific biochemical response, leading to the rejection of self-pollen.

The extensive studies in the last decade have generated a fairly clear picture of the molecular mechanism involved in pollen recognition of the SI system. However, the way SCR–SRK interaction is translated into self-pollen arrest remains largely unknown. One possible scenario (Kachroo *et al.*, 2002) is that the SCR–SRK signaling pathway results in release of calcium from the papillar cells to the pollen–papillae interface. Pollen maintains a calcium gradient. Uptake of calcium by the pollen grain changes the calcium gradient, leading to blockage of pollen germination. Another possibility is that the SCR–SRK interaction leads to activation of a physiological response in the papillae cells, which is analogous to the disease resistance response. Such a response would cause the papillae to secret components that could be detrimental to pollen.

2.3 Proteolytic processing of prohormones

Insulin was the first hormone found to be derived from a precursor (Steiner & Oyer, 1966). Later discoveries revealed that multiple peptide hormones can be produced from a common precursor (Mains *et al.*, 1977). Only 15 years ago the first prohormone processing enzyme (Kex2) was identified in yeast (Fuller *et al.*, 1989). It is now known that peptide signals are generally synthesized as larger inactive precursors that undergo limited proteolysis to yield active peptide signals. Such proteolytic processing is one of many mechanisms that regulate activity of peptide hormones.

Proteolytic processing of prohormones is catalyzed by five endopeptidases (commonly termed proteases): serine, cysteine, aspartic, metallo-, and threonine proteases (Barrett, 1998). Exopeptidases may also be involved in the process. For instance, consider the case of angiotensin II, the most important animal hormone in regulating blood pressure, and water and salt balance. Angiostensin II is derived from cleavage of its 55-kDa precursor by rennin (an aspartic protease) to generate the decapeptide angiotensin I, which is then processed by a carboxypeptidase to remove two carboxyterminal amino acids. Renin is the rate-limiting enzyme in the proteolytic cascade (Morris, 2003).

Most of the known peptide signals in plants are derived from large precursors, including systemins, RALFs, and PSKs. All of those peptide signals were initially identified using biochemical approaches based on their biological activities. It remains to be determined if CLV3, SCR, or ENOD40 also undergoes posttranslational proteolytic processing to generate a biologically active signal molecule.

To date, no protease has been identified in the maturation of the plant peptide signals. Proteolytic processing of prosystemin is thought to take place as a result of wounding, which brings prosystemin together with protease(s) from another cellular compartment. An aspartic protease is induced during the systemic wound response and could be involved in processing of prosystemin; however, no direct evidence exists to support that notion (Schaller & Ryan, 1996). Many prohormones in animals contain a pair of basic amino acids prior to and/or after the sequences of the mature hormones. Such dibasic residues are the preferred cleavage sites for prohormone-processing serine proteases. The *Arabidopsis* PSK precursors contain such dibasic residues near both sides of the PSK sequences (Yang *et al.*, 2001). Therefore, the PSK precursors may be cleaved at the dibasic sites by unknown serine proteases and further trimmed by exopeptidases to form mature PSKs. The *Arabidopsis* genome encodes over an estimated 550 putative proteases (Beers *et al.*, 2004). Future studies will likely lead to identification of many of them as prohormone maturation enzymes.

A gain-of-function genetic screen aimed at isolating components of the BRI1-mediated signaling pathway has resulted in the identification of an extracellular serine carboxypeptidase (BRS1) whose overexpresstion suppresses the *bri1* mutant phenotype (Li *et al.*, 2001). Serine carboxypeptidases belong to the serine protease family. BRS1 shares a sequence similarity to Kex1, which is involved in maturation of a-mating factor in yeast. The precursor of a-mating factor is processed by cleavage at a specific pair of basic amino acids by Kex2 (another serine protease) and followed by removal of dibasic amino acids at the C-terminus by Kex1. The authors proposed that BRS1 might be involved in processing a peptide signal that is an unidentified component of the brassinolide signaling pathway.

The disease resistance response in plants has been very well studied. A quick recognition of an invading pathogen is often the key for the plant to mount an effective defense response. This recognition triggers a battery of physiological responses including the hypersensitive response (HR), which is often accompanied by rapid cell death at the site of infection. HR, together with other defense mechanisms, contains and eventually eliminates the pathogen. In addition to such localized resistance, the plant develops mechanisms of systemic immunity following the initial pathogen attack. This systemic acquired resistance (SAR) occurs when a mobile signal is generated at the infected site and translocated throughout the rest of the plant. SAR is an enhanced resistance against a broad spectrum of pathogens.

To date, the identity of the mobile signal that activates SAR remains obscure. Possible candidates for long-range and/or short-range mobile SAR signals have included several small molecules: salicylate, H_2O_2, nitric oxide, and unidentified lipid molecules. Recently, the activation of an extracellular aspartic protease (CDR1) in *Arabidopsis* was found to generate an endogenous mobile signal that activates the SAR response (Xia *et al.*, 2004). The endogenous elicitor is associated with a protein fraction whose molecular size ranges from 3 to 10 kDa. The elicitor activity is inactivated by heating and after treatment with a digestive protease. The results suggest that the elicitor activity comes from a peptide signal. The study raises the possibility that SAR signaling is also mediated by a peptide signal in a manner analogous to the systemin-mediated systemic wound response.

In addition to proteolytic maturation, some proteases may be involved in inactivation of peptide signals. Since peptide signals generally act on cell surface, its degradation is not likely mediated through the widely used ubiquitin-mediated protein degradation pathway. Instead, extracellular proteases play an important role in the inactivation process. A tomato serine protease was initially identified as a systemin-binding protein (SBP50) (Schaller & Ryan, 1994). SBP50 is a membrane-associated protein with the ability to cleave systemin *in vitro*, which suggests that it could be involved in inactivating the systemin signaling pathway.

2.4 Technologies for discovering new peptide signals

As indicated from the above examples, peptide signals regulate a variety of important biological processes in plants. However, compared to the hundreds of peptide signals in animals, the number of known plant peptide signals is low. Growing evidence indicates that many more peptide signals in plants are yet to be discovered. For example, plants possess a large number of receptor-like kinases (RLKs). Six hundred ten RLKs have been identified through analysis of the *Arabidopsis* genome (Shiu & Bleecker, 2001). Among these, 216 contain extracellular LRR domains typical of peptide-binding motifs (Kobe & Deisenhofer, 1994, 1995). The receptors for CLV3, PSK, and systemin are all the members of LRR-containing RLKs. However, to date, ligands have only been identified for very few of these plant RLKs.

One of the hurdles in identifying peptides that function as signal molecules is their low abundance. For example, over 60 lb of tomato leaves was used in purifying the tomato systemin for sequence determination and characterization (Pearce *et al.*, 1991). Another hurdle is the lack of an assay for their biological activity. The discovery of the TomHypSys and TobHypSys systemins was facilitated by the availability of the alkalinization assay. Without a convenient assay, it would be very difficult to isolate a new peptide signal using a biochemical approach.

Two of the peptide signals, SCR and CLAVATA3, were identified through genetic approaches. Mutational analysis will continue to contribute to discovery of new peptide signals. However, most genes encoding peptide signals are likely small. This small target size reduces the chance to introduce a mutation in those genes. Even if some of them do encode large precursors, the active form of a peptide signal could be a short peptide such as PSK. Therefore, a large portion of mutations in those genes will be silent. As a result, a relatively large mutagenized population needs to be screened in order to isolate a loss-of-function mutation in a peptide signal-encoding gene.

Gain-of-function mutagenesis such as activation tagging (Weigel *et al.*, 2000) could be a powerful alternative genetic approach to isolate genes that encode peptide signals. Such an approach has led to the recent identification of an *Arabidopsis* gene (*DEVIL1*) encoding a novel 51-amino acid polypeptide whose gain-of-function mutation causes changes in a variety of developmental processes, including the leaf shape, plant stature, and silique development (Wen *et al.*, 2004). The *Arabidopsis* genome encodes 20 DVL1-like proteins, many of which appear to play a biological role similar to DVL1. The exact biological functions of the members of this gene family have yet to be determined. Besides, further studies are needed to prove that DVL1 and its homologues act as signal molecules.

The availability of whole genome sequences for a growing list of plant species as well as advances in functional genomics and proteomics technology provide a unique opportunity to identify new peptide signals in a high-throughput fashion. An *in silico* search of putative peptide signal-encoding genes is the first step toward this goal. However, the commonly used gene prediction algorithms predict genes on the basis of presence of a significant ORF of at least 100 amino acids (Harrison *et al.*, 2002). Similar gene prediction algorithms are also used to predict genes from the *Arabidopsis* and rice genomes (MacIntosh *et al.*, 2001; Goff *et al.*, 2002). Undoubtedly, a large number of genuine genes with small ORFs could not be predicted from those genomes. Therefore, many small ORFs that have been predicted as noncoding sequences could encode peptide signals.

A systemic analysis of small ORFs in *Arabidopsis* that encode putative secreted peptides has identified a family of 34 genes, many of which contain the RALF-like sequence (Olsen *et al.*, 2002). Biological functions of RALFs have not been well understood. The existence of the larger number of putative secreted RALF-like molecules suggests that many of them may be signals involved in a variety of biological pathways. Another search for CLV3 homologues has identified 42 *CLV3*-related (*CLE*) sequences in plants (Cock & McCormick, 2001). Among them, 24 are in the *Arabidopsis* genome. Speculatively, these *CLE* genes could encode the ligands for many RLKs. A similar search has identified a large number of putative genes that encode peptides homologous to SCR (Vanoosthuyse *et al.*, 2001).

One of the challenges of this approach is to determine that these *in silico* genes do encode polypeptides. The peptidomics approach may offer a way to aid the systematic discovery of peptides (Schulz-Knappe *et al.*, 2001). The approach takes advantage of high sensitivity of mass spectrometry technology in identification of proteins and improvements in separating and isolating peptides to systematically identify a complement of peptide molecules produced by particular tissues, organs, or whole organisms. Such an approach has recently led to the discovery of a large number of new animal neuropeptides and peptide hormones that had eluded identification through classical methods (Takahashi *et al.*, 1997; Sweedler *et al.*, 2000; Clynen *et al.*, 2001; Schrader & Schulz-Knappe, 2001; Svensson *et al.*, 2003). For instance, Svensson *et al.* (2003) simultaneously detected more than 550 endogenous neuropeptides from hypothalamic extracts, which include previously described and many novel neuropeptides. These structural peptidomics approaches can be explored to identify the major complement of peptides produced by plant cells.

A more challenging task is to determine whether a peptide really functions as a signal because many small proteins have other regulatory or nonregulatory functions. A method to address this question is to determine if a peptide binds to a receptor. Binding of a peptide to a cell surface receptor will provide strong evidence of function as a signal molecule. Several high-throughput protein–protein interaction technologies have been developed in recent years (Drewes & Bouwmeester, 2003). It should be feasible to apply them in generating interaction maps between peptide molecules and the putative receptor kinases encoded by the plant genomes. Such information will not only provide important clues as to whether a peptide functions as a signal but also facilitate identification of its receptors and downstream signaling components.

2.5 Concluding remarks

Intercellular communication mechanisms are believed to have evolved after plants and animals diverged from their common unicellular ancestor (Meyerowitz, 1999). The striking similarity in plant's and animal's signal perception and transduction pathways mediated by peptide signals indicates that those organisms have used a common set of original genes inherited from their common ancestor to build the signaling networks for their distinct biological processes.

Since the discovery of systemin, several peptides have been identified as signals in a variety of plant biological pathways including the defense response, cell proliferation and differentiation, and other developmental and reproductive processes. A peptide signal-mediated response could be part of a biological process regulated by other hormones. For instance, cytokinin, auxin, and PSK interplay to regulate proliferation of cultured cells. Similarly, systemins, jasmonate, and ethylene work together to activate an effective defense response.

Growing evidence indicates that peptide signals may be involved in many other developmental and physiological processes in plants. Advances in new technology will undoubtedly expedite discovery of additional peptide signals. Identification of other players involved in processing, transporting, relaying, amplifying, and integrating peptide signals will greatly increase our understanding of how functions of individual molecules have been integrated into a complex pathway and how individual pathways are integrated to develop a functional organism.

References

Alberts, B., Bray, D., Lewis, J., Raff, M., Roberts, K. & Watson, J.D. (1994) *Molecular Biology of the Cell*. Garland, New York.

Banting, F.G. & Best, C.H. (1922) The internal secretions of the pancreas. *J. Lab. Clin. Med.*, **5**, 251–266.

Barrett, A.J. (1998) Proteolytic enzymes: nomenclature and classification. In: *Handbook of Proteolytic Enzymes* (eds A.J. Barrett, N.D. Rawlings & J.F. Woessner), pp. 1–20. Academic Press, London.

Beers, E.P., Jones, A.M. & Dickerman, A.W. (2004) The S8 serine, C1A cysteine and A1 aspartic protease families in *Arabidopsis*. *Phytochemistry*, **65**, 43–58.

Bisseling, T. (1999) The role of plant peptides in intercellular signalling. *Curr. Opin. Plant Biol.*, **2**, 365–368.

Clark, S.E. (2001) Cell signaling at the shoot meristem. *Nat. Rev. Mol. Cell. Biol.*, **2**, 276–284.

Clark, S.E., Running, M.P. & Meyerowitz, E.M. (1993) CLAVATA1, a regulator of meristem and flower development in *Arabidopsis*. *Development*, **119**, 397–418.

Clark, S.E., Running, M.P. & Meyerowitz, E.M. (1995) *CLAVATA3* is a specific regulator of shoot and floral meristem development affecting the same processes as *CLAVATA1*. *Development*, **121**, 2057–2067.

Clark, S.E., Williams, R.W. & Meyerowitz, E.M. (1997) The *CLAVATA1* gene encodes a putative receptor kinase that controls shoot and floral meristem size in *Arabidopsis*. *Cell*, **89**, 575–585.

Clynen, E., Baggerman, G., Veelaert, D., Cerstiaens, A., Van der Horst, D., Harthoorn, L., Derua, R., Waelkens, E., De Loof, A. & Schoofs, L. (2001) Peptidomics of the pars intercerebralis–corpus cardiacum complex of the migratory locust, *Locusta migratoria*. *Eur. J. Biochem.*, **268**, 1929–1939.

Cock, J.M. & McCormick, S. (2001) A large family of genes that share homology with CLAVATA3. *Plant Physiol.*, **126**, 939–942.

Constabel, C.P., Yip, L. & Ryan, C.A. (1998) Prosystemin from potato, black nightshade, and bell pepper: primary structure and biological activity of predicted systemin polypeptides. *Plant Mol. Biol.*, **26**, 55–62.

Crespi, M.D., Jurkevitch, E., Poiret, M., d'Aubenton-Carafa, Y., Petrovics, G., Kondorosi, E. & Kondorosi, A. (1994) enod40, a gene expressed during nodule organogenesis, codes for a non-translatable RNA involved in plant growth. *EMBO J.*, **13**, 5099–5112.

Dixit, R. & Nasrallah, J.B. (2001) Recognizing self in the self-incompatibility response. *Plant Physiol.*, **125**, 105–108.

Dixit, R., Nasrallah, M.E. & Nasrallah, J.B. (2000) Post-transcriptional maturation of the S receptor kinase of *Brassica* correlates with co-expression of the *S*-locus glycoprotein in the stigmas of two *Brassica* strains and in transgenic tobacco plants. *Plant Physiol.*, **124**, 297–311.

Drewes, G. & Bouwmeester, T. (2003) Global approaches to protein–protein interactions. *Curr. Opin. Cell. Biol.*, **15**, 199–205.

Fletcher, J.C., Brand, U., Running, M.P., Simon, R. & Meyerowitz, E.M. (1999) Signaling of cell fate decisions by CLAVATA3 in *Arabidopsis* shoot meristems. *Science*, **283**, 1911–1914.

Fuller, R.S., Brake, A.S. & Thorner, J. (1989) Yeast prohormone processing enzyme (Kex2 gene product) is a Ca^{2+}-dependent serine protease. *Proc. Natl. Acad. Sci. U.S.A.*, **86**, 1434–1438.

Geurts, R. & Bisseling, T. (2002) Rhizobium nod factor perception and signalling. *Plant Cell*, **14** (Suppl.), S239–S249.

Goff, S.A., Ricke, D., Lan, T.H., Presting, G., Wang, R., Dunn, M., Glazebrook, J., Sessions, A., Oeller, P., Varma, H., Hadley, D., Hutchison, D., Martin, C., Katagiri, F., Lange, B.M., Moughamer, T., Xia, Y., Budworth, P., Zhong, J., Miguel, T., Paszkowski, U., Zhang, S., Colbert, M., Sun, W.L., Chen, L., Cooper, B., Park, S., Wood, T.C., Mao, L., Quail, P., Wing, R., Dean, R., Yu, Y., Zharkikh, A., Shen, R., Sahasrabudhe, S., Thomas, A., Cannings, R., Gutin, A., Pruss, D., Reid, J., Tavtigian, S., Mitchell, J., Eldredge, G., Scholl, T., Miller, R.M., Bhatnagar, S., Adey, N., Rubano, T., Tusneem, N., Robinson, R., Feldhaus, J., Macalma, T., Oliphant, A. & Briggs, S. (2002) A draft sequence of the rice genome (*Oryza sativa* L. ssp. *japonica*). *Science*, **296**, 92–100.

Habener, J.F. (1987) Gene structure and regulation. In: *Molecular Cloning of Hormone Genes* (ed. J.F. Habener), pp. 11–52. The Humana Press, Clifton, NJ.

Harrison, P.M., Kumar, A., Lang, N., Snyder, M. & Gerstein, M. (2002) A question of size: the eukaryotic proteome and the problems in defining it. *Nucleic Acids Res.*, **30**, 1083–1090.

Jeong, S., Trotochaud, A.E. & Clark, S.E. (1999) The *Arabidopsis* CLAVATA2 gene encodes a receptor-like protein required for the stability of the CLAVATA1 receptor-like kinase. *Plant Cell*, **11**, 1925–1934.

Johnson, R., Narvaez, J., An, G. & Ryan, C. (1989) Expression of proteinase inhibitors I and II in transgenic tobacco plants: effects on natural defense against *Manduca sexta* larvae. *Proc. Natl. Acad. Sci. U.S.A.*, **86**, 9871–9875.

Kachroo, A., Nasrallah, M.E. & Nasrallah, J.B. (2002) Self-incompatibility in the Brassicaceae: receptor-ligand signaling and cell-to-cell communication. *Plant Cell*, **14** (Suppl.), S227–S238.

Kachroo, A., Schopfer, C.R., Nasrallah, M.E. & Nasrallah, J.B. (2001) Allele-specific receptor-ligand interactions in *Brassica* self-incompatibility. *Science*, **293**, 1824–1826.

Kandasamy, M.K., Paolillo, D.J., Faraday, C.D., Nasrallah, J.B. & Nasrallah, M.E. (1989) The *S*-locus specific glycoproteins of *Brassica* accumulate in the cell wall of developing stigma papillae. *Dev. Biol.*, **134**, 462–472.

Kayes, J.M. & Clark, S.E. (1998) CLAVATA2, a regulator of meristem and organ development in *Arabidopsis*. *Development*, **125**, 3843–3851.

Kobe, B. & Deisenhofer, J. (1994) The leucine-rich repeat: a versatile binding motif. *Trends Biochem. Sci.*, **19**, 415–421.

Kobe, B. & Deisenhofer, J. (1995) A structural basis of the interactions between leucine-rich repeats and protein ligands. *Nature*, **374**, 183–186.

Kouchi, H., Takane, K.I., So, R.B., Ladha, J.K. & Reddy, P.M. (1999) Rice ENOD40: isolation and expression analysis in rice and transgenic soybean root nodules. *Plant J.*, **18**, 121–129.

Laux, T., Mayer, K.F., Berger, J. & Jurgens, G. (1996) The *WUSCHEL* gene is required for shoot and floral meristem integrity in *Arabidopsis*. *Development*, **122**, 87–96.

Li, J. & Chory, J. (1997) A putative leucine-rich repeat receptor kinase involved in brassinosteroid signal transduction [see comment]. *Cell*, **90**, 929–938.

Li, J., Lease, K.A., Tax, F.E. & Walker, J.C. (2001) BRS1, a serine carboxypeptidase, regulates BRI1 signaling in *Arabidopsis thaliana*. *Proc. Natl. Acad. Sci. U.S.A.*, **98**, 5916–5921.

Lindsey, K., Casson, S. & Chilley, P. (2002) Peptides: new signalling molecules in plants. *Trends Plant Sci.*, **7**, 78–83.

MacIntosh, G.C., Wilkerson, C. & Green, P.J. (2001) Identification and analysis of *Arabidopsis* expressed sequence tags characteristic of non-coding RNAs. *Plant Physiol.*, **127**, 765–776.

Mains, R.E., Eipper, B.A. & Ling, N. (1977) Common precursor to corticotropins and endorphins. *Proc. Natl. Acad. Sci. U.S.A.*, **74**, 3014–3018.

Matsubayashi, Y. & Sakagami, Y. (1996) Phytosulfokine, sulfated peptides that induce the proliferation of single mesophyll cells of *Asparagus officinalis* L. *Proc. Natl. Acad. Sci. U.S.A.*, **93**, 7623–7627.

Matsubayashi, Y. & Sakagami, Y. (2000) 120- and 160-kDa receptors for endogenous mitogenic peptide, phytosulfokine-alpha, in rice plasma membranes. *J. Biol. Chem.*, **275**, 15520–15525.

Matsubayashi, Y., Hanai, H., Hara, O. & Sakagami, Y. (1996) Active fragments and analogs of the plant growth factor, phytosulfokine: structure–activity relationships. *Biochem. Biophys. Res. Commun.*, **225**, 209–214.

Matsubayashi, Y., Morita, A., Matsunaga, E., Furuya, A., Hanai, N. & Sakagami, Y. (1999) Physiological relationships between auxin, cytokinin, and a peptide growth factor, phytosulfokine-a, in stimulation of asparagus cell proliferation. *Planta*, **207**, 559–565.

Matsubayashi, Y., Ogawa, M., Morita, A. & Sakagami, Y. (2002) An LRR receptor kinase involved in perception of a peptide plant hormone, phytosulfokine. *Science*, **296**, 1470–1472.

Matsubayashi, Y., Takagi, L. & Sakagami, Y. (1997) Phytosulfokine-alpha, a sulfated pentapeptide, stimulates the proliferation of rice cells by means of specific high- and low-affinity binding sites. *Proc. Natl. Acad. Sci. U.S.A.*, **94**, 13357–13362.

Matton, D., Nass, N., Clarke, A. & Newbigin, E. (1994) Self-incompatibility: how plants avoid ilegitimate offspring. *Proc. Natl. Acad. Sci. U.S.A.*, **91**, 1992–1997.

Mayer, K.F., Schoof, H., Haecker, A., Lenhard, M., Jurgens, G. & Laux, T. (1998) Role of WUSCHEL in regulating stem cell fate in the *Arabidopsis* shoot meristem. *Cell*, **95**, 805–815.

McGurl, B., Pearce, G., Orozco-Cardenas, M. & Ryan, C. (1992) Structure, expression, and antisense inhibition of the systemin precursor gene. *Science*, **255**, 1570–1573.

Meyerowitz, E.M. (1999) Plants, animals and the logic of development. *Trends Cell Biol.*, **9**, M65–M68.

Miklashevichs, E., Czaja, I., Rohrig, H., Schmidt, J., John, M., Schell, J. & Walden, R. (1996) Do peptides control plant growth and development? *Science*, **1**, 411.

Montoya, T., Nomura, T., Farrar, K., Kaneta, T., Yokota, T. & Bishop, G.J. (2002) Cloning the tomato curl3 gene highlights the putative dual role of the leucine-rich repeat receptor kinase tBRI1/SR160 in plant steroid hormone and peptide hormone signaling. *Plant Cell*, **14**, 3163–3176.

Morris, B.J. (2003) Renin: from 'pro' to promoter. *Bioessays*, **25**, 520–527.

Narvaez-Vasquez, J. & Ryan, C. (2004) The cellular localization of prosystemin: a functional role for phloem parenchyma in systemic wound signaling. *Planta*, **218**, 360–369.

Nasrallah, J., Kao, T.-H., Chen, C., Goldberg, M. & Nasrallah, M. (1987) Amino-acid sequence of glycoproteins encoded by three *S* alleles of the *S* locus of *Brassica oleracea*. *Nature*, **326**, 617–619.

Neves, S.R., Ram, P.T. & Iyengar, R. (2002) G protein pathways. *Science*, **296**, 1636–1639.

Olsen, A.N., Mundy, J. & Skriver, K. (2002) Peptomics, identification of novel cationic *Arabidopsis* peptides with conserved sequence motifs. *In Silico Biol.*, **2**, 441–451.

Pearce, G. & Ryan, C.A. (2003) Systemic signaling in tomato plants for defense against herbivores: isolation and characterization of three novel defense-signaling glycopeptide hormones coded in a single precursor gene. *J. Biol. Chem.*, **278**, 30044–30050.

Pearce, G., Johnson, S. & Ryan, C.A. (1993) Purification and characterization from tobacco (*Nicotiana tabacum*) leaves of six small, wound-inducible, proteinase isoinhibitors of the potato inhibitor ii family. *Plant Physiol.*, **102**, 639–644.

Pearce, G., Moura, D.S., Stratmann, J. & Ryan, C.A. (2001a) Production of multiple plant hormones from a single polyprotein precursor. *Nature*, **411**, 817–820.

Pearce, G., Moura, D.S., Stratmann, J. & Ryan, C.A. (2001b) RALF, a 5-kDa ubiquitous polypeptide in plants, arrests root growth and development. *Proc. Natl. Acad. Sci. U.S.A.*, **98**, 12843–12847.

Pearce, G., Strydom, D., Johnson, S. & Ryan, C.A. (1991) A polypeptide from tomato leaves induces wound-inducible proteinase-inhibitor proteins. *Science*, **253**, 895–898.

Pruitt, R.E. (1999) Complex sexual signals for the male gametophyte. *Curr. Opin. Plant Biol.*, **2**, 419–422.

Rohrig, H., Schmidt, J., Miklashevichs, E., Schell, J. & John, M. (2002) Soybean ENOD40 encodes two peptides that bind to sucrose synthase. *Proc. Natl. Acad. Sci. U.S.A.*, **99**, 1915–1920.

Ryan, C.A. & Pearce, G. (2003) Systemins: a functionally defined family of peptide signals that regulate defensive genes in Solanaceae species. *Proc. Natl. Acad. Sci. U.S.A.*, **100** (Suppl. 2), 14577–14580.

Schaller, A. & Ryan, C.A. (1994) Identification of a 50-kDa systemin-binding protein in tomato plasma membranes having Kex2p-like properties. *Proc. Natl. Acad. Sci. U.S.A.*, **91**, 11802–11806.

Schaller, A. & Ryan, C.A. (1996) Molecular cloning of a tomato leaf cDNA encoding an aspartic protease, a systemic wound response protein. *Plant Mol. Biol.*, **31**, 1073–1077.

Scheer, J.M. & Ryan, C.A. (2002) The systemin receptor SR160 from *Lycopersicon peruvianum* is a member of the LRR receptor kinase family [see comment]. *Proc. Natl. Acad. Sci. U.S.A.*, **99**, 9585–9590.

Scheer, J.M., Pearce, G. & Ryan, C.A. (2003) Generation of systemin signaling in tobacco by transformation with the tomato systemin receptor kinase gene. *Proc. Natl. Acad. Sci. U.S.A.*, **100**, 10114–10117.

Schlessinger, J. (2000) Cell signaling by receptor tyrosine kinases. *Cell*, **103**, 211–225.

Schopfer, C.R., Nasrallah, M.E. & Nasrallah, J.B. (1999) The male determinant of self-incompatibility in *Brassica*. *Science*, **286**, 1697–1700.

Schrader, M. & Schulz-Knappe, P. (2001) Peptidomics technologies for human body fluids. *Trends Biotechnol.*, **19**, S55–S60.

Schulz-Knappe, P., Zucht, H.D., Heine, G., Jurgens, M., Hess, R. & Schrader, M. (2001) Peptidomics: the comprehensive analysis of peptides in complex biological mixtures. *Comb. Chem. High Throughput Screen.*, **4**, 207–217.

Sharma, V.K., Carles, C. & Fletcher, J.C. (2003) Maintenance of stem cell populations in plants. *Proc. Natl. Acad. Sci. U.S.A.*, **100**, 11823–11829.

Shiu, S.H. & Bleecker, A.B. (2001) Receptor-like kinases from *Arabidopsis* form a monophyletic gene family related to animal receptor kinases. *Proc. Natl. Acad. Sci. U.S.A.*, **98**, 10763–10768.

Sousa, C., Johansson, C., Charon, C., Manyani, H., Sautter, C., Kondorosi, A. & Crespi, M. (2001) Translational and structural requirements of the early nodulin gene enod40, a short-open reading frame-containing RNA, for elicitation of a cell-specific growth response in the alfalfa root cortex. *Mol. Cell. Biol.*, **21**, 354–366.

Stein, J., Howlett, B., Boyes, D., Nasrallah, M. & Nasrallah, J. (1991) Molecular cloning of a putative receptor protein kinase gene encoded at the self-incompatibility locus of *Brassica oleracea*. *Proc. Natl. Acad. Sci. U.S.A.*, **88**, 8816–8820.

Steiner, D.F. & Oyer, P.E. (1966) The biosynthesis of insulin and a probable precursor of insulin by a human islet cell adenoma. *Proc. Natl. Acad. Sci. U.S.A.*, **57**, 473–480.

Stratmann, J., Scheer, J. & Ryan, C. A. (2000) Suramin inhibits initiation of defense signaling by systemin, chitosan, and a beta-glucan elicitor in suspension-cultured *Lycopersicon peruvianum* cells. *Proc. Natl. Acad. Sci. U.S.A.*, **97**, 8862–8867.

Svensson, M., Skold, K., Svenningsson, P. & Andren, P.E. (2003) Peptidomics-based discovery of novel neuropeptides. *J. Proteome Res.*, **2**, 213–219.

Sweedler, J.V., Li, L., Floyd, P. & Gilly, W. (2000) Mass spectrometric survey of peptides in cephalopods with an emphasis on the FMRFamide-related peptides. *J. Exp. Biol.*, **203** (Pt 23), 3565–3573.

Takahashi, T., Muneoka, Y., Lohmann, J., Lopez de Haro, M.S., Solleder, G., Bosch, T. C., David, C.N., Bode, H.R., Koizumi, O., Shimizu, H., Hatta, M., Fujisawa, T. & Sugiyama, T. (1997) Systematic isolation of peptide signal molecules regulating development in hydra: LWamide and PW families. *Proc. Natl. Acad. Sci. U.S.A.*, **94**, 1241–1246.

Takayama, S. & Isogai, A. (2003) Molecular mechanism of self-recognition in *Brassica* self-incompatibility. *J. Exp. Bot.*, **54**, 149–156.

Takayama, S., Shiba, H., Iwano, M., Shimosato, H., Che, F. S., Kai, N., Watanabe, M., Suzuki, G., Hinata, K. & Isogai, A. (2000) The pollen determinant of self-incompatibility in *Brassica campestris*. *Proc. Natl. Acad. Sci. U.S.A.*, **97**, 1920–1925.

Takayama, S., Shimosato, H., Shiba, H., Funato, M., Che, F.S., Watanabe, M., Iwano, M. & Isogai, A. (2001) Direct ligand-receptor complex interaction controls *Brassica* self-incompatibility. *Nature*, **413**, 534–538.

Trotochaud, A.E., Hao, T., Wu, G., Yang, Z. & Clark, S.E. (1999) The CLAVATA1 receptor-like kinase requires CLAVATA3 for its assembly into a signaling complex that includes KAPP and a Rho-related protein [see comment]. *Plant Cell*, **11**, 393–406.

Trotochaud, A.E., Jeong, S. & Clark, S.E. (2000) CLAVATA3, a multimeric ligand for the CLAVATA1 receptor-kinase. *Science*, **289**, 613–617.

van de Sande, K., Pawlowski, K., Czaja, I., Wieneke, U., Schell, J., Schmidt, J., Walden, R., Matvienko, M., Wellink, J., van Kammen, A., Franssen, H. & Bisseling, T. (1996) Modification of phytohormone response by a peptide encoded by ENOD40 of legumes and a nonlegume. *Science*, **273**, 370–373.

Vanoosthuyse, V., Miege, C., Dumas, C. & Cock, J. M. (2001) Two large *Arabidopsis thaliana* gene families are homologous to the *Brassica* gene superfamily that encodes pollen coat proteins and the male component of the self-incompatibility response. *Plant Mol. Biol.*, **46**, 17–34.

Weigel, D., Ahn, J.H., Blazquez, M.A., Borevitz, J.O., Christensen, S.K., Fankhauser, C., Ferrandiz, C., Kardailsky, I., Malancharuvil, E.J., Neff, M.M., Nguyen, J.T., Sato, S., Wang, Z.Y., Xia, Y., Dixon, R.A., Harrison, M.J., Lamb, C.J., Yanofsky, M.F. & Chory, J. (2000) Activation tagging in *Arabidopsis*. *Plant Physiol.*, **122**, 1003–1013.

Wen, J., Lease, K.A. & Walker, J.C. (2004) DVL, a novel class of small polypeptides: overexpression alters *Arabidopsis* development. *Plant J.*, **37**, 668–677.

Xia, Y., Suzuki, H., Borevitz, J., Blount, J., Guo, Z., Patel, K., Dixon, R.A. & Lamb, C. (2004) An extracellular aspartic protease functions in *Arabidopsis* disease resistance signaling. *EMBO J.*, **23**, 980–988.

Yang, H., Matsubayashi, Y., Nakamura, K. & Sakagami, Y. (1999) *Oryza sativa* PSK gene encodes a precursor of phytosulfokine-alpha, a sulfated peptide growth factor found in plants. *Proc. Natl. Acad. Sci. U.S.A.*, **96**, 13560–13565.

Yang, H., Matsubayashi, Y., Nakamura, K. & Sakagami, Y. (2001) Diversity of *Arabidopsis* genes encoding precursors for phytosulfokine, a peptide growth factor. *Plant Physiol.*, **127**, 842–851.

Yang, W.C., Katinakis, P., Hendriks, P., Smolders, A., de Vries, F., Spee, J., van Kammen, A., Bisseling, T. & Franssen, H. (1993) Characterization of GmENOD40, a gene showing novel patterns of cell-specific expression during soybean nodule development. *Plant J.*, 3, 573–585.

Zhang, Z.-P. & Baldwin, I. T. (1997) Transport of [2-^{14}C]jasmonic acid from leaves to roots mimics wound-induced changes in endogenous jasmonic acid pools in *Nicotiana sylvestris*. *Planta*, **203**, 436–441.

3 RNA as a signalling molecule
Patrice Dunoyer and Olivier Voinnet

3.1 Intercellular movement of plant mRNAs

3.1.1 Cell-to-cell movement of plant mRNAs

3.1.1.1 Plant plasmodesmata

In all multicellular organisms, the orchestration of key biological functions often involves intercellular communication. In animals, this process relies mainly on secreted signalling peptides interacting with specific cell surface receptors. Although plant cells can also communicate via secreted molecules (Bergey *et al.*, 1996; Fletcher *et al.*, 1999; McCarty & Chory, 2000; Clark, 2001; Matsubayashi *et al.*, 2001), they differ from their animal counterparts in that they are connected with each other via specific channels known as *plasmodesmata* (see also Chapter 5). Cell connections via plasmodesmata create a cytoplasmic continuum known as the *symplasm*. Plasmodesmata are used for transport of small molecules, such as ions, metabolites and hormones (Robards & Lucas, 1990), but these structures are also involved in the selective trafficking of macromolecules such as proteins and RNAs, providing the plant with a powerful communication system that allows control of many developmental and physiological processes.

Plasmodesmata can be simply defined as membrane-lined pores connecting two adjacent cells, which can arise during cytokinesis or between existing cell walls (Lucas, 1995; Crawford & Zambryski, 1999; Haywood *et al.*, 2002; Wu *et al.*, 2002). The two main structural components of plasmodesmata are an external membrane and a central desmotubule that derive from the plasma membrane and the endoplasmic reticulum, respectively, both of which are contiguous between connected cells. Plasmodesmata are typically defined by their size exclusion limit (or SEL), corresponding to the upper limit in the size of macromolecules that can freely diffuse from one cell to another. Several studies have identified two modes of cell-to-cell transport through plasmodesmata. Non-targeted movement appears to be based on passive diffusion and relies therefore solely on the intrinsic SEL of the plasmodesmata, which has been shown to vary greatly depending on the age of the tissues analysed. For instance, the progression of tobacco leaves from carbon sink tissues to carbon source tissues is accompanied by an important decrease in SEL that correlates with a change in plasmodesmata ultrastructure, from simple to branched. Non-targeted movement of macromolecules through plasmodesmata has been so far documented only for proteins (Imlau *et al.*, 1999; Oparka *et al.*, 1999; Crawford & Zambryski, 2000).

The second mode of transport, known as targeted movement, relies on specific interactions between plasmodesmata components and transported macromolecules, which leads to a dynamic increase in SEL. Our knowledge of the processes governing targeted movement through plasmodesmata has been greatly enhanced by the study of plant viruses (Lazarowitz, 1999). Successful infection by these pathogens implies an ability to move through the cell primary physical defence: the wall. To bypass this barrier, most plant viruses encode movement proteins (MPs) that actively target the plasmodesmata (Deom *et al.*, 1990) to increase their SEL (Wolf *et al.*, 1989; Oparka *et al.*, 1997), allowing movement of the viral ribonucleoprotein complex to the neighbouring cells. The possible mechanisms of viral-targeted transport in plants have been reviewed extensively in the past few years and will not be covered here. However, the existence of such mechanisms provided an important clue that, perhaps, endogenous ribonucleic complexes could also be transported through plasmodesmata.

3.1.1.2 Cell-to-cell movement of a transcription factor with its mRNA
Most plant organs originate post-embryonically from meristems, which correspond to groups of undifferentiated stem cells that are set aside during embryogenesis. The aerial parts of the plants are generated by the shoot apical meristem (SAM), which consists of three consecutive layers of tissue (respectively L1, L2 and L3, from the outer most layer to the inner most layer) that produce the epidermis, subepidermis and vasculature (see also Chapter 6). Extensive intercellular communication is required between these three layers to coordinate cell proliferation and organ differentiation. The demonstration that transcription factors could move together with their mRNA through different SAM layers was provided by studies of the KNOTTED1 (KN1) protein, a homeobox transcriptional regulator found in maize (Vollbrecht *et al.*, 1991; Reiser *et al.*, 2000). *In situ* hybridization and immunolocalization experiments revealed that the KN1 protein could be detected in the L1 of the maize SAM even though the *kn1* mRNA was apparently limited to L2 and L3 (Lucas *et al.*, 1995), therefore suggesting intercellular movement of KN1. This hypothesis was confirmed by microinjection of fluorescently labelled KN1 in maize and tobacco mesophyll cells. It was found that movement of KN1 into the surrounding cells was associated with an increase in plasmodesma SEL (Lucas *et al.*, 1995).

The interaction between KN1 and plasmodesmata, the resulting increase in SEL, and the capacity of the protein to mediate its own cell-to-cell transport were highly reminiscent of the standard properties of viral MPs. This suggested that KN1 could also promote the transport of nucleic acids, an idea that was subsequently confirmed experimentally. However, in contrast to the situation observed with viral MPs, movement of nucleic acids through KN1 was found to be sequence-specific, as it occurred only if sense *KN1* transcripts were involved. A phage display procedure involving plasmodesmata-enriched cell wall fractions was used to identify a KN1-derived peptide that strongly antagonized the increase in SEL normally potentiated by the full-length protein, presumably through competition for plasmodesmata-binding domains. Interestingly, this peptide completely abolished movement of the *KN1* sense mRNA while it only reduced movement of the protein (Kragler et al., 2000), suggesting that increased plasmodesmata SEL was not essential for KN1 transport (which probably occurred through passive diffusion, at least partly) but it was a prerequisite for the transport of the ribonucleoprotein KN1/*KN1* complexes.

A series of alanine scanning mutants was generated to investigate protein domains essential for KN1 transport through plasmodesmata. This study identified a potential nuclear localization signal (NLS) in the N-terminal part of the homeodomain, which, when mutated, significantly reduced the KN1 cell-to-cell movement ability. Since the antagonist peptide evoked did not span the region corresponding to the putative NLS, it may be that different domains in the KN1 protein have separate functions for its transport. For instance, the N-terminal part of KN1 could be involved in specific interactions of the protein with receptors located near the plasmodesmata pore, whereas the C-terminal part, containing the homeodomain, could be involved in targeting the protein to plasmodesmata and/or in modification of the SEL.

It is of note that although the ability of KN1 to trafflick its own mRNA has been unambiguously established, the biological relevance of this process is not entirely clear, since, *in vivo*, KN1 – but not its mRNA – is detected in the L1 layer of the SAM. Perhaps, the level of sensitivity of the *in situ* hybridization protocol used in this study was insufficient to detect the *kn1* mRNA. Quantitative RT-PCR on laser-microdissected tissue sample could help to resolve this issue.

The *Antirrhinum* MADS domain protein DEFICIENS was shown to move from the L2 into the L1 layer of floral meristems (Perbal et al., 1996). In *Arabidopsis*, the SHORTROOT (SHR) (Nakajima et al., 2001) and LEAFY (LFY) (Sessions et al., 2000) transcription factors have also been shown to function non-cell autonomously. For instance, the SHR mRNA is expressed in the stele of the root (Helariutta et al., 2000), yet the SHR protein is found both in the stele and in the surrounding endodermis, which is absent in *shr* mutants. Likewise, expression of *LFY* from the epidermis-specific *ML1* promoter was correlated with a gradient of LFY protein that extended over several internal cell layers beyond the epidermis (Sessions et al., 2000). So far, however, no evidence has been obtained to suggest that these transcription factors are able to mediate transport of nucleic acids, as seen in the case of KN1.

3.1.2 *Long-distance transport of plant mRNAs*

Long-distance transport of water, nutrients, hormones and other signals in plants relies on the plant vascular system, which comprises the xylem and the phloem. The xylem consists of rows of dead cells, whereas the phloem contains living, albeit enucleate cells called the sieve elements (SE) that are surrounded by nucleated companion cells (CCs). The thermodynamic driving force of the phloem sap relies on a pressure gradient created by the loading of sugars produced in source leaves and their subsequent unloading in sink tissue (Turgeon, 1996). The SE and surrounding CCs are connected by a considerable number of plasmodesmata that typically have a larger SEL compared to the ones found in non-vascular tissues of mature leaves

(Imlau *et al.*, 1999). This peculiarity is probably explained by the fact that extensive mRNA and protein communication between the SE and CCs is required to maintain the integrity of SE, which are not competent for protein synthesis because they lack a nucleus and Golgi apparatus.

Evidence for this kind of transport through SE–CCs plasmodesmata comes from studies of the spatial distribution of the sucrose transporter SUT1 mRNA (Kuhn *et al.*, 1997). *In situ* hybridization and immunogold labelling revealed that both SUT1 mRNA and protein are detected in SE although the SUT1 mRNA is produced exclusively in CCs (Kuhn *et al.*, 1997). The SUT1 mRNA has also been directly detected in the phloem sap (Ruiz-Medrano *et al.*, 1999) of cucurbits, providing the first experimental clue that movement of endogenous mRNAs through the CC–SE plamodesmata could allow their translocation into the phloem stream where they could potentially serve as long-distance signalling molecules.

This initial observation prompted the thorough analysis of the mRNA content of the cucurbit phloem sap (Sasaki *et al.*, 1998; Ruiz-Medrano *et al.*, 1999). These studies revealed a large population of at least 500 polyadenylated RNA molecules, several of which were subsequently confirmed to traffic over long distances, as discussed later on. One of these mRNAs, isolated from *Cucurbit maxima* (pumpkin), was found to encode an RNA-binding protein named CmPP16 (Ruiz-Medrano *et al.*, 1999), which shows functional similarity and a limited degree of sequence identity to the MP of red clover necrotic mosaic virus (RCNMV) (Xoconostle-Cazares *et al.*, 1999). CmPP16 was indeed detected with an antibody directed against the RCNMV-MP and exhibited a similar capacity to increase plasmodesmata SEL, to mediate its own cell-to-cell movement and to potentiate the intercellular trafficking of RNA in a non-sequence-specific manner (Xoconostle-Cazares *et al.*, 1999).

Heterografting experiments between pumpkin (rootstocks) and *Cucumis sativus* (cucumber scions) clearly demonstrated that both CmPP16 and its mRNA are indeed translocated over long distances through the phloem. The experiments exploited the polymorphism found between cucumber and pumpkin CmPP16 (both at the protein and mRNA level). Therefore, its detection in the scions could have only resulted from graft transmission. Altogether, these findings supported the idea that systemic transport of ribonucleoprotein complexes could participate in the integration of biological functions in plants.

The demonstration that phloem-mobile mRNA can play a role in developmental events was recently provided by grafting experiments conducted with tomato plants carrying the dominant gain-of-function mutation *Mouse ears* (*Me*), which affects leaf morphology (Kim *et al.*, 2001). The *Me* phenotype is caused by accumulation of fusion transcripts between the *PHOSPHATE-DEPENDENT PHOSPHO-FRUCTOKINASE* (*PFP*) gene, and a *KN1*-like homeobox gene, *LeT6* (Chen *et al.*, 1997). The resulting chimaeric *PFP-LeT6* mRNA was shown to be translocated in the phloem from *Me* stocks to wild-type (wt) tomato scions. Moreover, the graft-transmitted transcript accumulated in the SAM, resulting in phenotypic changes in the wt scions that were similar to those of the *Me* stocks (Kim *et al.*, 2001).

Implicit to the idea that phloem long-distance movement of mRNA could orchestrate crucial biological functions in plants is the notion that there must be control checkpoints ensuring, for instance, that macromolecules are not mistargeted. The presence of a finite number of mRNAs in cucurbit phloem sap suggests the existence of a selective recognition and transport mechanism able to discriminate between cell-autonomous and non-cell autonomous CC–SE transcripts. Moreover, the finding that only a subset of the mRNAs found in cucurbit phloem sap was detected in apices of heterografted plants (Ruiz-Medrano *et al.*, 1999) suggests that such discrimination process may also operate at the level of phloem exit. The mechanism and components of this elusive 'surveillance system' still remain to be identified.

3.2 Intercellular movement of viroids

3.2.1 What are viroids?

For decades, viruses were considered the smallest biological entity, a paradigm remaining undisputed until the discovery of viroids, a term coined by Diener to name the agent responsible for the Potato spindle tuber (PST) disease (Diener, 1971). Following from their discovery many studies provided compelling evidence that, far from being just a curiosity, viroids were in fact a novel class of subviral pathogens that accounted for some of the most devastating diseases of plants. For instance, the cadang–cadang viroid (CCCVd) has been directly implicated in the death of more than 20 million coconut trees in Southeast Asia.

On the basis of some of their autocatalytic properties, viroids are considered as possible relics of the RNA world (Diener, 2001). Isolation, cloning and sequencing procedures have led to the identification of more than 30 different species (Flores *et al.*, 2003; Pelchat *et al.*, 2003) that are classified in two families: the *Pospiviroidae*, whose type member is PSTVd; and the *Avsunviroidae*, represented by *avocado sun blotch viroid* (ASBVd). Viroids are single-stranded, non-coding and covalently closed RNA molecules, which adopt in most cases a quasi-rod-like secondary structure. Viroid genomes range in size from 246 to 401 nucleotides, which is approximately 10-fold smaller than the smallest known viral RNA. Their replication is either nuclear (*Pospiviroidae*) or chloroplastic (*Avsunviroidae*), and involves an asymmetric (*Pospiviroidae*) or symmetric (*Avsunviroidae*) rolling-circle mechanism with three catalytic steps that are briefly described below.

The first step involves rolling-circle transcription of the monomeric circular RNA by a cellular RNA polymerase to produce oligomeric strands. For PSTVd and presumably other members of the *Pospiviroidae*, the enzyme involved is the DNA-dependent RNA polymerase II (Muhlbach & Sanger, 1979; Rackwitz *et al.*, 1981; Schindler & Mülbach, 1992) whereas this reaction is performed by a nuclear-encoded homologue of a choloroplastic RNA polymerase in the case of *Avsunviroidae* (Flores & Semancik, 1982; Navarro *et al.*, 2000). The second step involves precise cleavage of the oligomeric transcripts to release linear monomeric forms of the viroid. For *Avsunviroidae*, this step is mediated by a hammerhead ribozyme embedded in both polarity strands of the viroid (Hutchins *et al.*, 1986; Hernandez *et al.*, 1992; Navarro & Flores, 1997). In the case of *Pospiviroidae*, it is thought that cleavage is mediated by a host-encoded RNase, which recognizes a specific conformation on the oligomeric viroid RNA, dictating cleavage specificity (Baumstark *et al.*, 1997; Liu & Symons, 1998). The last step of the replication cycle involves circularization of linear viroid monomers through RNA ligation. Whether this reaction is mediated by a host-encoded RNA ligase (Tabler *et al.*, 1992; Baumstark *et al.*, 1997) or by self-ligation (Lafontaine *et al.*, 1995; Cote *et al.*, 2001) is as yet undetermined (Flores *et al.*, 2000; Diener, 2001; and references therein).

3.2.2 Intercellular movement of viroids

In spite of the small size and non-coding nature of their genome, viroids are able to replicate, move from cell to cell, spread systemically, and cause symptoms on their hosts (Davies *et al.*, 1974; Hall *et al.*, 1974; Flores, 2001). Thus, these RNA molecules must rely entirely on cellular factors to carry out all the steps that are required for successful infection. They are therefore extremely valuable probes of the mechanisms and roles of endogenous RNA trafficking processes in plants.

3.2.2.1 Cell-to-cell movement of viroids

Cell-to-cell transport of PSTVd was analysed by microinjections of *in vitro* PSTVd RNA transcripts labelled with TOTO-1 iodide, a nucleotide-specific fluorescent dye (Glazer & Rye, 1992). When introduced into a symplastically connected mesophyll cell, rapid cell-to-cell movement of PSTVd occurred (Ding *et al.*, 1997). In contrast when similarly injected into symplastically isolated guard cells (Galatis, 1980; Wille & Lucas, 1984; Palevitz & Hepler, 1985), TOTO-labelled PSTVd remained in the injected cells, indicating that cell-to-cell movement of this viroid occurs through plasmodesmata. Interestingly, a 1400 nt long, heterologous RNA containing only vector sequences did not move unless it was fused to the PSTVd transcript (Ding *et al.*, 1997). This finding strongly suggested that PSTVd possesses a sequence or structural motif for plasmodesmata transport. Whether PSTVd 'piggy-backs' on cellular proteins that are transported between cells or interacts directly with plasmodesmata components remains to be determined.

3.2.2.2 Long-distance movement of viroids

At the whole plant level, PSTVd infection in tomato spreads in a pattern similar to photoassimilate transport, from source to sink tissues (Palukaitis, 1987; Zhu *et al.*, 2001), suggesting that PSTVd moves long distance through the phloem. Consistent with this hypothesis, *in situ* hybridization experiments have shown that PSTVd traffics through and replicates actively in the phloem during long-distance movement (Zhu *et al.*, 2001). Presumably, replication along the transport pathway provides more infectious PSTVd molecules either for further long-distance trafficking or for invasion of neighbouring, non-vascular cells. Analysis of mature floral organs indicated that PSTVd was detected in sepals but not in petals, stamens or ovaries. This observation was unexpected, since phloem connections exist in all floral parts and since ovaries, stamens and petals are strong sink tissues in which movement of plant mRNAs (Ruiz-Medrano *et al.*, 1999) and of the green fluorescent protein (GFP) had been reported (Imlau *et al.*, 1999). Absence of PSTVd in specific floral organs could result either from a failure of replication in these organs or from an inhibition of transport into those tissues. To distinguish between these possibilities, Zhu *et al.* (2002) used transgenic *Nicotiana benthamiana* that constitutively expressed the PSTVd cDNA under the control of the cauliflower mosaic virus (CaMV) 35S promoter (Hu *et al.*, 1997). *In situ* hybridization analysis using an RNA probe specific for the minus strand of PSTVd (diagnostic of its replication) showed that PSTVd could in fact replicate in all floral organs. Therefore its specific absence in ovaries, stamens and petals in mechanically inoculated plants was probably due to restriction of its trafficking in parts of the floral apex (Zhu *et al.*, 2002).

Restriction of long-distance viroid movement has also been indicated with hop stunt viroid (HSVd). HSVd-free plants could be regenerated by culture of an 0.2-mm shoot tip excised from infected plants, whereas plants generated from larger shoot tips contained HSVd, suggesting that HSVd is absent from the SAM of infected hop plants (Momma & Takahashi, 1983). Indeed, *in situ* hybridization failed to detect PSTVd in the SAM, even if the viroid was present in the vascular tissues (most likely the procambium and/or the protophloem) immediately below the SAM (Zhu *et al.*, 2001).

3.2.3 Cellular factors involved in viroid movement

3.2.3.1 Phloem Lectin 2

Only two proteins that may play a role in viroid RNA trafficking have been characterized so far. The first one is the phloem lectin 2 (PP2), one of the two most abundant proteins found in *C. maxima* phloem exudates (Beyenbach *et al.*, 1974). *In vitro*, this protein binds to HSVd RNA, PSTVd RNA as well as to other larger and less structured RNAs (including RNA with a polyA tail) but not to DNA (Gomez & Pallas, 2001; Owens *et al.*, 2001). Using genus-specific probes/antibodies and intergenic grafts, it was shown that PP2, but not its mRNA, is capable of long-distance trafficking through the phloem. PP2 was synthesized within the companion cells of the rootstocks, from which it moved into the SE and was subsequently transported in the phloem stream towards the scion tissues. Upon delivery PP2 was found to accumulate in CCs, indicating movement out of the scions' SE (Golecki *et al.*, 1999). Interestingly, the PP2 of *Cucurbitaceae* spp. has the capacity to modify plasmodesmata SEL and to move from cell to cell when it is microinjected into mesophyll cells of cucurbit cotyledons (Balachandran *et al.*, 1997). Altogether, these observations indicate that PP2 has the attributes that would be normally required for translocation of not only viroids but also plant RNAs. Further experiments *in vivo* are necessary to demonstrate this unequivocally.

3.2.3.2 VirP1

The second protein, named VirP1, was isolated using an RNA-ligand screening procedure from an expression library of *Lycopersicum esculentum* (Martinez de Alba *et al.*, 2003). VirP1 contains a bipartite nuclear localization signal, a bromodomain found in several eukaryotic transcription factors (Haynes *et al.*, 1992; Jeanmougin *et al.*, 1997) and an RNA-binding domain. It was shown that this protein interacts specifically *in vitro* with PSTVd RNA and immunoprecipitation from PSTVd-infected tissues confirmed that this complex does exist *in vivo*. Using the three-hybrid system, interaction domains between VirP1 and PSTVd RNA were mapped to an asymmetric internal loop called the *RY motif* (Gozmanova *et al.*, 2003; Maniataki *et al.*, 2003). This motif is contained within an approximately 80 nt long, bulged hairpin at the right-end side of the PSTVd RNA.

Sequence alterations in PSTVd (Hammond, 1994) had been described, that prevented the structural interactions in the terminally located RY motif. The resulting mutant, PSTVd-R+, was not infectious if it was mechanically inoculated onto tomato leaves, but it was able to replicate in crown galls developing on plants that had been injected with recombinant *Agrobacterium* carrying the PSTVd-R+ cDNA. Systemic infection occasionally occurred in those plants, because of the emergence of revertants. However, sequence analysis indicated the emergence of only one class of variant (apart from the wt PSTVd) in which the original base pair interaction at the RY motif was restored despite sequence alterations in the terminal right loop (Hammond, 1994). *In vivo* and *in vitro* binding assays have indeed shown that the R+ mutation impedes interaction of VirP1 with the PSTVd RNA, whereas this interaction is partially restored in variant 1 (Maniataki *et al.*, 2003). Taken together, these results strongly support the involvement of VirP1 in PSTVd RNA movement. Note that it was shown that VirP1 is expressed in petals (Martinez de Alba *et al.*, 2003), organs that have been found free of PSTVd (see above). Therefore, a reason other than the lack of VirP1 accounts for the absence of PSTVd unloading in petals.

At first, the nuclear localization signal of VirP1 and its homology with some transcription factors was difficult to reconcile with its role in PSTVd movement. However, PSTVd replication occurs in the nucleus and we have seen in the previous sections of this chapter that mutations of the NLS found in the maize transcription factor KN1 significantly inhibit transport of the *KN1* mRNA through plasmodesmata in mesophyll cells (Lucas *et al.*, 1995; reviewed in Wu *et al.*, 2002). This analogy could suggest that common mechanisms are involved in KN1-mediated trafficking of the KN1 mRNA and the VirP1-mediated trafficking of viroid RNAs and is consistent with the proposal that VirP1 may be broadly implicated in movement of endogenous transcripts. Based on this idea, it is of note that several viroids induce symptoms on their host that are reminiscent of the developmental anomalies usually elicited by mutations in bromodomain-containing proteins. Thus, it is possible that the developmental aberration of viroid-infected plants result, at least partly, from a competition between the binding of cellular RNA and of viroid RNA by VirP1.

3.3 Intercellular movement of RNA silencing

3.3.1 *Mechanism of RNA silencing*

This section highlights the most significant steps that contributed to our current understanding of the core reactions of RNA silencing in eukaryotes and, more specifically, in plants. The aim is to provide the reader with a basic knowledge of this process, and its biological relevance, together with key notions that are relevant to the non-cell autonomous nature of the phenomenon. More specific aspects about the mechanism and roles of RNA silencing can be found in recent reviews (Voinnet, 2001, 2002; Hutvagner & Zamore, 2002b; Bartel, 2004).

3.3.1.1 Co-suppression in petunia

Shortly after the discovery of the Ti plasmid of *Agrobacterium tumefaciens*, molecular biologists exploited transgenesis to modify agronomic traits of plants. However, for any given transgene construct, a variable fraction of the primary transformants did not express the product of the introduced gene, despite its stable integration into the plant genome. A now classical example of this phenomenon was provided more than 10 years ago by studies of transgenic petunias (Napoli *et al.*, 1990). These plants had been engineered to carry extra copies of the gene encoding the enzyme chalcone synthase (CHS), which is involved in synthesis of anthocyanins, the pigments that account for the purple colour of petals. However, although the aim of those experiments was to produce petunia plants with deep purple corolla, some transgenic lines had flowers with completely white petals, phenocopying a *CHS* knockout. Molecular analysis of these peculiar lines indicated that the white petal phenotype resulted from elimination of both the transgene and the corresponding endogenous CHS mRNAs, a phenomenon coined 'co-suppression'. Seminal nuclear run-on experiments showed that the reduction in CHS mRNA levels was at the steady state, rather than the transcriptional level (Napoli *et al.*, 1990; Vanderkrol *et al.*, 1993). Moreover, only RNAs identical in sequence to the introduced *CHS* transgene were targeted by this 'post-transcriptional gene silencing' (PTGS). Later on, many examples of co-suppression involving various transgene–transgene or transgene–endogenous gene combinations were described in plants and a related process called *quelling* was also reported in the filamentous fungus *Neurospora crassa* (Cogoni *et al.*, 1996; Vaucheret *et al.*, 1998).

3.3.1.2 Double-stranded RNA: trigger molecule of RNA silencing

In 1998, experiments carried out in *Caenorhabditis elegans* led to the discovery that exogenous delivery of double-stranded (ds)RNA into the body cavity of the worm induced a highly sequence-specific RNA degradation mechanism targeted against any cellular mRNA that shared sequence homology with the introduced dsRNA molecules (Fire *et al.*, 1998). There was little or no effect with either sense or antisense single-stranded (ss)RNA. This process, referred to as RNA interference or RNAi, was later found to occur in many other eukaryotes including fruitfly, mouse and human (Billy *et al.*, 2001; Elbashir *et al.*, 2001a,b). Obviously, the effects of RNAi were highly reminiscent of those of co-suppression and this prompted plant scientists to re-evaluate the potential of dsRNA as a possible trigger of the RNA degradation process involved. It was indeed found that a significant proportion of plants exhibiting co-suppression carried transgene loci that were organized in inverted repeats, with the potential to produce dsRNA-like transcripts through intramolecular base-pairing (Sijen *et al.*, 1996; Stam *et al.*, 1997). Moreover, transgenes that were deliberately engineered to produce dsRNA caused a very high incidence of co-suppression in tomato and *Arabidopsis* (Hamilton *et al.*, 1998; Chuang & Meyerowitz, 2000), leading to the suggestion that, as in animals, dsRNA was a critical molecule in the development of PTGS in plants.

These findings, however, did not explain how PTGS was triggered in transgenic lines expressing single copy, sense transgenes with no intrinsic potential to form dsRNA, such as the co-suppressed petunias described above. It was proposed that some single-stranded transcripts produced in these silenced lines had a distinctive, 'aberrant feature' that triggered their conversion into dsRNA through the action of an elusive plant-encoded RNA-dependent RNA polymerase (RdRp) (Dougherty & Parks, 1995; Wassenegger & Pelissier, 1998), a suggestion supported by the fact that RdRp activities had been indeed previously detected in plants (Schiebel *et al.*, 1993a,b, 1998). This hypothesis was subsequently confirmed by the identification of several genes that were specifically required for this 'sense transgene PTGS' in *Arabidopsis*. Among those genes, *SDE1/SGS2* was found to encode a putative RNA-dependent RNA polymerase and *SDE3* a putative RNA helicase (Mourrain *et al.*, 2000; Dalmay *et al.*, 2001). The current model is that the combined action of SDE1 and of SDE3 produces dsRNA *de novo* using transgene-derived, single-stranded transcripts as templates. The resulting dsRNA then triggers PTGS (Beclin *et al.*, 2002). A similar scenario probably accounts for the initial stages of quelling in *Neurospora*, since a SDE1 homologue, QDE1, was also found to be required for this process (Cogoni & Macino, 1999). At present, the presumed 'aberrant feature' that distinguishes RdRp templates from other cellular sense transcripts during the initiation of PTGS and quelling remains unclear. It may include the lack of a polyA tail or inadequate subcellular localization. Transgene dosage has been often positively correlated with the onset of co-suppression in plants (Pang *et al.*, 1996) and this can also be accommodated with the 'aberrant RNA model', since higher levels of transcription will increase the chances of such RNA being synthesized and detected in the cell.

3.3.1.3 *Short interfering (si)RNAs are the specificity determinants of RNA silencing*

Whether introduced exogenously, produced from inverted repeat constructs or synthesized by cellular RNA-dependent RNA polymerases (RdRps), the dsRNA trigger of RNAi and PTGS induces degradation of cytoplasmic RNAs that are identical in sequence. Therefore, both PTGS and RNAi must involve a *trans*-acting specificity determinant with a nucleic acid component, whose nature was first investigated in plants. Since several cases of co-suppression were triggered by sense transcripts and targeted against sense mRNAs, it was likely that the nucleic acid involved was an antisense RNA. However, many studies of co-suppressed lines failed to provide evidence for the involvement of a high-molecular-weight antisense RNA (Dougherty *et al.*, 1994; Kunz *et al.*, 1996). It was therefore reasoned that, perhaps, this elusive antisense molecule was in fact too short to be detected by conventional procedures.

To address this possibility, low-molecular-weight RNAs were selectively enriched from total RNA fractions of silenced plants and subjected to Northern analysis. It was found that a discrete species of 21–24 nt long RNAs with antisense sequence of the silenced genes invariably accumulated in the co-suppressed lines tested (Hamilton & Baulcombe, 1999). Subsequent studies in *Drosophila* and *C. elegans* demonstrated the involvement of similar molecules in the RNAi process (Tuschl *et al.*, 1999; Parrish *et al.*, 2000).

Interestingly, 21–24 nt long RNAs in the sense orientation were found in equal abundance in silenced plants, suggesting that these small RNAs accumulated as duplexes that could be derived from the dsRNA that triggers PTGS/RNAi. This idea was indeed confirmed by the demonstration that long dsRNA is converted into 21–24 nt RNA duplexes in *Drosophila* embryo extract (Zamore *et al.*, 2000). These duplexes have 5′ terminal phosphate and 2 nt long overhanging 3′ ends that are characteristic of the product of RNaseIII-like enzymes (Elbashir *et al.*, 2001c). Analysis of the *Drosophila* genome identified three genes that could potentially encode such proteins. However, only one of these proteins, Dicer, was able to produce 21–24 nt RNAs when incubated with long dsRNA in embryo extracts (Bernstein *et al.*, 2001). Defining features of Dicer, in addition to an RNaseIII domain, include two dsRNA-binding domains, an RNA helicase domain and a putative protein–protein interaction domain known as PAZ (PIWI/ARGONAUTE/ZWILLE) (Cerutti *et al.*, 2000). In agreement with a key role for Dicer in RNA silencing, genes encoding Dicer-like enzymes have been identified in the genome of all eukaryotes in which PTGS/RNAi has been experimentally verified (Hutvagner & Zamore, 2002b). Moreover, some organisms such as *Arabidopsis* and *Drosophila* have several Dicers that may serve specific functions (Finnegan *et al.*, 2003). For instance, in silenced plants, siRNAs with sequence of the silenced gene are in two discrete classes of 21 nt and 24 nt that are the likely products of separate Dicers and appear to be functionally distinct (Hamilton *et al.*, 2002; Tang *et al.*, 2003; Xie *et al.*, 2004).

Altogether, these findings support the previously made suggestion that 21–24 nt RNA duplexes could provide sequence specificity to the machinery that degrades homologous mRNAs in RNAi/PTGS. In agreement with this suggestion, chemically synthesized small RNAs with features of Dicer cleavage products were sufficient to mediate RNAi in *Drosophila* embryos and in human cells (Elbashir *et al.*, 2001a,b). Consequently, the silencing-related small RNAs were coined 'small interfering' or siRNAs.

3.3.1.4 RNA-induced silencing complex RISC

Fractionation experiments in *Drosophila* embryo extracts indicated that the nuclease that generates the siRNAs (Dicer) could be separated from an activity that accounts, on its own, for the degradation of RNAs that are identical in sequence to the siRNAs (Hammond *et al.*, 2000). This activity belongs to a multi-subunit complex referred to as the 'RNA-induced silencing complex' (RISC), which co-purifies with siRNAs. RISC activities have been found in other animals, including *C. elegans* and human (Hutvagner & Zamore, 2002a; Caudy *et al.*, 2003). Biochemical characterization of the *Drosophila* RISC revealed that it contains several proteins, among which is Argonaute 2 (Ago2), a homologue of the translation initiation factor eIF2C (Hammond *et al.*, 2001). Interestingly, Ago2 contains a PAZ domain that is potentially involved in mediating Dicer–RISC interaction. Upon this interaction, it is thought that the Dicer-generated siRNAs are incorporated into RISC and thereby guide degradation of homologous RNAs in a sequence-specific manner. This scenario also most likely applies in plants, since a RISC activity has been detected in wheat germ extracts (Tang *et al.*, 2003). The endonucleolytic cleavage of RISC targets occurs at the centre of the siRNA/RNA hybrid (11th nucleotide) (Elbashir *et al.*, 2001c) and is mediated by Ago2 in *Drosophila* (Liu *et al.*, 2004; Meister *et al.*, 2004).

3.3.1.5 Transitive RNA silencing

As explained above, cellular RdRps play a key role in the initiation of sense transgene PTGS in plants and of quelling in *N. crassa*, presumably through recognition of transgene-derived aberrant RNAs and their subsequent conversion into double-stranded molecules. In addition, cellular RdRps may also play important functions in amplifying already established silencing. For instance, it was shown, in *C. elegans*, that RNAi triggered by siRNAs corresponding to one part of a transcript leads to production of secondary siRNAs that are homologous in sequence to regions located outside of the initially targeted portion of this transcript (Sijen *et al.*, 2001). This process leading to *de novo* synthesis of siRNAs is referred to as 'transitivity' and requires the action of RFF1, a putative RdRp homologue of *C. elegans* (Sijen *et al.*, 2001). It was proposed that the primary siRNAs that trigger RNAi prime the activity of RRF1 such that it converts siRNA-homologous transcripts into dsRNA that are subsequently cleaved by Dicer into secondary siRNAs. The idea of this 'primer-dependent' transitivity was supported by the fact that only secondary siRNAs located 5′ of the initially targeted region were apparently synthesized (Sijen *et al.*, 2001). Thus, by generating more siRNAs than originally used to trigger RNAi, the RFF1-mediated transitivity may account for the very high potency of RNAi in *C. elegans* (Fire *et al.*, 1998).

Transitivity had also been reported in transgenic tobacco exhibiting PTGS and in *Arabidopsis* infected with recombinant viruses or expressing constitutively an inverted repeat transgene (Voinnet *et al.*, 1998; Vaistij *et al.*, 2002; Himber *et al.*, 2003). Although, as in *C. elegans*, this process was dependent upon the activity of an RdRp (SDE1), occurrence of secondary siRNAs was from both the 5′ and 3′ part of the originally targeted transcript (Voinnet *et al.*, 1998; Vaistij *et al.*, 2002). To account for this phenomenon, it is possible that 5′ transitivity is primer-dependent but that 3′ transitivity is primer-independent in plants. For instance, it has been suggested that the 3′ cleavage products generated by the action of RISC could be perceived as 'aberrant' and used directly as templates by SDE1 (Hutvagner & Zamore, 2002b), in a manner similar to the events that initiates co-suppression and quelling from sense transcripts (see above section). In any case, and as seen in *C. elegans*, the involvement of transitivity in RNA silencing in plants provides a powerful system for the amplification of molecules that are actively involved in RNA silencing, namely, the dsRNA and the siRNAs that derive from it.

3.3.1.6 Biological functions of RNA silencing in plants

Most plant viruses have an RNA genome whose replication involves synthesis of cytoplasmic, double-stranded intermediates that are potent inducers of RNA silencing, as shown by the high accumulation of virus-derived siRNAs in infected tissues (Hamilton & Baulcombe, 1999; Szittya *et al.*, 2002; Lakatos *et al.*, 2004). Thus, silencing has evolved as an antiviral mechanism in plants, a proposal consistent with the findings that most, if not all, plant viruses encode proteins that inhibit various steps of the silencing mechanism (Voinnet, 2001). The existence of these 'silencing suppressors', extremely diverse in sequence and structure, provides compelling evidence for an ongoing silencing-based arms race between plants and viruses (Voinnet, 2001). A second defensive role of RNA silencing is in the control of transposable elements and maintenance of genome integrity. For instance, 24 nt siRNAs derived from the atSIN1 elements are associated with methylation of the corresponding DNA that impact on RNA levels of this element (Hamilton *et al.*, 2002; Zilberman *et al.*, 2003; Xie *et al.*, 2004).

Recent findings in both animals and plants have highlighted an important role of RNA silencing in patterning endogenous gene expression. For instance, reduction of Dicer activity has dramatic developmental consequences in mice and *Arabidopsis* (Jacobsen *et al.*, 1999; Bernstein *et al.*, 2003). In many organisms, abundant endogenous, single-stranded 21–24 nt long RNAs are processed by Dicer-like enzymes from short stem-loop precursor transcripts that are encoded in intergenic regions. These microRNAs (miRNAs) are perfectly or partially complementary to the coding region or the 3' UTR of cellular transcripts that they regulate at the level of stability or translation (for review: Carrington & Ambros, 2003; Bartel, 2004). Analyses in plants suggest that endonucleolytic cleavage is the prevalent mode of action of miRNAs (Llave *et al.*, 2002; Bartel & Bartel, 2003). Like siRNAs, they appear to be incorporated into an RISC (Tang *et al.*, 2003) upon their processing by a Dicer-like enzyme called DCL-1 (Jacobsen *et al.*, 1999; Papp *et al.*, 2003; Xie *et al.*, 2003, 2004). Many transcription factors that have been crucially implicated in plant development have been predicted or experimentally characterized as miRNA targets in plants and it is currently thought that at least some miRNAs play a crucial role in clearing the cellular content of specific transcription factors to orchestrate the emergence of cell lineages that are crucial for organ formation in plants (Aukerman & Sakai, 2003; Bartel & Bartel, 2003; Palatnik *et al.*, 2003; Chen, 2004).

3.3.2 The discovery of systemic RNA silencing

The first hint of the non-cell-autonomous nature of PTGS in plants was provided by studies of transgenic tobacco lines exhibiting spontaneous silencing of a highly expressed nitrate reductase transgene (*Nia*). Co-suppression was manifested by the occurrence of cholorosis due to perturbation in nitrogen availablility, which provided a visual phenotype of the silenced state (Palauqui & Vaucheret, 1995). Detailed analysis of the spatial and temporal pattern of the chlorotic phenotype in those plants showed that development of silencing was a dynamic process (Palauqui *et al.*, 1996). It was first apparent within one or several expanding clusters of cells that were stochastically distributed in mature leaves. These silencing foci eventually reached a vein from which the co-suppressed state was transmitted to the new growth, following a pattern reminiscent of the phloem translocation (Palauqui *et al.*, 1996). The existence of a signal molecule for RNA silencing was confirmed in elegant experiments involving grafting of non-silenced transgenic plants onto the *Nia* co-suppressed plants (Palauqui *et al.*, 1997). Grafting resulted in 100% transmission of the silenced state from the rootsocks to transgenic scions that expressed an initially nonsilenced *Nia* transgene. These experiments provided a clear demonstration that a silencing signal emitted from the rootstocks was able to travel over long distances through the vascular system and could trigger *de novo* silencing of a homologous transgene in remote tissues of the plant. Interestingly, systemic patterns similar to those of the *Nia* silenced plants had been observed with other tobacco-based silencing systems that involved co-suppression of S-adenosyl-L-methionine synthetase and chitinase genes (Boerjan *et al.*, 1994; Kunz *et al.*, 1996). This suggested that systemic RNA silencing was not a peculiarity of the *Nia* transgenic system.

Transgenic *N. benthamiana* plants expressing the GFP provided a powerful experimental system to confirm that non-cell-autonomous silencing was indeed a broadly applicable principle in plants (Voinnet & Baulcombe, 1997). Infiltration of a culture of recombinant *Agrobacterium* was used to allow a local and transient production of GFP RNA within a cluster of mesophyll cells in one or two leaves of the GFP transgenic plants. This treatment triggered first the loss of GFP fluorescence within the infiltrated region of the leaf as a result of local PTGS of both the stably integrated and ectopic GFP transgenes. Phenotypically, silencing was manifested as the appearance of red fluorescent tissue under ultraviolet (UV) illumination, owing to chlorophyll autofluorescence. This red (i.e. silenced) tissue was restricted to the *Agrobacterium* infiltrated area, whereas the surrounding non-silenced tissue remained green under UV light. Although silencing had been triggered locally, it moved to remote parts of the plants, which eventually became uniformly silenced for GFP (Voinnet & Baulcombe, 1997).

A remarkable aspect of the long-distance signalling elicited in the *Nia* and GFP systems was its nucleotide sequence specificity, both in terms of initiation and propagation. Hence, graft transmission of *Nia* silencing was not observed if the scions were from transgenic plants expressing a transgene that was divergent in sequence from the co-suppressed *Nia* transgene in the rootstocks (Palauqui *et al.*, 1997). In addition, systemic silencing of GFP did not take place unless the *Agrobacterium*-infiltrated cultures carried a transgene that was identical in sequence to the stably integrated GFP transgene (Voinnet & Baulcombe, 1997). Therefore, to account for its nucleotide sequence-specific effects, it was proposed that the systemic signal for RNA silencing must have a nucleic acid component (Palauqui *et al.*, 1997; Voinnet & Baulcombe, 1997).

3.3.3 Initiation of systemic RNA silencing

3.3.3.1 Spontaneous activation of systemic RNA silencing

Systemic silencing was originally discovered in plants exhibiting co-suppression in a spontaneous manner. In the case of the *Nia* tobacco transformants described above, however, only a subset of the lines exhibited this property (see Plate 3.1A, B, following page 146). These plants were referred to as 'Class-II plants', in contrast to the 'Class-I plants', which were unable to trigger *Nia* silencing *de novo* unless they were grafted onto silenced scions of Class II, which were competent to send a systemic silencing signal (Palauqui & Vaucheret, 1998). Molecular analyses indicated that the difference between Class-I and Class-II lines was not due to transgene copy number *per se* but, rather, to an ill-defined transgene feature that potentiated the capacity of spontaneous triggering through accumulation of a particular transgene product (Palauqui & Vaucheret, 1998). Since the construct integrated into the genome of the Class-II plants was designed to produce sense transcripts, it is very likely that spontaneous silencing was elicited via an SDE1-dependent pathway that could sense and convert aberrant transgene transcripts into dsRNA molecules (see Section 3.3.1.2). Possibly, the transgene constructs in Class-II plants were more prone to produce the aberrant RNA than those of Class-I plants. Significantly, transgene dosage was found to be pivotal in the initiation of silencing in Class-II plants, since analysis of five independent lines carrying each a single copy of the *Nia* transgene revealed that only homozygous (as opposed to hemizygous) plants displayed the spontaneous silencing phenotype (Palauqui & Vaucheret, 1998). Assuming that the sense transgene constructs carried by the Class-II plants were prone to the production of only a low level of aberrant transcripts, the duplication of the corresponding loci may have enhanced the chances of such transcripts being detected in the cell and subsequently processed into dsRNA. Therefore, both qualitative and quantitative features of the *Nia* loci/mRNA influenced the onset of spontaneous silencing.

A detailed analysis revealed that, under standard greenhouse growth conditions, spontaneous co-suppression was triggered at various frequencies (ranging from 5 to 42%) between individual Class-II lines (Palauqui & Vaucheret, 1995). However, for all lines, triggering consistently occurred during a phenocritical period ranging from 15 days post-germination to flowering. In addition, the incidence of co-suppression between individuals from each line was increased if plants were grown *in vitro* prior to their transfer in greenhouse. Similar observations were made with co-suppressed lines of tobacco plants expressing a nitrite reductase (*Nii*) transgene (Palauqui & Vaucheret, 1995). These findings indicate that the physiological state of the plant (i.e. transition from vegetative to reproductive growth) and environmental factors exert a critical influence upon activation of spontaneous systemic silencing. In addition, triggering of *Nia* and *Nii* silencing was found to occur exclusively in mature leaves, indicating that the intrinsic physiological status of those organs also played an important role in the initiation of spontaneous silencing (Palauqui & Vaucheret, 1995). In this regard, it is significant that many cells in mature leaves (unlike those in young developing tissues) undergo the process of endoreduplication, a somatic polyploidysation resulting from repeated rounds of DNA replication in the absence of intervening mitoses (Sugimoto-Shirasu & Roberts, 2003). It is conceivable that the duplication of chromosome sets in those cells may have contributed to increase in the pool of aberrant RNAs evoked above. The well-established random nature of the endoreduplication process in leaves (Sugimoto-Shirasu & Roberts, 2003) would also explain the stochastic distribution of silencing foci observed in source leaves of the Class-II plants (Palauqui et al., 1996).

3.3.3.2 Exogenously induced systemic silencing

Unlike the Class-II *Nia* plants, the transgenic GFP lines described previously did not have the intrinsic capacity to spontaneously activate systemic silencing (Ruiz et al., 1998). Initiation and propagation of silencing occurred only if the plants were provided with excess doses of the *GFP* transgene via *Agrobacterium*-mediated transient expression of ectopic constructs (Voinnet & Baulcombe, 1997). A biolistic procedure was developed as an alternative method to deliver the silencing triggers in those plants and was exploited to show that systemic GFP silencing could be initiated from as little as 8–10 randomly bombarded cells within one single leaf (Voinnet et al., 1998). Upon bombardment, GFP silencing progressively radiated around those cells and formed macroscopically detectable red fluorescent foci resembling the chlorotic spots observed on mature leaves of the Class-II *Nia* co-suppressed plants (Voinnet et al., 1998). Silencing eventually reached a vein from which it was transmitted to remote parts of the plants that, eventually, became uniformly silenced for GFP. Biolistic delivery of *Nia* DNA constructs was also shown to trigger localized and systemic silencing in non-silenced Class-II tobacco plants (Palauqui & Balzergue, 1999).

The use of this technique allowed the rapid assessment of the requirements for triggering systemic silencing in both transgenic systems. Various DNA constructs were tested (homologous to the *Nia* and *GFP* coding sequence, respectively), ranging from plasmids containing promoter-driven and full-length cDNA to gel-purified PCR-amplified fragments. The results in both systems were remarkably similar (Voinnet et al., 1998; Palauqui & Balzergue, 1999) and indicated that sense, antisense and promoterless constructs could trigger systemic silencing, although to varying degrees. For instance, promoterless full-length cDNA constructs were consistently less efficient than equivalent constructs with a 35S promoter, which was also observed with biolistically induced systemic silencing of chitinase transgenes in tobacco. Both in the *GFP* and *Nia* systems, the proportion of plants exhibiting silencing upon bombardment also increased as the length of homology between bombarded and stably integrated sequences was higher.

More recently, the same *GFP* system and a novel *GUS*-based silencing reporter system were used to investigate the effect of bombarding homologous RNA, rather than DNA (Klahre et al., 2002). It was found that full-length sense and antisense RNA could trigger systemic silencing at a low frequency. The frequency was greatly enhanced (up to 75%) if the sense and antisense RNAs were pre-annealed to form dsRNA prior to bombardment. Shorter-than-full-length dsRNA was less efficient in triggering systemic silencing. Importantly, this study also established that a single species of synthetic siRNA duplex was nearly as competent in triggering systemic silencing as was the full-length dsRNA, whereas there was no effect if the individual siRNA strands were bombarded on their own (Klahre *et al.*, 2002).

Collectively, these results indicate that virtually any type of nucleic acid has the potential to trigger systemic silencing, provided it bares sequence homology with the mRNA of the target transgene. Importantly, the second set of experiments (Klahre *et al.*, 2002) shows that RNA molecules are directly capable of inducing this phenomenon, excluding the possible complications that are linked to transcription of the delivered DNA, or the delivered DNA itself. Therefore, it is likely that RNA is the trigger molecule of systemic silencing, which is consistent with the observation that the presence of a promoter significantly enhanced silencing initiation by DNA-based constructs. Moreover, the fact that the two key RNA components of the silencing pathway (i.e. dsRNA and the siRNAs that derive from this molecule; see Sections 3.3.1.2 and 3.3.1.3) were potent and direct triggers of systemic silencing suggests that the other RNA molecules tested in those experiments (sense and antisense RNAs) were probably ultimately converted into ds- and siRNAs, presumably through the action of a cellular RdRp akin to SDE1. It remains unclear how promoterless DNA plasmid constructs or PCR-amplified DNA fragments triggered systemic RNA silencing (Brigneti *et al.*, 1998; Palauqui & Balzergue, 1999), but spurious transcription from endogenous promoters upon their integration into the genome could be involved.

3.3.4 Propagation of systemic RNA silencing

3.3.4.1 Long-distance movement of RNA silencing

Long-distance transport occurs in the phloem. Long-distance movement of molecules in plants can occur through the xylem and the phloem conduits, both of which are restored upon graft junctions. However, only the phloem distributes the photoassimilates throughout the plant, following a specific pattern whereby mature, photosynthetically autonomous organs export – but do not import – carbohydrates. These tissues are said to be 'source' of phloem. Conversely, import of carbohydrates occurs in the new growth (meristems, primordia, young leaves), which has not yet reached photosynthetic autonomy and is therefore a 'sink' of phloem. The intermediate situation is the sink-to-source transition, which typically occurs in leaves in a basipetal direction down the leaf axis towards the petiole.

Detailed analysis of the pattern and dynamics of silencing activated upon *Agrobacterium* leaf infiltration of the GFP transgenic *N. benthamiana* revealed that at 20 days post-infiltration of lower leaves, silencing was strongest in systemic, young developing leaves (sink leaves) and was very pronounced in the shoot tips, although the meristematic regions were still green fluorescent (Voinnet *et al.*, 1998). There was also silencing in upper leaves that were already expanding at the time of infiltration (sink–source transition leaves), but it was fainter and less extensive than in the new growth. In contrast, the leaves immediately above and below the infiltrated leaves (source leaves) remained fully green fluorescent. At 30 days post-infiltration, the stem and roots below the infiltrated leaves also showed GFP silencing, thus indicating that the movement of the silencing signal was bidirectional in the plant. Overall, this spatial distribution was consistent with movement occurring through the phloem. The development of silencing in leaves was also similar to the unloading of a phloem-transported dye through Class I, II and III veins of *N. benthamiana* leaves (Roberts *et al.*, 1997; Voinnet *et al.*, 1998). Further support for phloem transport of the signal came from experiments in which the GFP plants were infiltrated in just one single leaf. These experiments differed from those described above in which the infiltration involved several leaves on both sides of the plant. At 1 month post-infiltration, GFP silencing in the stem was restricted to the side of the originally infiltrated leaf. Shoots that had emerged from the silenced portion of the stem were silenced, while those emerging from the non-silenced half were not, a spatial distribution again similar to the movement of phloem-translocated dyes in *N. benthamiana* (Roberts *et al.*, 1997; Voinnet *et al.*, 1998). Spontaneous *Nia* co-suppression in *N. tabaccum* plants of Class II also occasionally started on one single leaf situated at the bottom of the plant. In this case, the first leaves to be affected by chlorosis were also all positioned on the same side of the plant (Palauqui *et al.*, 1996).

The fact that systemic spread of RNA silencing is strongly influenced by source–sink relationships between silenced and non-silenced organs explains why the efficiency of graft transmission of silencing of a chitinase transgene in tobacco was highly dependent upon the grafting method used (Crete *et al.*, 2001). Thus, reciprocal grafts involving fully developed sections of plants or exchanges of mature tissue plugs were unsuccessful for silencing transmission, even if the former method has been routinely used to indicate the spread of systemic acquired resistance (see, for instance, Crete *et al.*, 2001). It appears, from this and other studies, that the top-grafting method in which a young vegetative shoot scion acts as a strong sink for the transport of silencing signals from source leaves constitutes the most efficient approach for transmission of systemic silencing, at least in Solanacaeous species.

The speed of translocation of the silencing signal from leaf to vasculature was assessed in the GFP transgenic system by removal of the infiltrated leaf 1, 2, 3, 4 or 5 days after infiltration of the *A. tumefaciens* strain (Voinnet *et al.*, 1998). There was systemic loss of GFP fluorescence in 10% of the plants if the infiltrated leaf was removed 2 days post-infiltration. A progressively higher proportion of plants exhibited systemic silencing when the infiltrated leaf was removed 3 days post-infiltration or later. Similar values were obtained with bombardment experiments carried out in the Class-II *Nia* tobacco plants (Palauqui & Balzergue, 1999). These observations indicate that 2–3 days are sufficient for the signal to accumulate and translocate into the phloem long-distance transport stream.

Molecular requirements for the long-distance transport and systemic effects of RNA silencing. Triple-grafting experiments in which an intermediate section of non-transgenic tobacco was inserted between silenced *Nia* rootstocks, and non-silenced

Nia scions of Class I or Class II indicated that the presence of the *Nia* transgene in the middle section was not required for transmission of the silenced state to the top scion (Palauqui *et al.*, 1997). In fact, up to 20 cm of wt plant segment could be interspaced without affecting the efficiency or the rate of silencing transmission (Plate 3.1C). However, as this intermediate section carried the endogenous copy of the *Nia* gene, it was still possible that long-distance transmission of the signal was dependent upon homologous DNA or RNA. The fact that triple-grafting experiments conducted with the GFP system gave similar results ruled out this possibility, as the intermediate wt section had no GFP sequence at all (Voinnet *et al.*, 1998). Therefore, these results demonstrated that the silencing signal could move over long distances through cells in which there is no corresponding nuclear gene.

Grafting experiments were also used to address the molecular requirements for the suppression effect of the phloem-transported signal in the *Nia* tobacco plants (Palauqui *et al.*, 1997). It was found that wt scions grafted onto co-suppressed Class-II plants were unable to undergo silencing (as assessed by normal accumulation of the endogenous *Nia* transcript and the lack of chlorosis), despite the fact that the same tissues were clearly competent to transport the signal over long distances (Plate 3.1B, C). Silencing occurred only if the scion carried a *Nia* transgene (Class-I and Class-II plants). However, the presence of the target transgene *per se* was not sufficient for silencing to occur because a *Nia* transgene that had been transcriptionally silent was not competent to perceive the silencing signal (Palauqui *et al.*, 1997). Further experiments established that non-transgenic mutant plants in which the *Nia* mRNA levels exceeded those normally found in wt plants (Class-III plants) also exhibited a silencing phenotype when grafted onto silenced stocks of Class-II plants (Palauqui & Vaucheret, 1998). Thus, transgenes appear to be dispensable for the perception of the signal and the mRNA degradation process that follows (Plate 3.1B). However, high levels of target transcripts, whether of endogenous or transgenic origin, seem to be absolutely required, at least in the *Nia* system.

3.3.4.2 *Cell-to-cell movement of RNA silencing*

Cell-to cell movement upon phloem unloading. Upon phloem unloading, silencing usually spreads across the whole leaf lamina, indicating cell-to-cell movement of the silencing signal. The pathway for this cell-to-cell movement was elucidated by observations made in the *N. benthamiana* inducible GFP silencing system. Hence, there was no GFP silencing in the stomata guard cells of leaves that were already expanding at the time of silencing initiation (Voinnet *et al.*, 1998). Stomata guard cells progressively lose plasmodesmatal connections with neighbouring cells as the age of the leaf increase; therefore, they would have been symplastically isolated before the signal had moved into those mature leaves. However, in leaves developing after the signal had spread to the apical growing point, GFP was uniformly silenced, even in the stomata guard cells. Thus, guard cells are competent for gene silencing, provided that the signal invades leaves early in their development, before their symplastic isolation (Voinnet *et al.*, 1998). Therefore, the most straightforward interpretation of these observations is that the signal moves from cell to cell *via* the symplasm through the plasmodesmata.

A particularly striking aspect of silencing movement in tissues that are competent for signal perception and mRNA degradation is the extent and uniformity of the silenced phenotype. This aspect is even more apparent in bombardment experiments in which complete systemic silencing in the new growth can be elicited with a limited quantity of trigger nucleic acid within a few cells of a single, mature leaf (Voinnet *et al.*, 1998; Palauqui & Balzergue, 1999; Klahre *et al.*, 2002). Moreover, transmission of silencing becomes rapidly independent of the initially bombarded leaves, as shown by the leaf detachment experiments described in Section 3.4.1.1. Taken together, these observations indicate that movement of silencing involves a relay-amplification process, whereby secondary signal molecules are synthesized as silencing moves away from its sites of initiation. This property is most clearly evidenced upon phloem unloading of the signal, as silencing initially restricted around the veins progressively invades the entire leaf lamina.

As explained previously, RdRps have been implicated in the amplification of silencing effector molecules, both in plants and in *C. elegans* (Dalmay *et al.*, 2000; Mourrain *et al.*, 2000; Sijen *et al.*, 2001). An early indication of the potential involvement of RdRps in the extensive nature of silencing movement in plants came from biolistic experiments carried out in the GFP transgenic *N. benthamiana* plants (Voinnet *et al.*, 1998). In those experiments, the DNA constructs used to initiate systemic silencing were PCR-amplified fragments with sequence of the 5′ part of the GFP cDNA. It was shown that silencing developing in systemic leaves was targeted not only against the region of the GFP transcript corresponding to the bombarded DNA, but also against regions located outside of the sequence used to trigger RNA silencing. This 'transitive silencing' is diagnostic of the activity of RdRps (Sijen *et al.*, 2001; Vaistij *et al.*, 2002) and this prompted the evaluation of the contribution of SDE1 – the *Arabidopsis* RdRp required for initiation of sense transgene silencing (Dalmay *et al.*, 2000; Mourrain *et al.*, 2000) – to the extent of silencing cell-to-cell movement.

To recreate in *Arabidopsis* the phloem unloading and subsequent cell-to-cell spread of silencing observed in *N. benthaminana* leaves, a system was set up that involved *Arabidopsis* plants carrying a constitutively expressed GFP transgene (Himber *et al.*, 2003). A second set of plants was from an isogenic GFP line carrying, in addition to the GFP transgene, a null mutation in the gene for SDE1. Both types of plants were supertransformed with a second construct in which a panhandled transgene with GFP sequences was cloned under the control of the phloem companion cell-specific AtSUC2 promoter. Thus, silencing of the GFP transgene was triggered specifically in the vasculature of those plants, allowing a precise assessment of its movement into the neighbouring mesophyll cells in both wt and *sde1* backgrounds (see Plate 3.2A, following page 146). All of the transformants with wt copies of *SDE1* exhibited a uniformly red (i.e. GFP silenced; Plate 3.2B) phenotype, whereas 80% of the *sde1* mutant plants showed a silencing phenotype that was centred around the vascular network, affecting a near constant number of

10–15 cells on both sides of the veins (Himber *et al.*, 2003) (Plate 3.2C–E). These results indicated that the extent of silencing movement outside of the vasculature was directly dependent upon the activity of SDE1, which was presumably involved in production of secondary signal molecules. They also suggested that the primary signal molecules generated from the veins could move over 10–15 cells in the absence of relay-amplification by SDE1. Similar experiments carried out with the *sde3* mutant showed that movement of GFP silencing outside of the veins was more extensive than in the *sde1* mutant, but less pronounced than in a wt background, indicating that SDE3 also contributes to the extensive cell-to-cell movement, but to a lesser degree than SDE1 (Himber *et al.*, 2003) (see Plate 3.3A–D, following page 146). This was in accordance with the previous demonstration that, unlike in the *sde1* mutant, initiation of sense-transgene silencing in plants deficient for the SDE3 RNA helicase was not completely abolished, presumably because of residual SDE1 activity (Dalmay *et al.*, 2000, 2001).

Cell-to cell movement of silencing in mature leaves. We saw in a previous section that spontaneous triggering in Class-II *Nia* plants is first manifested by the emergence of chlorotic spots on mature leaves, which are also observed upon bombardment of *Nia*-homologous DNA in single cells (Palauqui *et al.*, 1996; Palauqui & Balzergue, 1999). Likewise, biolistic delivery of GFP DNA in cells of mature leaves of GFP transgenic *N. benthamiana* leads to the rapid appearance of red fluorescent (i.e. silenced) foci that are similar in size to the chlorotic spots of the *Nia* plants (Voinnet *et al.*, 1998; Klahre *et al.*, 2002). Interestingly, in both systems, the diameter of these clusters of silenced cells does not increase further throughout time, indicating that silencing moves over a limited number of cells from its site of initiation in mature leaves, a phenomenon referred to as *localized movement of silencing*. To gain insight into this phenomenon, the agroinfiltration procedure was exploited in the GFP plants (Himber *et al.*, 2003). In addition to the GFP sequences that trigger silencing, the T-DNA used in those experiments carried a GUS transgene, whose open reading frame (ORF) was interrupted by an intron. The presence of this intron meant that GUS could be produced only within plant cells, upon T-DNA transfer (see Plate 3.4A, following page 146).

The infiltrated patch rapidly became uniformly red fluorescent, indicating GFP silencing had taken place. GUS histochemical blue staining was used to image precisely the cells that had received the T-DNA, and therefore the cells in which silencing had been triggered. GFP silencing clearly extended outside of the stained patch, forming a border of red fluorescent tissue that followed exactly the shape of the infiltrated patch itself (Plate 3.4B–D). Microscopic inspection and cell counting revealed that the width of this red border was remarkably constant and affected 10–15 cells, which was strikingly reminiscent of the extent of silencing movement observed near the veins of the *sde1 Arabidopsis* mutant. In further experiments, a cork borer was used to remove most of the inner part of an infiltrated area immediately after delivery of the *Agrobacterium*. This treatment did not modify the development and extent of localized silencing at the edge of the infiltrated patch, indicating that the cells located in the outmost periphery of the delivered area are sufficient to produce the local signal (Himber *et al.*, 2003). Taking also into account the bombardment data, these results suggest that short-distance movement of silencing over 10–15 adjacent cells can be elicited from one single cell. Based on the fact that those experiments involved tissues in which SDE1 was presumably active, it is surprising that silencing did not expand further by means of relay-amplification. This situation is similarly encountered with the limited increase in size of primary viral lesions on inoculated leaves, before systemic movement of the pathogen. With both silencing and virus movement, the size restriction of primary foci is most likely explained by the source status of the delivered organs (Santa Cruz *et al.*, 1998).

Molecular requirements for cell-to-cell movement of RNA silencing. The molecular requirements for cell-to-cell movement of RNA silencing were studied in both *N. benthamiana* and *Arabidopsis* (Himber *et al.*, 2003). In the former system, co-delivery of *Agrobacterium* cultures producing viral-encoded silencing suppressors was used to alter the silencing phenotype normally elicited in the infiltrated patch by GFP constructs. Molecular analysis of RNA extracted from such silenced patches shows that the ectopic and transgenic GFP mRNAs were below detection limit and that there was high accumulation of both 21 nt and 24 nt species of GFP siRNAs. In addition, the border of red fluorescent tissue, characteristic of the cell-to-cell movement of silencing from the infiltrated region to the surrounding cells, was readily detected. However, when those experiments were repeated in the presence of the P19 suppressor protein of tomato bushy stunt virus (TBSV), silencing in the patch was completely abolished: GFP siRNAs of both size classes were undetectable and the levels of GFP and GFP mRNA were high. As expected from the absence of silencing within those tissues, the development of the red border was also abolished. Co-treatments with the P1 suppressor of rice yellow mottle virus (RYMV), in contrast, did not prevent the onset of GFP silencing within the infiltrated tissue, as indicated by the complete lack of GFP mRNA. However, although the levels of 21 nt long GFP siRNAs were high in those tissues, the 24 nt long siRNAs was below detection limit. Interestingly, and despite the absence of the 24 nt siRNAs, cell-to-cell movement of GFP silencing was as extensive as it was in leaves that had been infiltrated with GFP alone (Himber *et al.*, 2003). These experiments indicated that the occurrence of the 21 nt siRNA at the source of silencing initiation was sufficient for cell-to-cell movement of silencing to take place at the edge of the patch. Note, however, that these experiments did not rule out the possibility that, although not necessary, the 24 nt siRNAs may also be sufficient for short-distance spread to take place.

As discussed in Section 3.4.1.2, high levels of *Nia* transcripts were mandatory for the development of co-suppression upon graft transmission of the *Nia* silencing signal into the scions. To test if, similarly, the cell-to-cell movement of GFP silencing from an infiltrated patch to the surrounding cells required the presence of the GFP mRNA in recipient cells, a second set of agroinfiltration experiments was carried out (Himber *et al.*, 2003). These experiments involved sequential infiltrations of the GFP *Agrobacterium* strain in leaves of wt *N. benthamiana* such that the second infiltrated patch overlapped with the first one. The two infiltrations were carried out at a 5-day interval. The rationale was that if a localized GFP silencing signal had moved from the first infiltrated patch to adjacent, non-transgenic cells, GFP expression from the second overlapping patch would be prevented at the edge of the first one. In the absence of cell-to-cell signalling from the first patch, however, expression of GFP from the second patch would coincide with the edge of the first patch (see Plate 3.5, following page 146). The outcome of those experiments showed that localized signalling indeed occurred in non-transgenic leaves and that its extent was the same as in leaves of the GFP transgenic *N. benthamiana*. Moreover, as in the GFP transgenic *N. benthamiana*, this process was abolished if the first patch had received a P19 co-treatment (Himber *et al.*, 2003). Therefore, unlike systemic movement of *Nia* silencing, the cell-to-cell movement of GFP silencing did not require prior transcription of a homologous transgene in recipient tissues.

Molecular analysis of the cell-to-cell movement of GFP silencing outside of the *Arabidopsis* vasculature was also carried out in wt, *sde1* and *sde3* mutant backgrounds (Himber *et al.*, 2003). In these plants, the pan-handled transgene construct used to trigger GFP silencing in the phloem companion cells corresponded to the 5' part of the GFP cDNA ('GF'). In all three genetic backgrounds there were comparable levels of 21 nt and 24 nt 'GF' siRNAs, despite very dissimilar silencing movement phenotypes. Thus, accumulation of the primary 'GF' siRNAs produced into the phloem was not correlated with the extent of silencing movement in the lamina. 'P' siRNAs originating from the non-overlapping part between the silencing trigger and the GFP transgene were then analysed. These secondary siRNAs are diagnostic of transitivity. It was found that in contrast to the 'GF' siRNAs, accumulation of the 'P' siRNAs was strictly correlated to the extent of silencing movement. Hence, high levels of 'P' siRNAs in wt plants were linked to full silencing, lower levels of 'P' siRNAs in the *sde3* mutant correlated with intermediate movement of silencing whereas the complete lack of 'P' siRNAs in the *sde1* mutant was associated with the vein-centred movement of silencing over 10–15 cells only. Interestingly, the secondary 'P' siRNAs were found to be exclusively of the 21 nt size class, indicating that the dsRNA product of transitivity is processed by a specific Dicer that generates 21 nt siRNAs only (Himber *et al.*, 2003). Thus, extensive movement is linked to the recruitment, by SDE1 and SDE3, of GFP mRNA in recipient cells for the production of 21 nt long secondary siRNAs. By contrast, short-distance silencing movement is independent of SDE1 and SDE3, does not require the presence of GFP mRNA in recipient cells and can occur in the absence of the 24 nt siRNA at the site of initiation.

3.3.5 *Maintenance of systemic silencing*

Grafting involving the *Nia* transgenic plants showed that tissues of Class-I, Class-II and Class-III plants were all competent to perceive the systemic silencing signal and activate the sequence-specific degradation of *Nia* transcripts. However, de-grafting experiments indicated that Class-I and Class-III plants were unable to sustain production of the systemic signal, as they progressively lost the silenced phenotype in the new growth and they were unable to transmit silencing into new *Nia* scions (Palauqui & Vaucheret, 1998). Only plants of Class II could sustain signal production when de-grafted (see Plate 3.6B, following page 146), a property called 'maintenance'. Interestingly, these plants were also shown to trigger co-suppression spontaneously, suggesting a link between maintenance and spontaneous activation of RNA silencing. This was indeed confirmed by the demonstration that the homozygous state of the *Nia* transgene in Class-II plants was strictly required for maintenance, as was shown for spontaneous initiation (Palauqui & Vaucheret, 1998) (Plate 3.6B). It is important to note that hemizygous Class-II plants were still competent to perceive the systemic silencing signal when grafted onto silenced rootstocks, indicating that the RNA degradation that is triggered upon signal delivery is distinct from the process of maintenance. Therefore, long-distance transport can occur in the absence of maintenance.

Maintenance of silencing has also been observed with the GFP transgenic *N. benthamiana*, as tissues taken from silenced scions were able to sustain signal production when used as rootstocks (Voinnet *et al.*, 1998). Analysis of silenced and non-silenced tissues in those plants indicated that entry of the systemic signal into cells was correlated with *de novo* methylation of the GFP transgene coding region (Jones *et al.*, 1999). Analysis carried out in systemically silenced GUS tobacco plants provided a similar conclusion (Mallory *et al.*, 2001). Work in *Arabidopsis* has also linked methylation of transgene coding region to PTGS (Elmayan *et al.*, 1998; Mourrain *et al.*, 2000). Moreover, a recent study of plants carrying a mutation in the gene for the major DNA methyltransferase MET1 revealed that this protein was necessary for maintenance rather than initiation of PTGS, as the new growth of *met1* plants exhibiting PTGS of a GUS transgene progressively lost the silencing phenotype (Morel *et al.*, 2000). Altogether, these results support the idea that methylation of the coding region plays an active role in the process of maintenance.

As seen previously in this chapter, spontaneous systemic silencing in the *Nia* Class-II plants has been proposed to result from production of a specific transgene product, referred to as the *aberrant RNA*. Since spontaneous initiation and maintenance of systemic silencing have similar requirements, and given the above results, methylation could be either a cause or a consequence of aberrant RNA production. The first scenario is supported by results obtained with the fungus *N. crassa*, where DNA methylation induced by DNA repeats at specific loci compromises elongation of transcription from these loci, leading to the production of truncated (and therefore aberrant) mRNAs (Rountree & Selker, 1997).

3.3.6 *What is the nucleic acid component of the silencing signal?*

The nucleotide sequence-specificity of the effects of systemic silencing indicates that the signal(s) involved have a nucleic acid component. As detailed in Section 3.3.8.1, signalling of silencing is elicited not only by transgenes, but also by viruses with no DNA phases in their replication cycles. Therefore, the most likely molecule for the nucleic acid component of the signal(s) is RNA.

3.3.6.1 *Cell-to-cell movement and phloem transport of silencing involve separate mechanisms and, most likely, separate signals*

Before considering the possible nature of the RNA involved in signalling, it is worth taking into consideration a series of observations that strongly suggest the existence of two separate mechanisms and, presumably, two separate signalling molecules, which account for cell-to-cell and long-distance movement of silencing in plants.

The first evidence comes from agroinfiltration experiments that were carried out in the presence of viral suppressors of RNA silencing (Himber *et al.*, 2003). For instance, co-treatments of GFP transgenic *N. benthamiana* leaves with the GFP silencing trigger and the AC2 suppressor of African Cassava mosaic virus led to 100% systemic silencing, whereas localized silencing was never observed at the margin of the infiltrated zones. Conversely, patches co-treated with the P1 protein of RYMV had a normal localized silencing phenotype (as explained in Section 3.4.2.3) but systemic silencing was abolished in those plants (Himber *et al.*, 2003). The second evidence is provided by analysis of the three classes of *Nia* transgenic plants. Indeed, bombardment of *Nia*-homologous DNA led to the development of chlorotic spots (through localized movement of *Nia* silencing from the bombarded cells) that were identical in diameter in all three classes of plants (Palauqui & Balzergue, 1999). Localized silencing occurring to the same extent was also observed when leaves of wt plants were bombarded. However, only in Class-II plants was the silencing able to move systemically (Palauqui & Balzergue, 1999). The third evidence is from experiments involving cadmium. Indeed, treatments of GUS silenced tobacco and GFP silenced *N. benthamiana* with non-toxic concentrations of cadmium prevented phloem transport, but not cell-to-cell spread of silencing targeted against GUS and GFP, respectively (Ueki & Citovsky, 2001). Altogether, these observations are consistent with the involvement of at least two distinct RNA species in cell-to-cell and long-distance transport of silencing. The possible nature of each RNA species is therefore discussed in separate sections.

3.3.6.2 *Possible nature of the RNA species involved in cell-to-cell movement of silencing*

Cell-to-cell movement of RNA silencing can be separated into short-distance (i.e. localized) and extensive processes. Short-distance movement of RNA silencing occurs over 10–15 cells and can be initiated from one single cell. This movement is independent of the presence of homologous transcripts in the recipient cells, does not require the activity of SDE1 and SDE3 and does not require the accumulation of 24 nt siRNAs at the site of silencing activation. This makes the involvement of target mRNA unlikely and excludes participation of the predicted products of transitivity (i.e. *de novo* synthesized dsRNA and secondary siRNAs). Therefore, primary siRNAs of the 21 nt size class are good candidates for the short-distance signal molecule (Himber *et al.*, 2003).

Extensive cell-to-cell movement is dependent upon SDE1 and SDE3, which use homologous transcripts as templates to produce *de novo* secondary siRNAs that are exclusively of the 21 nt size class, the proposed nucleic acid component of the short-distance signal (Himber *et al.*, 2003). In principle, limited and extensive cell-to-cell movement of silencing could be mediated by distinct mechanisms involving separate molecules. However, the difference could be more simply explained in terms of a single movement process with varying intensities. In their model, Himber *et al.* (2003) proposed that local initiation of silencing would produce 21 nt and 24 nt primary siRNAs. The primary 21 nt siRNA would move to 10–15 adjacent cells, independently of the presence of homologous transcripts in those cells. This initial wave of movement could then have two possible outcomes. First, primary 21 nt siRNAs could initiate synthesis of secondary 21 nt siRNAs through the action of SDE1 and SDE3 using homologous transcripts as templates. As proposed for primary 21 nt siRNAs, the newly synthesized 21 nt siRNAs could then move over a further distance of 10 ± 15 cells in which the same SDE1/SDE3-mediated process would be initiated. Such reiterated short-distance signalling events would then eventually translate into extensive movement. The second possible outcome would be that silencing does not move any further because of a lack or inability of homologous transcripts to act as templates for SDE1 and SDE3. This would preclude further production of 21 nt siRNAs and movement would stop (Himber *et al.*, 2003).

3.3.6.3 *No specific RNA species has been correlated with long-distance transport of silencing in plants*

The nature of the nucleic acid component of the signal involved in systemic silencing remains highly controversial. On the one hand, the use of several silencing suppressors indicated a tight correlation between the occurrence of the 24 nt long GFP siRNA in infiltrated tissues and the onset of systemic silencing in GFP transgenic *N. benthamiana* (Hamilton *et al.*, 2002). However, experiments involving stable expression of the HcPro suppressor of silencing in GUS-silenced tobacco lines showed that this protein eliminated the production of both 21 nt and 24 nt siRNAs, yet it did not prevent graft transmission of the silenced state from HcPro rootstocks to non-HcPro scions (Mallory *et al.*, 2001). The presence of HcPro in the scions did, however, prevent perception of the silencing signal in those tissues. Additional use of HcPro in conjunction with other silencing systems involving an inverted repeat construct and a transgene encoding a replicating virus, respectively, further suggested that long-distance silencing in those systems was not correlated with accumulation of any small RNA species, larger RNA or long dsRNA molecules (Mallory *et al.*, 2003). The picture is even more complicated by the fact that different classes of transgenes are likely to produce different patterns of systemic silencing, if any, as exemplified here by the very distinct response of the Class-I, Class-II and Class-III plants in systemic silencing of *Nia*. Therefore, it may well be that there is not one single nucleic acid species that serves as the long-distance signal. Rather, potentially any RNA intermediate in the silencing pathway that is able to move systemically could act as the systemic RNA silencing signal (Mallory *et al.*, 2003).

3.3.7 Plant factors required for movement of RNA silencing

To date, no plant factors that are required for silencing movement have been identified. However, the demonstrated sensitivity of long-distance silencing transport to non-toxic concentrations of cadmium (Ueki & Citovsky, 2001) could pave the way to the identification of such factors. Indeed, it has recently been shown that a tobacco glycine-rich protein, cdiGRP, is specifically induced by low cadmium concentration and accumulates in the cell walls of vascular tissues (Ueki & Citovsky, 2002). Constitutive expression of cdiGRP was shown to inhibit long-distance spread of turnip vein-clearing tobamovirus (TVCV) by enhancing callose deposits in the vasculature. It remains to be addressed whether enhanced cdiGRP expression also inhibits systemic movement of RNA silencing. If that is the case, cdiGRP could provide a powerful handle towards identification of plant proteins that regulate transport of silencing.

Factors required for silencing cell-to-cell movement in *Arabidopsis* are currently being investigated through a genetic approach developed in our laboratory. The principle of this approach is the same as described in Plate 3.2, except that the target of silencing in this system is an endogenous mRNA (*SULPHUR*) rather than the GFP mRNA. Phloem-triggered silencing of *SULPHUR* causes the appearance of yellow chlorosis that expands 10–15 cells away from the vein network (see Plate 3.7, following page 146). These plants have been mutagenized in order to retrieve individuals in which the vein-centred silencing phenotype is compromised. Some of these individuals should carry mutations that impair or enhance the short-distance movement of *SULPHUR* silencing. We have indeed retrieved several classes of such mutants and they are currently being characterized.

3.3.8 Biological functions of non-cell autonomous RNA silencing in plants

3.3.8.1 Antiviral defence

Obviously, plants did not elaborate such sophisticated signalling systems for the purpose of long-distance silencing of transgenes. The link with antiviral defence first became apparent from the striking similarities between the timing and pathways of systemic silencing and virus movement in plants (Santa Cruz *et al.*, 1996, 1998; Voinnet *et al.*, 1998). Based on the discovery that viruses were potent triggers of a PTGS-like response within infected cells (see Section 3.3.1.6), it was speculated that non-cell autonomous silencing could represent the systemic arm of this response, whereby transmission of a virus-induced-silencing signal ahead of the infection front would prime silencing in cells that are yet to be infected. Consequently, movement of the pathogen into those cells would be delayed or precluded (Voinnet *et al.*, 1998). This hypothesis received support from experiments in which silencing was activated in upper leaves of wt tobacco plants by inoculating lower leaves with movement defective mutants of PVX (Voinnet *et al.*, 2000). Systemic spread of silencing could be monitored *in planta*, because the modified PVX contained fragments of endogenous genes. Thus, the silencing signal generated by localized virus replication was primed not only against the viral genome but also against the corresponding endogenous transcripts, therefore generating a visual systemic silencing phenotype. Systemic silencing was apparent only when replication competent PVX was used as inoculum (Voinnet et al., 2000). Because PVX has an RNA genome that is replicated via RNA intermediates, this result is consistent with the idea that the observed signal, or at least part of it, has an RNA component.

Further support for an antiviral role of RNA silencing came from the finding of viral-encoded proteins that are specifically directed against systemic, as opposed to intracellular RNA, silencing. For instance, systemic silencing from PVX could be achieved only if the ORF for the 25-kDa movement protein (P25) was deleted in the viral genome, whereas silencing within the inoculated region remained unaffected by the presence or absence of P25 (Voinnet et al., 2000). Likewise, the 2b protein of cucumber mosaic virus was found to have a poor effect on intracellular silencing triggered by a transgene. However, triple-grafting experiments demonstrated unambiguously that 2b was a very potent inhibitor of the physical long-distance trafficking of silencing (Guo & Ding, 2002). A detailed analysis of the contribution of silencing suppression to systemic viral infection was provided in a recent study of the P19 protein of cymbidium ringspot virus (CymRSV) (Havelda et al., 2003). Previous work had already established that the P19-defective CymRSV accumulates to wt levels in single cells, suggesting that P19 targets a non-cell autonomous step of RNA silencing (Silhavy et al., 2002). Using a combination of in situ hybridization and immunohistochemistry, the authors have now convincingly demonstrated that the lack of P19 does not alter the phloem-dependent movement or the replication of CymRSV in and around the vascular bundles of systemic leaves (Havelda et al., 2003). However, it prevents further viral invasion of the leaf lamina, which, although virus free, exhibits nucleotide sequence-specific resistance to secondary infection with either CymRSV or an unrelated recombinant virus carrying sequence identity to CymRSV (Szittya et al., 2002). Therefore, P19 most likely prevents the onset or movement of a mobile virus induced silencing signal. Upon phloem unloading of the pathogen, the signal primes the destruction of viral RNAs ahead of the infection front. As observed for P19-deficient CymRSV, deletion of P25 caused PVX to be confined within the vasculature of systemically infected leaves. Thus, the movement-promoting function of P25 may also be explained in terms of preventing the synthesis or spread of a virus-induced silencing signal ahead of the virus. However, P25 has also been shown to facilitate cell-to-cell trafficking of PVX by increasing plasmodesmata size-exclusion limit in infected cells (Angell et al., 1996). Therefore, PVX movement might depend on both physical gating of the viral genome and inhibition of an antiviral signal.

3.3.8.2 A role in non-cell autonomous regulation of gene expression?
Since plant miRNAs resemble siRNAs, both biochemically and functionally (Bartel & Bartel, 2003), the possibility of their movement appears to be a legitimate question in light of the data summarized in this chapter. Limited silencing movement

over a nearly constant number of cells could participate in non-cell autonomous regulation of gene expression through miRNA trafficking. For instance, movement of miRNAs over a set number of cells could generate gradients of gene expression in meristems and primordia. As explained before, we are currently screening for *Arabidopsis* mutants with altered patterns of short-distance spread of *SULPHUR* and GFP silencing. If cell-to-cell movement of RNA silencing has developmental functions, we predict that some of these mutants will exhibit aberrant or modified morphology.

Note added in proof: A recent report describes the isolation of siRNAs and several miRNAs in cucurbit phloem sap, suggesting that at least some of those molecules may have the potential to move over long distances. This work describes a protein that specifically trafficks single-stranded small RNAs between cells (Yoo *et al.*, 2004).

References

Angell, S.M., Davies, C. & Baulcombe, D.C. (1996) Cell-to-cell movement of potato virus X is associated with a change in the size-exclusion limit of plasmodesmata in trichome cells of *Nicotiana clevelandii*. *Virology*, **216**, 197–201.

Aukerman, M.J. & Sakai, H. (2003) Regulation of flowering time and floral organ identity by a MicroRNA and its APETALA2-like target genes. *Plant Cell*, **15**, 2730–2741.

Balachandran, S., Xiang, Y., Schobert, C., Thompson, G.A. & Lucas, W.J. (1997) Phloem sap proteins from cucurbita maxima and *Ricinus communis* have the capacity to traffic cell to cell through plasmodesmata. *Proc. Natl. Acad. Sci. U.S.A.*, **94**, 14150–14155.

Bartel, B. & Bartel, D.P. (2003) MicroRNAs: at the root of plant development? *Plant Physiol.*, **132**, 709–717.

Bartel, D.P. (2004) MicroRNAs: genomics, biogenesis, mechanism, and function. *Cell*, **116**, 281–297.

Baumstark, T., Schroder, A.R. & Riesner, D. (1997) Viroid processing: switch from cleavage to ligation is driven by a change from a tetraloop to a loop E conformation. *EMBO J.*, **16**, 599–610.

Beclin, C., Boutet, S., Waterhouse, P. & Vaucheret, H. (2002) A branched pathway for transgene-induced RNA silencing in plants. *Curr. Biol.*, **12**, 684–688.

Bergey, D.R., Howe, G.A. & Ryan, C.A. (1996) Polypeptide signaling for plant defensive genes exhibits analogies to defense signaling in animals. *Proc. Natl. Acad. Sci. U.S.A.*, **93**, 12053–12058.

Bernstein, E., Caudy, A.A., Hammond, S.M. & Hannon, G.J. (2001) Role for a bidentate ribonuclease in the initiation step of RNA interference. *Nature*, **409**, 363–366.

Bernstein, E., Kim, S.Y., Carmell, M.A., Murchison, E.P., Alcorn, H., Li, M.Z., Mills, A.A., Elledge, S.J., Anderson, K.V. & Hannon, G.J. (2003) Dicer is essential for mouse development. *Nat. Genet.*, **35**, 215–217.

Beyenbach, J., Weber, C. & Kleinig, H. (1974) Sieve-tube proteins from *Cucurbita maxima*. *Planta*, **119**, 113–124.

Billy, E., Brondani, V., Zhang, H.D., Muller, U. & Filipowicz, W. (2001) Specific interference with gene expression induced by long, double-stranded RNA in mouse embryonal teratocarcinoma cell lines. *Proc. Natl. Acad. Sci. U.S.A.*, **98**, 14428–14433.

Boerjan, W., Bauw, G., Van Montagu, M. & Inzé, D. (1994) Distinct phenotypes generated by overexpression and supression of S-adenosyl-L-methionine synthetase reveal developmental patterns of gene silencing in tobacco. *Plant Cell*, **6**, 1401–1414.

Brigneti, G., Voinnet, O., Li, W.X., Ji, L.H., Ding, S.W. & Baulcombe, D.C. (1998) Viral pathogenicity determinants are suppressors of transgene silencing in *Nicotiana benthamiana*. *EMBO J.*, **17**, 6739–6746.

Carrington, J.C. & Ambros, V. (2003). Role of microRNAs in plant and animal development. *Science*, **301**, 336–338.

Caudy, A.A., Ketting, R.F., Hammond, S.M., Denli, A.M., Bathoorn, A.M., Tops, B.B., Silva, J.M., Myers, M.M., Hannon, G.J. & Plasterk, R.H. (2003) A micrococcal nuclease homologue in RNAi effector complexes. *Nature*, **425**, 411–414.

Cerutti, L., Mian, N. & Bateman, A. (2000) Domains in gene silencing and cell differentiation proteins: the novel PAZ domain and redefinition of the Piwi domain. *Trends Biochem. Sci.*, **25**, 481–482.

Chen, J.J., Janssen, B.J., Williams, A. & Sinha, N. (1997) A gene fusion at a homeobox locus: alterations in leaf shape and implications for morphological evolution. *Plant Cell*, **9**, 1289–1304.

Chen, X. (2004) A microRNA as a translational repressor of APETALA2 in *Arabidopsis* flower development. *Science*, **303**, 2022–2025.

Chuang, C.-H. & Meyerowitz, E.M. (2000) Specific and heritable genetic interference by double-stranded RNA in *Arabidopsis thaliana*. *Proc. Natl. Acad. Sci. U.S.A.*, **97**, 4985–4990.

Clark, S.E. (2001) Cell signalling at the shoot meristem. *Nat. Rev. Mol. Cell. Biol.*, **2**, 276–284.

Cogoni, C. & Macino, G. (1999) Gene silencing in *Neurospora crassa* requires a protein homologous to RNA-dependent RNA polymerase. *Nature*, **399**, 166–169.

Cogoni, C., Irelan, J.T., Schumacher, M., Schmidhauser, T.J., Selker, E.U. & Macino, G. (1996) Transgene silencing of the *Al-1* gene in vegetative cells of *Neurospora* is mediated by a cytoplasmic effector and does not depend on DNA–DNA interactions or DNA methylation. *EMBO J.*, **15**, 3153–3163.

Cote, F., Levesque, D. & Perreault, J.P. (2001) Natural $2',5'$-phosphodiester bonds found at the ligation sites of peach latent mosaic viroid. *J. Virol.*, **75**, 19–25.

Crawford, K.M. & Zambryski, P.C. (1999) Plasmodesmata signaling: many roles, sophisticated statutes. *Curr. Opin. Plant Biol.*, **2**, 382–387.

Crawford, K.M. & Zambryski, P.C. (2000) Subcellular localization determines the availability of non-targeted proteins to plasmodesmatal transport. *Curr. Biol.*, **10**, 1032–1040.

Crete, P., Leuenberger, S., Iglesias, V.A., Suarez, V., Schob, H., Holtorf, H., van Eeden, S. & Meins, F. (2001) Graft transmission of induced and spontaneous post-transcriptional silencing of chitinase genes. *Plant J.*, **28**, 493–501.

Dalmay, T., Hamilton, A.J., Rudd, S., Angell, S. & Baulcombe, D.C. (2000) An RNA-dependent RNA polymerase gene in *Arabidopsis* is required for posttranscriptional gene silencing mediated by a transgene but not by a virus. *Cell*, **101**, 543–553.

Dalmay, T.D., Horsefield, R., Braunstein, T.H. & Baulcombe, D.C. (2001) SDE3 encodes an RNA helicase required for post-transcriptional gene silencing in *Arabidopsis*. *EMBO J.*, **20**, 2069–2078.

Davies, J.W., Kaesberg, P. & Diener, T.O. (1974) Potato spindle tuber viroid, XII: An investigation of viroid RNA as a messenger for protein synthesis. *Virology*, **61**, 281–286.

Deom, C.M., Schubert, K.R., Wolf, S., Holt, C.A., Lucas, W.J. & Beachy, R.N. (1990) Molecular characterization and biological function of the movement protein of tobacco mosaic virus in transgenic plants. *Proc. Natl. Acad. Sci. U.S.A.*, **87**, 3284–3288.

Diener, T.O. (1971) Potato spindle tuber 'virus', IV: A replicating, low molecular weight RNA. *Virology*, **45**, 411–428.

Diener, T.O. (2001) The viroid: biological oddity or evolutionary fossil? *Adv. Virus Res.*, **57**, 137–184.

Ding, B., Kwon, M.-O., Hammond, R. & Owens, R. (1997) Cell-to-cell movement of potato spindle tuber viroid. *Plant J.*, **12**, 931–936.

Dougherty, W.G. & Parks, T.D. (1995) Transgenes and gene suppression – telling us something new? *Curr. Opin. Cell Biol.*, **7**, 399–405.

Dougherty, W.G., Lindbo, J.A., Smith, H.A., Parks, T.D., Swaney, S. & Proebsting, W.M. (1994) RNA-mediated virus resistance in transgenic plants: exploitation of a cellular pathway possibly involved in RNA degradation. *Mol. Plant Microbe Interact.*, **7**, 544–552.

Elbashir, S.M., Harborth, J., Lendeckel, W., Yalcin, A., Weber, K. & Tuschl, T. (2001a) Duplexes of 21-nucleotide RNAs mediate RNA interference in cultured mammalian cells. *Nature*, **411**, 494–498.

Elbashir, S.M., Lendeckel, W. & Tuschl, T. (2001b) RNA interference is mediated by 21- and 22-nucleotide RNAs. *Genes Dev.*, **15**, 188–200.

Elbashir, S.M., Martinez, J., Patkaniowska, A., Lendeckel, W. & Tuschl, T. (2001c) Functional anatomy of siRNAs for mediating efficient RNAi in *Drosophila melanogaster* embryo lysate. *EMBO J.*, **20**, 6877–6888.

Elmayan, T., Balzergue, S., Beon, F., Bourdon, V., Daubremet, J., Guenet, Y., Mourrain, P., Palauqui, J.C., Vernhettes, S., Vialle, T., Wostrikoff, K. & Vaucheret, H. (1998) *Arabidopsis* mutants impaired in cosuppression. *Plant Cell*, **10**, 1747–1757.

Finnegan, E.J., Margis, R. & Waterhouse, P.M. (2003) Posttranscriptional gene silencing is not compromised in the *Arabidopsis* CARPEL FACTORY (DICER-LIKE1) mutant, a homolog of Dicer-1 from *Drosophila*. *Curr. Biol.*, **13**, 236–240.

Fire, A., Xu, S., Montgomery, M.K., Kostas, S.A., Driver, S.E. & Mello, C.C. (1998) Potent and specific genetic interference by double-stranded RNA in *Caenorhabditis elegans*. *Nature*, **391**, 806–811.

Fletcher, J.C., Brand, U., Running, M.P., Simon, R. & Meyerowitz, E.M. (1999) Signaling of cell fate decisions by CLAVATA3 in *Arabidopsis* shoot meristems. *Science*, **283**, 1911–1914.

Flores, R. (2001) A naked plant-specific RNA ten-fold smaller than the smallest known viral RNA: the viroid. *C. R. Acad. Sci. III*, **324**, 943–952.

Flores, R. & Semancik, J.S. (1982) Properties of a cell-free system for synthesis of citrus exocortis viroid. *Proc. Natl. Acad. Sci. U.S.A.*, **79**, 6285–6288.

Flores, R., Daros, J.A. & Hernandez, C. (2000) Avsunviroidae family: viroids containing hammerhead ribozymes. *Adv. Virus Res.*, **55**, 271–323.

Flores, R., Randles, J.W. & Owens, R.A. (2003) Classification. In: *Viroids* (eds A. Hadidi, R. Flores, J.W. Randles & J.S. Semancik), pp. 71–75. CSIRO Publishing, Collingwood, VIC, Australia.

Galatis, B. (1980) Microtubules and guard-cell morphogenesis in *Zea mays* L. *J. Cell Sci.*, **45**, 211–244.

Glazer, A.N. & Rye, H.S. (1992) Stable dye–DNA intercalation complexes as reagents for high-sensitivity fluorescence detection. *Nature*, **359**, 859–861.

Golecki, B., Schulz, A. & Thompson, G.A. (1999) Translocation of structural P proteins in the phloem. *Plant Cell*, **11**, 127–140.

Gomez, G. & Pallas, V. (2001) Identification of an in vitro ribonucleoprotein complex between a viroid RNA and a phloem protein from cucumber plants. *Mol. Plant Microbe Interact.*, **14**, 910–913.

Gozmanova, M., Denti, M.A., Minkov, I.N., Tsagris, M. & Tabler, M. (2003) Characterization of the RNA motif responsible for the specific interaction of potato spindle tuber viroid RNA (PSTVd) and the tomato protein Virp1. *Nucleic Acids Res.*, **31**, 5534–5543.

Guo, H.S. & Ding, S.W. (2002) A viral protein inhibits the long range signaling activity of the gene silencing signal. *EMBO J.*, **21**, 398–407.

Hall, T.C., Wepprich, R.K., Davies, J.W., Weathers, L.G. & Semancik, J.S. (1974) Functional distinctions between the ribonucleic acids from citrus exocortis viroid and plant viruses: cell-free translation and aminoacylation reactions. *Virology*, **61**, 486–492.

Hamilton, A.J. & Baulcombe, D.C. (1999) A species of small antisense RNA in post-transcriptional gene silencing in plants. *Science*, **286**, 950–952.

Hamilton, A.J., Brown, S., Han, Y.H., Ishizuka, M., Lowe, A., Solis, A.G.A. & Grierson, D. (1998) A transgene with repeated DNA causes high frequency, post-transcriptional suppression of ACC-oxidase gene expression in tomato. *Plant J.*, **15**, 737–746.

Hamilton, A.J., Voinnet, O., Chappell, L. & Baulcombe, D.C. (2002) Two classes of short interfering RNA in RNA silencing. *EMBO J.*, **21**, 4671–4679.

Hammond, R.W. (1994) Agrobacterium-mediated inoculation of PSTVd cDNAs onto tomato reveals the biological effect of apparently lethal mutations. *Virology*, **201**, 36–45.

Hammond, S.M., Bernstein, E., Beach, D. & Hannon, G. (2000) An RNA-directed nuclease mediates post-transcriptional gene silencing in *Drosophila* cell extracts. *Nature*, **404**, 293–296.

Hammond, S.M., Boettcher, S., Caudy, A.A., Kobayashi, R. & Hannon, G.J. (2001) Argonaute2, a link between genetic and biochemical analyses of RNAi. *Science*, **293**, 1146–1150.

Havelda, Z., Hornyik, C., Crescenzi, A. & Burgyan, J. (2003) In situ characterization of cymbidium ringspot tombusvirus infection-induced posttranscriptional gene silencing in *Nicotiana benthamiana*. *J. Virol.*, **77**, 6082–6086.

Haynes, S.R., Dollard, C., Winston, F., Beck, S., Trowsdale, J. & Dawid, I.B. (1992) The bromodomain: a conserved sequence found in human, *Drosophila* and yeast proteins. *Nucleic Acids Res.*, **20**, 2603.

Haywood, V., Kragler, F. & Lucas, W.J. (2002) Plasmodesmata: pathways for protein and ribonucleoprotein signaling. *Plant Cell*, **14** (Suppl.), S303–S325.

Helariutta, Y., Fukaki, H., Wysocka-Diller, J., Nakajima, K., Jung, J., Sena, G., Hauser, M.T. & Benfey, P.N. (2000) The SHORT-ROOT gene controls radial patterning of the *Arabidopsis* root through radial signaling. *Cell*, **101**, 555–567.

Hernandez, C., Daros, J.A., Elena, S.F., Moya, A. & Flores, R. (1992) The strands of both polarities of a small circular RNA from carnation self-cleave in vitro through alternative double- and single-hammerhead structures. *Nucleic Acids Res.*, **20**, 6323–6329.

Himber, C., Dunoyer, P., Moissiard, G., Ritzenthaler, C. & Voinnet, O. (2003) Transitivity-dependent and -independent cell-to-cell movement of RNA silencing. *EMBO J.*, **22**, 4523–4533.

Hu, Y., Feldstein, P.A., Hammond, J., Hammond, R.W., Bottino, P.J. & Owens, R.A. (1997) Destabilization of potato spindle tuber viroid by mutations in the left terminal loop. *J. Gen. Virol.*, **78** (Pt 6), 1199–1206.

Hutchins, C.J., Rathjen, P.D., Forster, A.C. & Symons, R.H. (1986) Self-cleavage of plus and minus RNA transcripts of avocado sunblotch viroid. *Nucleic Acids Res.*, **14**, 3627–3640.

Hutvagner, G. & Zamore, P.D. (2002a) A microRNA in a multiple-turnover RNAi enzyme complex. *Science*, **297**, 2056–2060.

Hutvagner, G. & Zamore, P.D. (2002b) RNAi: nature abhors a double-strand. *Curr. Opin. Genet. Dev.*, **12**, 225–232.

Imlau, A., Truernit, E. & Sauer, N. (1999) Cell-to-cell and long distance trafficking of the green fluorescent protein in the phloem and symplastic unloading of the protein into sink tissues. *Plant Cell*, **11**, 309–322.

Jacobsen, S.E., Running, M.P. & Meyerowitz, E.M. (1999) Disruption of an RNA helicase/RNAse III gene in *Arabidopsis* causes unregulated cell division in floral meristems. *Development*, **126**, 5231–5243.

Jeanmougin, F., Wurtz, J.M., Le Douarin, B., Chambon, P. & Losson, R. (1997) The bromodomain revisited. *Trends Biochem. Sci.*, **22**, 151–153.

Jones, L., Hamilton, A.J., Voinnet, O., Thomas, C.L., Maule, A.J. & Baulcombe, D.C. (1999) RNA–DNA interactions and DNA methylation in post-transcriptional gene silencing. *Plant Cell*, **11**, 2291–2302.

Kim, M., Canio, W., Kessler, S. & Sinha, N. (2001) Developmental changes due to long-distance movement of a homeobox fusion transcript in tomato. *Science*, **293**, 287–289.

Klahre, U., Crete, P., Leuenberger, S.A., Iglesias, V.A. & Meins, F.J. (2002) High molecular weight RNAs and small interfering RNAs induce systemic posttranscriptional gene silencing in plants. *Proc. Natl. Acad. Sci. U.S.A.*, **99**, 11981–11986.

Kragler, F., Monzer, J., Xoconostle-Cazares, B. & Lucas, W.J. (2000) Peptide antagonists of the plasmodesmal macromolecular trafficking pathway. *EMBO J.*, **19**, 2856–2868.

Kuhn, C., Franceschi, V.R., Schulz, A., Lemoine, R. & Frommer, W.B. (1997) Macromolecular trafficking indicated by localization and turnover of sucrose transporters in enucleate sieve elements. *Science*, **275**, 1298–1300.

Kunz, C., Hanspeter, S., Stam, M., Kooter, J.M. & Meins, F.J. (1996) Developmentally regulated silencing and reactivation of tobacco chitinase transgene expression. *Plant J.*, **10**, 437–450.

Lafontaine, D., Beaudry, D., Marquis, P. & Perreault, J.P. (1995) Intra- and intermolecular nonenzymatic ligations occur within transcripts derived from the peach latent mosaic viroid. *Virology*, **212**, 705–709.

Lakatos, L., Szittya, G., Silhavy, D. & Burgyan, J. (2004) Molecular mechanism of RNA silencing suppression mediated by p19 protein of tombusviruses. *EMBO J.*, **23**, 876–884.

Lazarowitz, S.G. (1999) Probing plant cell structure and function with viral movement proteins. *Curr. Opin. Plant Biol.*, **2**, 332–338.

Liu, J.D., Carmell, M.A., Rivas, F.V., Marsden, C.E., Thomson, J.M., Song, J.J., Hammond, S.M., Joschua-Tor, L. & Hannon, E.T. (2004) Argonaute 2 is the catalytic engine of mammalian RNAi. *Science*, **305**, 1437–1441.

Liu, Y.H. & Symons, R.H. (1998) Specific RNA self-cleavage in coconut cadang cadang viroid: potential for a role in rolling circle replication. *RNA*, **4**, 418–429.

Llave, C., Xie, Z., Kasschau, K.D. & Carrington, J.C. (2002) Cleavage of Scarecrow-like mRNA targets directed by a class of *Arabidopsis* miRNA. *Science*, **297**, 2053–2056.

Lucas, W.J. (1995) Plasmodesmata: intercellular channels for macromolecular transport in plants. *Curr. Opin. Cell Biol.*, **7**, 673–680.

Lucas, W.J., Bouché-Pillon, S., Jackson, D.P., Nguyen, L., Baker, L., Ding, B. & Hake, S. (1995) Selective trafficking of KNOTTED1 homeodomain proteins and its mRNA through plasmodesmata. *Science*, **270**, 1980–1983.

Mallory, A.C., Ely, L., Smith, T.H., Marathe, R., Anandalakshmi, R., Fagard, M., Vaucheret, H., Pruss, G., Bowman, L. & Vance, V.B. (2001) HC-Pro suppression of transgene silencing eliminates the small RNAs but not transgene methylation or the mobile signal. *Plant Cell*, **13**, 571–583.

Mallory, A.C., Mlotshwa, S., Bowman, L.H. & Vance, V.B. (2003) The capacity of transgenic tobacco to send a systemic RNA silencing signal depends on the nature of the inducing transgene locus. *Plant J.*, **35**, 82–92.

Maniataki, E., Martinez de Alba, A.E., Gesser, R.S., Tabler, M. & Tsagris, M. (2003) Viroid RNA systemic spread may depend on the interaction of a 71-nucleotide bulged hairpin with the host protein VirP1. *RNA*, **9**, 346–354.

Martinez de Alba, A.E., Sagesser, R., Tabler, M. & Tsagris, M. (2003) A bromodomain-containing protein from tomato specifically binds potato spindle tuber viroid RNA in vitro and in vivo. *J. Virol.*, **77**, 9685–9694.

Matsubayashi, Y., Yang, H. & Sakagami, Y. (2001) Peptide signals and their receptors in higher plants. *Trends Plant Sci.*, **6**, 573–577.

McCarty, D.R. & Chory, J. (2000) Conservation and innovation in plant signaling pathways. *Cell*, **103**, 201–209.

Meister, G. & Tuschl, T. (2004) Mechanisms of gene silencing by double stranded RNA. *Nature*, **431**, 343–349.

Momma, T. & Takahashi, T. (1983) Cytopathology of shoot apical meristem of hop plants infected with hop stunt viroid. *Phytopath. Z.*, **106**, 272–280.

Morel, J.B., Mourrain, P., Beclin, C. & Vaucheret, H. (2000) DNA methylation and chromatin structure affect transcriptional and post-transcriptional transgene silencing in *Arabidopsis*. *Curr. Biol.*, **10**, 1591–1594.

Mourrain, P., Beclin, C., Elmayan, T., Feuerbach, F., Godon, C., Morel, J.-B., Jouette, D., Lacombe, A.-M., Nikic, S., Picault, N., Remoue, K., Sanial, M., Vo, T.-A. & Vaucheret, H. (2000) *Arabidopsis SGS2* and *SGS3* genes are required for posttranscriptional gene silencing and natural virus resistance. *Cell*, **101**, 533–542.

Muhlbach, H.P. & Sanger, H.L. (1979) Viroid replication is inhibited by alpha-amanitin. *Nature*, **278**, 185–188.

Nakajima, K., Sena, G., Nawy, T. & Benfey, P.N. (2001) Intercellular movement of the putative transcription factor SHR in root patterning. *Nature*, **413**, 307–311.

Napoli, C., Lemieux, C. & Jorgensen, R.A. (1990) Introduction of a chimeric chalcone synthase gene into Petunia results in reversible co-suppression of homologous genes *in trans*. *Plant Cell*, **2**, 279–289.

Navarro, B. & Flores, R. (1997) Chrysanthemum chlorotic mottle viroid: unusual structural properties of a subgroup of self-cleaving viroids with hammerhead ribozymes. *Proc. Natl. Acad. Sci. U.S.A.*, **94**, 11262–11267.

Navarro, J.A., Vera, A. & Flores, R. (2000) A chloroplastic RNA polymerase resistant to tagetitoxin is involved in replication of avocado sunblotch viroid. *Virology*, **268**, 218–225.

Oparka, K.J., Prior, D.A.M., Santa Cruz, S., Padgett, H.S. & Beachy, R.N. (1997) Gating of epidermal plasmodesmata is restricted to the leading edge of expanding infection sites of tobacco mosaic virus (TMV). *Plant J.*, **12**, 781–789.

Oparka, K.J., Roberts, A.G., Boevink, P., Santa Cruz, S., Roberts, I., Pradel, K.S., Imlau, A., Kotlizky, G., Sauer, N. & Epel, B. (1999) Simple, but not branched plasmodesmata allow the nonspecific trafficking of proteins in developing tobacco leaves. *Cell*, **97**, 743–754.

Owens, R.A., Blackburn, M. & Ding, B. (2001) Possible involvement of the phloem lectin in long-distance viroid movement. *Mol. Plant Microbe Interact.*, **14**, 905–909.

Palatnik, J.F., Allen, E., Wu, X., Schommer, C., Schwab, R., Carrington, J.C. & Weigel, D. (2003) Control of leaf morphogenesis by microRNAs. *Nature*, **425**, 257–263.

Palauqui, J.-C. & Balzergue, S. (1999) Activation of systematic acquired silencing by localised introduction of DNA. *Curr. Biol.*, **9**, 59–66.

Palauqui, J.-C. & Vaucheret, H. (1995) Field trial analysis of nitrate reductase co-suppression – a comparative-study of 38 combinations of transgene loci. *Plant Mol. Biol.*, **29**, 149–159.

Palauqui, J.-C. & Vaucheret, H. (1998) Transgenes are dispensable for the RNA degradation step of cosuppression. *Proc. Natl. Acad. Sci. U.S.A.*, **95**, 9675–9680.

Palauqui, J.C., Elmayan, T., Deborne, F.D., Crete, P., Charles, C. & Vaucheret, H. (1996) Frequencies, timing and spatial patterns of co-suppression of nitrate reductase and nitrite reductase in transgenic tobacco plants. *Plant Physiol.*, **112**, 1447–1456.

Palauqui, J.-C., Elmayan, T., Pollien, J.-M. & Vaucheret, H. (1997) Systemic acquired silencing: transgene-specific post-transcriptional silencing is transmitted by grafting from silenced stocks to non-silenced scions. *EMBO J.*, **16**, 4738–4745.

Palevitz, B.A. & Hepler, P.K. (1985) Changes in dye coupling of stomatal cells of *Allium* and *Commelina* demonstrated by microinjection of lucifer yellow. *Planta*, **164**, 473–479.

Palukaitis, P. (1987) Potato spindle tuber viroid: investigation of the long-distance, intra-plant transport route. *Virology*, **158**, 239–241.

Pang, S.Z., Jan, F.J., Carney, K., Stout, J., Tricoli, D.M., Quemada, H.D. & Gonsalves, D. (1996) Post-transcriptional transgene silencing and consequent tospovirus resistance in transgenic lettuce are affected by transgene dosage and plant development. *Plant J.*, **9**, 899–909.

Papp, I., Mette, M.F., Aufsatz, W., Daxinger, L., Schauer, S.E., Ray, A., van der Winden, J., Matzke, M. & Matzke, A.J. (2003) Evidence for nuclear processing of plant micro RNA and short interfering RNA precursors. *Plant Physiol.*, **132**, 1382–1390.

Parrish, S., Xu, S., Mello, C. & Fire, A. (2000) Functional anatomy of a dsRNA trigger: differential requirements for the two trigger strands in RNA interference. *Mol. Cell*, **6**, 1077–1087.

Pelchat, M., Rocheleau, L., Perreault, J. & Perreault, J.P. (2003) SubViral RNA: a database of the smallest known auto-replicable RNA species. *Nucleic Acids Res.*, **31**, 444–445.

Perbal, M.C., Haughn, G., Saedler, H. & Schwarz-Sommer, Z. (1996) Non-cell-autonomous function of the *Antirrhinum* floral homeotic proteins DEFICIENS and GLOBOSA is exerted by their polar cell-to-cell trafficking. *Development*, **122**, 3433–3441.

Rackwitz, H.R., Rohde, W. & Sanger, H.L. (1981) DNA-dependent RNA polymerase II of plant origin transcribes viroid RNA into full-length copies. *Nature*, **291**, 297–301.

Reiser, L., Sanchez-Baracaldo, P. & Hake, S. (2000) Knots in the family tree: evolutionary relationships and functions of knox homeobox genes. *Plant Mol. Biol.*, **42**, 151–166.

Robards, A.W. & Lucas, W.J. (1990) Plasmodesmata. *Annu. Rev. Plant Physiol. Plant Mol. Biol.*, **41**, 369–419.

Roberts, A.G., Santa Cruz, S., Roberts, I.M., Prior, D.A.M., Turgeon, R. & Oparka, K.J. (1997) Phloem unloading in sink leaves of *Nicotiana benthamiana*: comparison of a fluorescent solute with a fluorescent virus. *Plant Cell*, 9, 1381–1396.

Rountree, M.R. & Selker, E.U. (1997) DNA methylation inhibits elongation but not initiation of transcription in *Neurospora crassa*. *Genes Dev.*, **11**, 2383–2395.

Ruiz, M.T., Voinnet, O. & Baulcombe, D.C. (1998) Initiation and maintenance of virus-induced gene silencing. *Plant Cell*, **10**, 937–946.

Ruiz-Medrano, R., Xoconostle-Cazares, B. & Lucas, W.J. (1999) Phloem long-distance transport of CmNACP mRNA: implications for supracellular regulation in plants. *Development*, **126**, 4405–4419.

Santa Cruz, S., Chapman, S., Roberts, A.G., Roberts, I.M., Prior, D.A.M. & Oparka, K.J. (1996) Assembly and movement of a plant-virus carrying a green fluorescent protein overcoat. *Proc. Natl. Acad. Sci. U.S.A.*, **93**, 6286–6290.

Santa Cruz, S., Roberts, A.G., Prior, D.A.M., Chapman, S. & Oparka, K.J. (1998) Cell-to-cell and phloem-mediated transport of potato virus X: the role of virions. *Plant Cell*, **10**, 495–510.

Sasaki, T., Chino, M., Hayashi, H. & Fujiwara, T. (1998) Detection of several mRNA species in rice phloem sap. *Plant Cell Physiol.*, **39**, 895–897.

Schiebel, W., Haas, B., Marinkovic, S., Klanner, A. & Sanger, H.L. (1993a) RNA-directed RNA polymerase from tomato leaves, I: Purification and physical properties. *J. Biol. Chem.*, **268**, 11851–11857.

Schiebel, W., Haas, B., Marinkovic, S., Klanner, A. & Sanger, H.L. (1993b) RNA-directed RNA polymerase from tomato leaves, II: Catalytic *in vitro* properties. *J. Biol. Chem.*, **268**, 11858–11867.

Schiebel, W., Pelissier, T., Reidel, L., Thalmeir, S., Schiebel, R., Kempe, D., Lottspeich, F., Sanger, H.L. & Wassenegger, M. (1998) Isolation of an RNA-directed RNA polymerase-specific cDNA clone from tomato. *Plant Cell*, **10**, 2087–2102.

Schindler, I.M. & Mülbach, H.P. (1992) Involvement of nuclear-dependent RNA polymerases in potato spindle tuber viroid replication: a reevaluation. *Plant Sci.*, **84**, 221–229.

Sessions, A., Yanofsky, M.F. & Weigel, D. (2000) Cell–cell signaling and movement by the floral transcription factors LEAFY and APETALA1. *Science*, **289**, 779–782.

Sijen, T., Fleenor, J., Simmer, F., Thijssen, K.L., Parrish, S., Timmons, L., Plasterk, R.H.A. & Fire, A. (2001) On the role of RNA amplification in dsRNA-triggered gene silencing. *Cell*, **107**, 465–476.

Sijen, T., Wellink, J., Hiriart, J.-B. & Van Kammen, A. (1996) RNA-mediated virus resistance: role of repeated transgenes and delineation of targetted regions. *Plant Cell*, **8**, 2277–2294.

Silhavy, D., Molnar, A., Lucioli, A., Szittya, G., Hornyik, C., Tavazza, M. & Burgyan, J. (2002) A viral protein suppresses RNA silencing and binds silencing-generated, 21- to 25-nucleotide double-stranded RNAs. *EMBO J.*, **21**, 3070–3080.

Stam, M., de Bruin, R., Kenter, S., van der Hoorn, R.A.L., van Blokland, R., Mol, J.N.M. & Kooter, J.M. (1997) Post-transcriptional silencing of chalcone synthase in Petunia by inverted transgene repeats. *Plant J.*, **12**, 63–82.

Sugimoto-Shirasu, K. & Roberts, K. (2003) 'Big it up': endoreduplication and cell-size control in plants. *Curr. Opin. Plant Biol.*, **6**, 544–553.

Szittya, G., Molnar, A., Silhavy, D., Hornyik, C. & Burgyan, J. (2002) Short defective interfering RNAs of tombusviruses are not targeted but trigger post-transcriptional gene silencing against their helper virus. *Plant Cell*, **14**, 359–372.

Tabler, M., Tzortzakaki, S. & Tsagris, M. (1992) Processing of linear longer-than-unit-length potato spindle tuber viroid RNAs into infectious monomeric circular molecules by a G-specific endoribonuclease. *Virology*, **190**, 746–753.

Tang, G., Reinhart, B.J., Bartel, D.P. & Zamore, P.D. (2003) A biochemical framework for RNA silencing in plants. *Genes Dev.*, **17**, 49–63.

Turgeon, R. (1996) Phloem loading and plasmodesmata. *Trends Plant Sci.*, **1**, 418–423.

Tuschl, T., Zamore, P.D., Lehmann, R., Bartel, D.P. & Sharp, P.A. (1999) Targeted mRNA degradation by double-stranded RNA in vitro. *Genes Dev.*, **13**, 3191–3197.

Ueki, S. & Citovsky, V. (2001) Inhibition of systemic onset of postranscriptional gene silencing by non-toxic concentrations of cadmium. *Plant J.*, **28**, 283–291.

Ueki, S. & Citovsky, V. (2002) The systemic movement of a tobamovirus is inhibited by a cadmium-ion-induced glycine-rich protein. *Nat. Cell Biol.*, **4**, 478–486.

Vaistij, F.E., Jones, L. & Baulcombe, D.C. (2002) Spreading of RNA targeting and DNA methylation in RNA silencing requires transcription of the target gene and a putative RNA-dependent RNA polymerase. *Plant Cell*, **14**, 857–867.

Vanderkrol, A.R., Brunelle, A., Tsuchimoto, S. & Chua, N.H. (1993) Functional-analysis of petunia floral homeotic mads box gene *pmads1*. *Genes Dev.*, **7**, 1214–1228.

Vaucheret, H., Beclin, C., Elmayan, T., Feuerbach, F., Godon, C., Morel, J.-B., Mourrain, P., Palauqui, J.-C. & Vernhettes, S. (1998) Transgene-induced gene silencing in plants. *Plant J.*, **16**, 651–659.

Voinnet, O. (2001) RNA silencing as a plant immune system against viruses. *Trends Genet.*, **17**, 449–459.

Voinnet, O. (2002) RNA silencing: small RNAs as ubiquitous regulators of gene expression. *Curr. Opin. Plant Biol.*, **5**, 444–451.

Voinnet, O. & Baulcombe, D.C. (1997) Systemic signalling in gene silencing. *Nature*, **389**, 553.

Voinnet, O., Lederer, C. & Baulcombe, D.C. (2000) A viral movement protein prevents systemic spread of the gene silencing signal in *Nicotiana benthamiana*. *Cell*, **103**, 157–167.

Voinnet, O., Vain, P., Angell, S. & Baulcombe, D.C. (1998) Systemic spread of sequence-specific transgene RNA degradation is initiated by localised introduction of ectopic promoterless DNA. *Cell*, **95**, 177–187.

Vollbrecht, E., Veit, B., Sinha, N. & Hake, S. (1991) The developmental gene Knotted-1 is a member of a maize homeobox gene family. *Nature*, **350**, 241–243.

Wassenegger, M. & Pelissier, T. (1998) A model for RNA-mediated gene silencing in higher plants. *Plant Mol. Biol.*, **37**, 349–362.

Wille, A.C. & Lucas, W.J. (1984) Ultrastructural and histochemical studies on guard cells. *Planta*, **160**, 129–142.

Wolf, S., Deom, C.M., Beachy, R.N. & Lucas, W.J. (1989) Movement protein of tobacco mosaic virus modifies plasmodesmatal size exclusion limit. *Science*, **246**, 377–379.

Wu, X., Weigel, D. & Wigge, P.A. (2002) Signaling in plants by intercellular RNA and protein movement. *Genes Dev.*, **16**, 151–158.

Xie, Z., Johansen, L.K., Gustafson, A.M., Kasschau, K.D., Lellis, A.D., Zilberman, D., Jacobsen, S.E. & Carrington, J.C. (2004) Genetic and functional diversification of small rna pathways in plants. *PLoS Biol.*, **2**, E104.

Xie, Z., Kasschau, K.D. & Carrington, J.C. (2003) Negative feedback regulation of Dicer-Like1 in *Arabidopsis* by microRNA-guided mRNA degradation. *Curr. Biol.*, **13**, 784–789.

Xoconostle-Cazares, B., Yu, X., RuizMedrano, R., Wang, H.L., Monzer, J., Yoo, B.C., McFarland, K.C., Franceschi, V.R. & Lucas, W.J. (1999) Plant paralog to viral movement protein that potentiates transport of mRNA into the phloem. *Science*, **283**, 94–98.

Yoo, B.C., Kragler, F., Varkonyi-Casic, E., Haywood, V., Archer-Evans, S., Lee, Y.M., Lough, T.J. & Lucas, W.J. (2004) A systemic small RNA signaling system in plants. *Plant Cell*, **16**, 1979–2000.

Zamore, P.D., Tuschl, T., Sharp, P.A. & Bartel, D.P. (2000) RNAi: double-stranded RNA directs the ATP-dependent cleavage of mRNA at 21 to 23 nucleotide intervals. *Cell*, **101**, 25–33.

Zhu, Y., Qi, Y., Xun, Y., Owens, R. & Ding, B. (2002) Movement of potato spindle tuber viroid reveals regulatory points of phloem-mediated RNA traffic. *Plant Physiol.*, **130**, 138–146.

Zhu, Y.L., Green, L., Woo, Y.M., Owens, R. & Ding, B. (2001) Cellular basis of potato spindle tuber viroid systemic movement. *Virology*, **279**, 69–77.

Zilberman, D., Cao, X. & Jacobsen, S.E. (2003) ARGONAUTE4 control of locus-specific siRNA accumulation and DNA and histone methylation. *Science*, **299**, 716–719.

4 The plant extracellular matrix and signalling
Andrew J. Fleming

4.1 Introduction

Plant cells are fixed relative to another via a cellulose-based cell wall. As a consequence, cell migration during the ontogeny of a plant is extremely limited, yet cells in spatially separated parts of an organ must differentiate in a coordinated fashion if a fully functional plant is to be produced. This coordination of differentiation could be based on cell lineage (i.e. if all cells underwent a precise pattern of growth and division, then it would be possible to reliably generate a functioning organism). However, the vast majority of data indicate that cell-lineage-based mechanisms of differentiation are rare in plants (Kessler *et al.*, 2002) and the general consensus is that the patterns of differentiation observed in plants are based on extensive networks of intercellular communication. Significant progress has been made in this area over the last few years. Indeed, there has been a veritable boom in the interest and advances in our understanding of how plant cells communicate with each other. As will be seen from even a brief perusal of the other chapters in this book, plants employ a plethora of different signalling mechanisms, and a major challenge for the future is not simply to decipher these signals, but also to understand how the different signals are integrated. Added to the complexity of the basic signalling network is the further complication that plants are sessile organisms that must sense and respond to their environment. Thus, the precise timing and position of events of differentiation are frequently influenced by environmental factors, ranging from temperature and light to a massive spectrum of potential pathogens and herbivores. These responses may occur both locally at the point of stimulation and at a distance. Thus, the signalling schemes employed by plants must somehow combine a consistency that maintains species-specific forms of growth and differentiation with an adaptability that can encompass the myriad external factors that can potentially influence the precise form and timing of growth and differentiation.

Much classical work on intercellular communication in plants has concentrated on traditional hormones (e.g. auxin, cytokinin, gibberellin, abscisic acid and ethylene). However, the last decade has seen intense interest (and progress) in novel signalling components. Thus, recent research has led to exciting discoveries on the function of peptides and RNA in intercellular communication in plants, and many of the following chapters of this book reflect this progress. By necessity, this means that less emphasis is given to some of the classical hormones implicated in cell-to-cell signalling. Readers are directed towards a number of excellent reviews that cover this area (e.g. Finkelstein *et al.*, 2002; Hutchison & Kieber, 2002; Wang *et al.*,

2002). However, even the novel mechanisms and classical plant hormones that are the focus of several of the following chapters do not encompass all the likely players involved in intercellular communication. Indeed, the cast of characters is likely to increase as research in this highly active area progresses. Recent examples include the implication of novel carotenoid derivatives in a root-sourced long-distance signalling mechanism that mediates the classically described repression of axillary bud growth by auxin (Beveridge *et al.*, 2000, Sorefan *et al.*, 2003) and the identification of novel peptide phytosulphokine growth factors (Yang *et al.*, 1999).

However, by concentrating on 'success' stories and novel signalling molecules, there is the danger that we forget alternative mechanisms of intercellular communication for which there is, as yet, incomplete evidence as to their significance. Some of these alternative mechanisms were regarded as novel in their time and it is instructive to view the progress that has (or has not) been made in consolidating their potential role in the complex procedure by which plant cells communicate with each other. In this context, much hard work has been invested in the potential role that the plant cell wall might play in intercellular communication and the remainder of this chapter will focus on examining the evidence that the cell wall functions as a key component in plant signalling. As will be seen, although the evidence is sometimes not totally conclusive, the various strands of data from the wide ranging experiments into the roles that the cell wall could play in cell-to-cell signalling provide sufficient indication that it does indeed play an important role in plant intercellular communication. However, it will also be seen that there is still much work to be done.

4.2 The cell wall and signalling

Plant cells are surrounded by a relatively rigid cellulose-based cell wall. A number of authors have suggested that this complex structure of polysaccharides, proteins and lipids be more precisely termed the plant extracellular matrix (ECM) (Roberts, 1994). This provides both a more vivid vision of the complex macromolecular interconnections surrounding and joining plant cells, and draws attention to a possible comparison with the animal ECM. Since a large and convincing body of data indicates that the ECM plays a key role in intercellular communication in animals, the ECM could also, by analogy, be a key element in plant cell signalling. In this chapter, the terms *cell wall* and *ECM* will be used interchangeably simply because the majority of researchers will be more familiar with the former term.

This chapter will also tend to concentrate on the potential role of the cell wall in communication events associated with endogenous processes of growth and development, rather than on plant response to pathogens and herbivores. However, very often the analysis of signalling associated with plant–microbe interactions provides insights into endogenous developmental signalling events and these aspects will be referred to when necessary.

Some of the initial interest in the cell wall as a source of signals arose on conceptual grounds. It is clear that the plant cell wall is an intricate composite of a large number of extremely complex molecules (McNeil *et al.*, 1984). Such complexity could engender a large supply of information. Such information might, for instance, be used to distinguish one cell from another and to inform neighbouring cells of their tissue context. Since virtually all cells in the plant are surrounded by a cell wall, it could act as a universal mediator of signalling events. Thus, the idea that cell wall polymers (or breakdown products) play a role in intercellular communication is intellectually enticing. Certainly, a significant body of data from research on animal systems has revealed how the extracellular matrix can influence signalling, acting as a source and modulator of signalling compounds (e.g. Perrimon & Bernfield, 2001). It seems reasonable to suppose that the plant cell wall could function in an analogous way. However, in addition to the conceptual similarity of the plant and animal ECM, it is also clear that the plant cell wall has some unique characteristics. Chief among these is its rigidity. This biophysical specialisation of the plant cell wall has encouraged some researchers to explore the possibility that the mechanical properties of the plant ECM might allow it to be part of a biophysical-based system of intercellular signalling, with the cell wall acting as the conduit (and modulator) of the forces involved. This concept is discussed further towards the end of this chapter. To start with, however, let us examine the biochemical structure of the cell wall and consider the evidence that at least some of these chemical moieties act as intercellular signals.

4.3 The cell wall as a potential source of chemical signals

Central to wall structure is a polymer of (1-4)-b-D-glucan (cellulose) that is coated with a heteropolymer of xyloglucan. This is structurally similar to cellulose but is substituted at positions along the backbone with mono-, di- and trisaccharide side chains. These side chains always contain xylose and may contain galactose and fucose residues (Fig. 4.1). Cellulose and xyloglucan are thought to constitute a matrix that provides the cell wall with resistance to tensile stress. A second matrix within the cell wall is based on pectin. This is an extraordinarily complex matrix whose essential structure is based on polygalacturonic acid. This may be a homopolymer (homogalacturonan) or can be substituted at particular sites with complex side chains to form rhamnogalacturonan II. Other pectins include rhamnogalacturonan I (a polymer containing a backbone of alternating galacturonan and rhamnose units) and arabinogalactan (containing a backbone polymer of arabinose). This complex pectin matrix is thought to function biophysically as a gel to contain compressive forces that build up within the cell wall.

In addition to these purely polysaccharide components, the cell wall also contains protein and lipids. Thus, many cell walls contain an extensin-based protein matrix and also an array of proteins and enzymes thought to function in cell wall assembly. In addition, most epidermal cells possess a coating of lipid-derived cutins and waxes that perform primarily a protective role to restrict water loss and to defend the plant against attack. Finally, there is a spectrum of complex molecules that may consist of

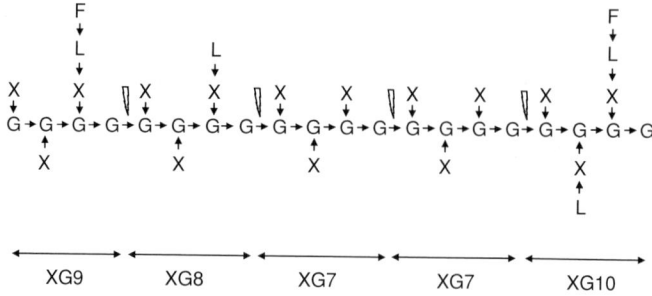

Figure 4.1 Structure of xyloglucan to demonstrate the generation of potential signalling compounds (based on Aldington & Fry, 1993). Enzymatic hydrolysis of xyloglucan tends to occur at free positions along the glucan backbone of the molecule, as indicated in the diagram. Depending on the pattern of side chains attached to the glucan polymer, such hydrolysis leads to the generation of xyloglucan fragments (XGs) of various specific sizes (e.g. XG7, XG8, XG9 and XG10, as indicated). Some of these fragments have been implicated in growth control.

combinations of lipid or carbohydrate-modified proteins. Among the most complex of these are the arabinogalactan proteins, which will be discussed later.

This brief summary can only provide a glimpse of the complexity and variety of cell wall chemistry and architecture, elements of which are still being discovered (e.g. Vincken *et al.*, 2003). Nevertheless, it forms a basis for examining the idea that components derived from the cell wall function in signalling.

4.3.1 *Polysaccharide signals*

Enzymatic breakdown of the xyloglucan matrix leads to the release of fragments of different sizes. Since hydrolysis tends to occur at the non-substituted glucose residues along the glucan backbone, these fragments can be characterised by the number of sugar residues that they contain, e.g. XG7, XG8, XG9 and XG10 (see Fig. 4.1; Aldington & Fry, 1993). A number of studies have shown that such xyloglucan fragments can influence tissue growth and that the effectiveness of the fragment depends on its size. For example, York *et al.* (1984) showed that a particular xyloglucan derivative (XG9) at a low concentration (1 nM) could block the promotion of growth induced by auxin in excised pea stem segments, whereas higher concentrations (1 mM) of the same fragment induced growth (McDougall & Fry, 1988). Similar reports have been made by other groups (e.g. Hoson & Masuda, 1991). These observations led to the proposal that xyloglucan fragments derived from the cell wall might play a role as an endogenous regulator or mediator of growth. Moreover, since such fragments would be free within the ECM, it is easy to conceive that they might move within the apoplast to coordinate growth both at a local level and at a distance. However, despite the convincing demonstrations that xyloglucan derivatives can influence tissue growth, a number of questions as to their physiological significance remain. In particular, it is unclear to what extent such fragments exist in normally growing tissue and to what extent they actually move from cell to cell (i.e. whether they function in communicating or coordinating growth) (Aldington & Fry, 1993).

A recent paper from Takeda *et al.* (2002) may shed an important light on this issue. These authors provide data demonstrating that small xyloglucan polymers supplied exogenously to plant tissue can become incorporated into the cell wall matrix via the activity of the enzyme xyloglucan transglycosylase (XET). This incorporation leads to an alteration in the growth characteristics of the wall such that it becomes more extensible. This would tend to increase the potential for extension growth. It may be that incorporation of xyloglucan polymers into the cell wall matrix reflects the normal mechanism by which XET activity synthesises xyloglucan. This would certainly influence plant cell growth characteristics but would not necessarily imply that small xyloglucan polymers normally act as intercellular communicators to regulate the growth characteristics of neighbouring cells. In this scenario, the modulation of extension by the exogenous supply of xyloglucan polymers to *in vitro* cultured tissue might mimic or disrupt the endogenous process of xyloglucan biosynthesis, but might not accurately reflect an endogenous mechanism for the coordination of tissue growth.

Data also exists that fragments from the pectin matrix can act in intercellular signalling. As with xyloglucans, most of these experiments have been performed with preparations of cell walls obtained by enzymatic or chemical treatments that lead to the release of fragments of differing polymer size. The best characterised of these putative signals are oligomers of galacturonic acid (OGAs), which can be defined by their degree of polymerisation (DP). For example, fragments of polygalaturonic acid generated by pectinase have been reported to block the growth-promoting activity of auxin in excised pea segments (Branca *et al.*, 1988), with OGAs of DP 10–17 being most effective. These experiments indicated a requirement for an OGA concentration in the micromolar range. However, as with xyloglucans, it is still unclear as to whether such free OGAs occur to a significant level in unwounded tissue and whether their mobility in the apoplast is sufficient to allow them to function in intercellular communication to coordinate growth *in vivo*.

An added interest in OGAs and other pectin-derived fragments arose from reports that they influence developmental processes, most notably flowering and morphogenesis. This was most dramatically demonstrated in experiments in which pectin-derived material was shown to induce floral morphogenesis from cultured strips of tobacco tissue (Tran Thanh Van *et al.*, 1985). Further investigation of these initial observations indicated a very complex situation in which the influence of pectin-derived fragments on morphogenesis was heavily dependent on the hormonal regime required to maintain the plant explants used in these experiments (Eberhard *et al.*, 1989). As a consequence, research and interest in the potential role of pectin-derived fragments as developmental signals appears to have dwindled.

To summarise, the physiological role of oligosaccharides in controlling plant growth processes (as well as their role in defence responses) remains debatable (Ryan & Farmer, 1991; Aldington & Fry, 1993). It is clear that responses can be observed under *in vitro* conditions, but it is unclear to what extent this reflects normal processes of growth control. In particular, the vast majority of experiments have been performed with oligosaccharides derived from chemical or enzymatic preparations. It is often unclear how these preparations relate to endogenously generated oligosaccharides. In addition, the mobility of polysaccharide fragments appears to be rather limited in intact tissue, suggesting that any intercellular signalling role is likely to be of a paracrine nature. Furthermore, receptors for these ligands have not yet been identified.

An initial impulse for the interest in the potential role of cell wall polysaccharides in signalling came from the chemical complexity (thus high information content) of this matrix. However, this very complexity provides a major challenge to future work in this area. Our ignorance on basic elements of cell wall biosynthesis and organisation (as reflected in the continuing evolution of models of cell wall structure and biosynthesis (e.g. Vincken *et al.*, 2003; Dhugga *et al.*, 2004)) represents a significant block to progress in this area. Novel techniques for the *in vivo* analysis of the plant cell wall are being developed and have already led to re-assessments of current models of cell wall structure (e.g. Wilson *et al.*, 2000; Lerouxel *et al.*, 2002; Moille *et al.*, 2003), but further progress in this direction is essential. However, even if advances are made in this area, the metabolic flexibility exhibited by plants, which allows mutation in one particular polysaccharide biosynthetic pathway to be compensated for by the generation of novel cell wall components, which can (more or less) functionally compensate for any deleterious outcome of a single genetic lesion (e.g. Zablackis *et al.*, 1996), makes the genetic and biochemical analysis of polysaccharide-derived compounds extremely challenging.

4.3.2 *Arabinogalactan proteins as signals*

In addition to pure polysaccharides, the plant cell wall contains proteins and lipids that may contain carbohydrate modifications. With respect to intercellular signalling, one family of proteoglycan-based molecules has aroused intense interest – the arabinogalactan proteins (AGPs) (Majewska-Sawska & Nothnagel, 2000).

AGPs are highly complex molecules that contain a protein core modified to generate a final structure that may consist of greater than 90% carbohydrate (Fig. 4.2A). According to cDNA sequences, AGPs generally possess a C-terminal hydrophobic domain and biochemical analysis has revealed that this can be processed to generate a glycerolphosphatidylinositol (GPI) lipid anchor (Oxley & Bacic, 1999; Svetek *et al.*, 1999) by which some AGPs could be tethered to a membrane. The site and mechanism of this processing in plants is unclear, but the presence of such a lipid anchor adds complexity to an already complex molecule and provides intriguing possibilities for the proposed functions of AGPs in intercellular signalling, as will be discussed below.

Initial interest in the potential signalling function of AGPs was aroused by the finding that epitopes of specific AGP-associated carbohydrates occurred in tissue-specific patterns. For example, Roberts and colleagues demonstrated that an AGP recognised by the antibody JIM13 is initially expressed exclusively in metaxylem

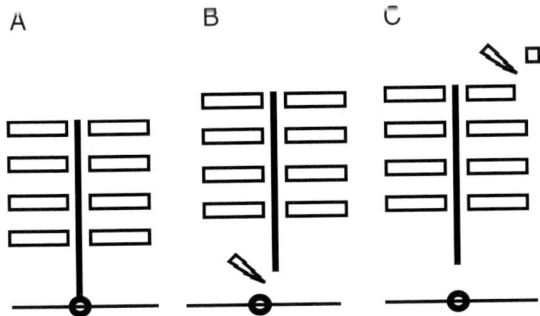

Figure 4.2 Arabinogalactan protein (AGP) structure and potential for signal generation. (A) A typical AGP consists of a protein backbone that is extensively modified by the attachment of complex carbohydrates. The AGP may be anchored to a membrane via a lipid anchor. (B) The AGP may be detached from the lipid anchor via action of a phospholipase, generating a free AGP with signalling potential. (C) Further processing of AGPs containing N-acetylglucosamine oligomers in the carbohydrate side chains may occur via action of chitinases. Both the modified AGP and the N-acetylglucosamine oligomers released could act as signals.

cells above the quiescent centre and subsequently appears in files of cells more proximal to the root tip (Dolan et al., 1995). Other AGP epitopes were found to show tissue-specific patterns in flowers of peas (Pennel & Roberts, 1990), developing carrot roots (Knox et al., 1991) as well as in pistils (Du et al., 1996), and germinating pollen tubes (Li et al., 1995). Such cell-specific patterns were also found in *in vitro* cultured plant tissues, most notably those associated with somatic embryogenesis. For example, analysis of carrot suspension cultures revealed that cells that possessed embryogenic potential were distinguishable from their non-embryogenic counterparts by the presence of a particular AGP epitope recognised by the antibody JIM8 (McCabe et al., 1997). Such correlative observations led to the hypothesis that AGPs might act as markers for particular developmental fates and, moreover, that they might actually function as cell-wall-based signals to communicate cell identity between neighbouring cells and/or to influence tissue growth. There are some functional data to support this hypothesis.

Firstly, the supply of free AGPs to *in vitro* systems has been shown to influence tissue growth and differentiation. Thus, AGPs added to suspension cultures were found to inhibit cell division (Thompson & Knox, 1998) and exogenous AGPs were shown to influence the embryogenic potential of suspension culture cells (Toonan et al., 1997; Van Hengel et al., 2001). These last data are reminiscent of the findings of McCabe et al. (1997), who showed that non-embryogenic cells secreted a compound into the medium that facilitated embryogenesis from embryogenic cells. Although the nature of this compound was not demonstrated, correlative data suggested that the factor might be an AGP.

Further data supporting a functional role for AGPs in differentiation has come from analysis of another *in vitro* system, xylogenesis in *Zinnia elegans*. Mesophyll cells isolated from leaves of *Zinnia* can be induced by appropriate hormonal manipulations to undergo terminal differentiation to form tracheary elements (Fukada & Komamine, 1980). During this process, a number of proteins and carbohydrates are secreted into the medium (Stacey *et al.*, 1995) and specific changes in gene expression occur (Milioni *et al.*, 2002). Motose *et al.* (2001) reported that a secreted compound facilitates tracheary element differentiation and that an active fraction identified through their analysis contained AGP. Whether AGP is the active principle in this fraction has to be substantiated, but these data are again indicative of AGPs playing a functional role in differentiation.

The identification of genes encoding the core proteins of AGPs has opened the door to the use of molecular genetic tools to further investigate AGP function. The first data from these approaches are now being published and have proved informative. For example, an AGP has been described in cucumber whose expression is gibberellin-induced (Park *et al.*, 2003). Ectopic expression of this gene in transgenic tobacco plants led to increased AGP accumulation and increased internode elongation, suggesting that the influence of gibberellin on plant growth might be, at least in part, mediated via AGPs. However, the interpretation of gene function based purely on ectopic expression can be problematic and more definitive data can be obtained from experiments in which gene expression is abrogated. In this context, the first reported phenotypes of plants with mutations in AGP-encoding genes are intriguing, and at the same time somewhat surprising. Van Hengel and Roberts (2003) reported on the characterisation of an insertional mutant in an *Arabidopsis* gene encoding AtAGP30, leading to the loss of expression of this specific AGP. Plant growth and development was apparently normal, but *in vitro* root regeneration was suppressed, providing a link to previous data implicating AGPs in aspects of root growth (Van Hengel & Roberts, 2002). In addition, the mutant seeds showed an altered response to abscisic acid, a hormone correlated with various elements of seed development and germination. How exactly loss of AtAGP30 function might lead to the observed phenotypes is not yet clear, but the data provide some of the first molecular genetic data on AGP function.

In another report on loss of AGP function, Shi *et al.* (2003) were investigating *Arabidopsis* mutants altered in their response to salt stress. One of the genes identified by this approach (*SOS5*) encoded a protein that contained AGP-like domains but was distinct from previously characterised AGP sequences. Under normal growth condition, the mutant appeared normal, but after salt stress the root tips became swollen and cellular organisation was disrupted. The authors proposed that the SOS5 AGP-like protein plays a role in cell wall architecture and adhesion and that mutation of the gene leads to at least some cells becoming prone to abnormal expansion. In particular, it is possible that the SOS5 protein (and other AGP-like proteins) is involved in linking the plasma membrane and cell wall. This idea is interesting, especially bearing in mind the recent data showing that AGPs can interact with wall-associated kinases (WAKs) (Gens *et al.*, 2000). WAKs (which will be examined in more detail later in this chapter) have been proposed as potential bridges between the plasma membrane and the cell wall and to be involved in the transduction of information from the ECM (Anderson et al., 2001). The possibility of a complex of WAKs and AGPs at the plasma membrane/cell wall interface is intriguing.

A further unexpected link between AGPs and intercellular signalling has come from the finding that they can act as substrates for chitinases (Van Hengel et al., 2001). Chitinases are enzymes that catalyse the hydrolysis of linkages between b-(1-4) linked polymers of N-acetylglucosamine. Such polymers are found in the ECM of most fungi and the generally accepted view is that chitinases act as part of a defence mechanism for the plant (Mauch et al., 1988). Surprisingly, investigation of a temperature-sensitive mutant line of a carrot embryogenic culture revealed that the mutation (which led to decreased embryogenic potential) could be rescued by the addition of chitinase to the medium (De Jong et al., 1992). Moreover, embryogenic cultures themselves secreted a chitinase that could be purified from the medium and which could rescue mutant cultures in cross-feeding experiments. Further experiments also indicated that chitinases influenced somatic embryogenesis (Van Hengel et al., 1998). A puzzle at the time of these experiments was that there was no known endogenous substrate for chitinases in plant tissue. However, recent data show that at least some AGPs contain N-acetylglucosamine and that these AGPs can act as a substrate for cleavage by chitinases, presumably leading to the release of N-acetylglucosamine-containing oligosaccharides (Van Hengel et al., 2001). Moreover, the pretreatment of AGPs with chitinase influenced the potency of the AGPs in an assay designed to test their efficacy in promoting embryogenesis. Interestingly, it has been shown that a class of lipooligosaccharides (nod factors) can also rescue defects in somatic embryogenesis (De Jong et al., 1993). These molecules contain an oligosaccharide backbone of four to five b-(1-4)-linked residues of N-acetylglucosamine attached to a C16 or C18 fatty acid and play a key role in signalling between *Rhizobium* species and plant roots destined for nodulation (Lerouge et al., 1990)

Taken together, these data provide a tantalising link between three previously disparate lines of investigation, i.e. AGPs, chitinases and oligosaccharides. Thus, AGPs could act as the endogenous substrates for developmentally regulated chitinases whose action leads to the generation of modified AGPs and oligosaccharides, both of which might act as signals involved in growth and development.

A final piece to the puzzle of AGPs is the discovery that the C-terminal part of the some AGPs can undergo modification to generate a lipid anchor (Youl et al., 1998). Such lipid anchors would attach the AGP to the plasma membrane but could be susceptible to attack by phospolipases to generate a free, presumably diffusible AGP. Phospholipases have been shown to be involved in many signalling pathways, thus (theoretically) providing a link between classically defined signalling mechanisms and a novel AGP-based mechanism of intercellular communication. In such a speculative pathway, specific AGPs might be initially linked to the plasma membrane until cleavage by phospholipase to release the proteoglycan element of the AGP. This moiety might itself be subject to further processing by chitinases to generate signals, as described above (Figs. 4.2B and 4.2C). It should be stressed that such a model is speculative, but it provides a model for testing in future work.

To summarise, an increasing body of evidence supports the hypothesis that AGPs function in intercellular communication. The patterns of AGP expression, the observed outcome of exogenous AGPs in various systems and the novel linkages that can now be made with data from research on chitinases and oligosaccharide signalling all provide strong indications of a functional role for AGPs in intercellular signalling. However, a significant amount of data have been obtained from the analysis of *in vitro* cultured tissue (somatic embryos, suspension cultures, *Zinnia* mesophyll cells), raising the question as to what extent AGPs function in the intact plant as signalling molecules. The analysis of AGP function using molecular genetic approaches has begun to shed light on this important issue and it can be expected that definitive evidence on the physiological role of these fascinating molecules will be described in the near future.

4.3.3 Cutin and signalling

Aerial organs of plants are normally formed as discrete primordia from meristems distributed throughout the plant. However, many plants at points in their development generate structures that are the result of the fusion of initially independently formed organs. Such post-genital fusion is especially common in the generation of floral structures (Verbeke, 1992). The observation that only some plant organs undergo fusion whereas the majority do not suggests that a signalling process is involved. For example, experiments in which young primordia were physically forced into close contact did not lead to fusion; i.e., the process is not just a consequence of spatially restricted growth of neighbouring organs. Initial work on carpel fusion in *Catheranthus roseus* indicated that the organs involved secreted a substance that promoted subsequent fusion (Siegel & Verbeke, 1989). Thus, insertion of thin agar slices between the presumptively fused adaxial sides of carpels led to the accumulation within the agar of a water-soluble substance that, when applied to the abaxial side of carpels, induced elements of epidermal cell de-differentiation that were normally observed only during carpel fusion. A range of experiments using membranes of varying porosity and placed at different regions of the developing carpels indicated that the signalling process was a two-way affair, i.e., that both carpels involved in a fusion event needed to secrete a substance if successful fusion were to occur. The nature of this substance has not yet been characterised.

More recently, a number of mutants of *Arabidopsis* have been identified that display abnormal organ fusion, e.g. post-genital fusion of leaves and stem. Molecular genetic analysis of these mutants has revealed that the disrupted genes encode enzymes that potentially play a role in production of the plant cuticle.

The cuticle is a complex matrix consisting mainly of lipid-based cutin and wax (Kolattukudy, 1980). These components are secreted onto the epidermal surface where they play an essentially protective function, e.g. to limit loss of water from the plant and to reflect irradiance. However, mutants in which the cuticle appears to be compromised frequently display morphogenic abnormalities, most notably in the occurrence of abnormal organ fusions. For example, the *LACERATA* gene of *Arabidopisis* encodes a cytochrome P450 monooxygenase which catalyses fatty acid hydroxylation (Wellesen *et al.*, 2001) and the *FIDDLEHEAD* gene encodes a ketoacyl synthase (Pruitt *et al.*, 2000). Both enzymes could potentially play a role in cutin synthesis and both mutants are characterised by organ fusions. The potential importance of cutin in organ fusion was further substantiated by the work of Sieber *et al.* (2000). These authors over-expressed a gene encoding a cutinase in *Arabidopsis* and observed that this led to a number of organ fusion events and an abnormal structure of the plant cuticle.

Although these data indicate the importance of cutin in influencing organ fusion, they do not necessarily imply that cutin (or cutin-derivatives) acts as a signal. It could be that the role of the cuticle is permissive rather than instructive. In this interpretation, fusion of primary cell walls would be a default pathway that is normally prevented only by the presence of an intact cuticle. Any factor leading to loss of cuticle structure would thus lead to organ fusion. How earlier data indicating a positive factor inducing organ fusion fit into this interpretation is unclear. One possibility is that during normal post-genital fusion events a positive diffusible signal is involved, which leads to either the induction of cutin breakdown in the target tissue or to the suppression of cutin biosynthesis. In such a scenario, the mutants in which uncontrolled post-genital fusion events are observed would mimic the normal process but might not necessarily identify the normal control mechanism.

Further evidence pointing to an important (but possibly permissive) role of cutin and wax in intercellular communication comes from the analysis of patterning events in the leaf epidermis, in particular of stomata and trichome formation. Stomatal density is normally negatively correlated with the CO_2 level to which a plant is exposed (Gray *et al.*, 2002). In the *HIC* mutant of *Arabidopsis*, this linkage is broken and a decrease in stomatal density does not occur in response to elevated CO_2 concentration. At ambient CO_2 levels, stomatal density in the *HIC* genetic background is similar to that observed in wild-type plants. Molecular genetic analysis revealed that the *HIC* gene encodes a ketoacyl synthase similar to that identified by analysis of the *FIDDLEHEAD* mutation, i.e., an enzyme potentially involved in cutin biosynthesis. Organ fusion events are not observed in the *HIC* mutant background and the cuticle appears normal. However, it is interesting that a component of the *FIDDLEHEAD* mutant phenotype is an altered density of trichomes on the leaves (Pruitt *et al.*, 2000). Both trichomes and stomata are epidermal-derived structures that are generated in specific patterns (see Chapter 9). Moreover, analysis of *Arabidopsis* mutants has indicated that stomatal patterning is disrupted by the altered expression of a secreted protease, giving rise to the idea that a diffusible peptide generated with the cell wall might act as intercellular signal influencing stomatal differentiation (Berger & Altmann, 2000). Viewing the analysis of the *HIC* mutant in the context of these data, it is possible that (again) the cuticle is influencing intercellular signalling in a permissive fashion rather than an instructive one, i.e., that cutin structure influences the diffusion of an extracellular signal. This idea is strengthened by the observation that at least some mutants with altered wax biosynthesis also show altered stomatal density (Gray *et al.*, 2002). In this scenario, the general cuticle structure would restrict or influence the diffusion of a CO_2-associated extracellular signal. The identity of this signal is unknown.

An interesting extension of the work on the *HIC* mutant is the finding that the exposure of a relatively mature leaf to a high CO_2 concentration leads to a lower stomatal density in younger primordia that are not exposed to elevated CO_2 levels (Lake *et al.*, 2001). These data indicate that leaves exposed to altered CO_2 levels generate a signal that acts at a distance to regulate stomatal density in developing leaf primordia. As with the *HIC* mutant, the nature of the intercellular signal is unknown and there is no reason to implicate the cell wall in this instance. Nevertheless, it provides another example of the increasing spectrum of intercellular signalling pathways that are being uncovered in plants. The application of molecular genetic tools to these problems promises to reveal the nature of these novel signalling entities.

To summarise, a number of lines of evidence indicate that the structure of the cuticle can influence intercellular communication in plants. However, the majority of these data suggest that this action is essentially permissive. An interesting question for the future is to investigate whether there is an endogenous regulation of cuticle permeability with respect to the extracellular signals that are hypothesised to flow through this component of the ECM; i.e., do the observed outcomes of disruption of cuticle synthesis and architecture on organ fusion and epidermal patterning reflect an endogenous mechanism by which intercellular signalling is normally modulated by the plant? This seems to be especially pertinent to the developmentally regulated process of organ fusion observed in many floral structures. With respect to actual signals derived from the cuticle, the evidence is weaker. As with other components of the ECM, the chemical complexity of the cuticle, coupled with the limits of our understanding of its synthesis and breakdown, represents significant hurdles to further progress in this area. Novel approaches to the identification of mutants in cuticle structure promise to provide important advances in this area (Tanaka *et al.*, 2004).

4.3.4 Uncharacterised cell wall determinants involved in signalling

The survey of ECM-associated intercellular signalling mechanisms described above has focused on chemical components that, although sometimes not fully characterised, at least can be placed into reasonably specific chemical groupings (e.g. OGAs, xyloglucans, AGPs). However, in addition to these 'known' signals, there are many data in the literature that indicate or imply the presence of an intercellular signalling process but do not enable any precise identification of the actual signalling component itself. For example, sequence analysis has revealed that plants express proteins with similarity to receptors for tumour-necrosis factor peptide hormones that influence a variety of developmental processes in animals (Becraft *et al.*, 1996; Tanaka *et al.*, 2002; Gifford *et al.*, 2003). Moreover, mutations in at least some of these putative receptors lead to phenotypes consistent with disruption of an intercellular signalling mechanism. Thus mutation of the *CR4* gene in maize leads to altered leaf epidermal identity (Becraft *et al.*, 1996) and the *ARC4* gene in *Arabidopsis* is involved in cellular organisation during growth of the ovule integument (Gifford *et al.*, 2003). Although it seems likely that peptide factors are involved in these signalling processes, the factors have not yet been identified. Similarly, the finding that mutations in subtilisin-like proteases influence epidermal differentiation and stomatal patterning also strongly indicates the presence of as yet uncharacterised peptide based signalling components that are processed within the ECM (Berger & Altmann, 2000) (see also Chapter 2).

These examples have arisen from the analysis of higher plants, but work with other experimental systems has also implicated the cell wall as a key element in determining cellular differentiation. In particular, research using the marine alga *Fucus* has led to novel insights.

Fertilised eggs of *Fucus* undergo a series of conserved cell divisions to generate apical thallus (shoot-like) and basal rhizoid (root-like) organs. This process occurs in the free-living zygote and the early events of cell growth and division can easily be followed under the microscope. This is in contrast to the situation in angiosperms where such events occur hidden deep within the surrounding sporophytic tissue. The *Fucus* system has thus been used as experimental model to gain insight into the earliest events of plant embryogenesis (Brownlee & Berger, 1995). Using novel tools of laser ablation, Brownlee and colleagues undertook a series of elegant experiments to decipher the outcome on embryo development of the selected removal of particular cell types (Berger *et al.*, 1994). One of the most informative experiments involved the destruction of the rhizoid protoplast to leave a presumptive thallus cell subtended by fragments of the rhizoid wall. A variety of growth patterns were observed, but if thallus-cell growth led (fortuitously) to contact with the remnants of the rhizoid cell wall, the thallus cell underwent a transdifferentiation to form a rhizoid-like structure. The conclusion from this experiment is that the rhizoid cell wall must contain some type of information that specifies rhizoid cell fate. The molecular character of this information is as yet unknown. Nevertheless, the experiments clearly indicate the potential of cell wall epitopes as determinants of cell fate and suggest that such cell wall determinants might partake in a cross-talk between adjacent cells, leading to the appropriate differentiation of each cell type. Such two-way cellular discussions are common in animal systems and are characterised by the Notch/delta system involving the proteolytic cleavage and sensing of signals within the ECM (Artavanis-Tsakonas *et al.*, 1999; Fortini, 2001). Obvious counterparts to Notch/delta components have not yet been reported in plants.

To summarise, a number of intriguing experiments and observations indicate the presence of many as yet uncharacterised chemical signals in plants. For some of these there are hints and leads as to their identity. For example, the genomic data indicating the presence of novel peptide-type signalling components provide a strong synergy with biochemical and molecular genetic data that have led to the identity of peptide-based signals (Fletcher *et al.*, 1999; Yang *et al.*, 1999) (see Chapter 2). Although the precise linkage between signal processing system, signal factor and specific receptor has generally not yet been achieved, rapid and significant progress in this area can be expected. With respect to other data (such as those implicating cell wall epitopes as determinants of differentiation), progress may be rapid if molecules already implicated in cell-to-cell signalling turn out to be the causal agent (e.g., AGPs). However, if these signals are, for example, novel polysaccharide components of the cell wall, their identification and characterisation may present a major challenge. It seems safe to state that we have certainly not exhausted the list of chemical signals involved in plant intercellular communication.

4.4 The cell wall and biophysical signalling

The plant cell wall forms a semi-rigid case around the protoplast and connects cells together to form a tissue. Because of turgor pressure, the cell wall is generally under tension, leading to a pattern of physical stress both around individual cells and connecting neighbouring cells. Altered tissue growth or altered biophysical characteristics of the cell wall are likely to lead to alteration in this stress pattern. If individual cells can sense and respond to changes of stress pattern around them, then the cell wall could act as a conduit for a physical-stressed-based transfer of information within a tissue. These ideas, and the data to support them, are most advanced in animal systems.

Animal cells (like plant cells) are also embedded in an ECM. This is generally not as rigid as a cellulose-based matrix, but it can nevertheless generate a physical stress on the cell. Moreover, different types of matrixes lead to and are associated with specific differentiation pathways. Although there are clearly chemical-based interactions between the cell and the ECM, models have been proposed in which the mechanical stresses generated around cells are transmitted across the plasma membrane to the cytoskeleton to generate cell-specific responses (Ingber, 2003). A key element in these models is that the tensional forces generated within the cytoskeleton and the surrounding ECM constitute a balanced system (termed tensional integrity). Such tensegrity systems are characterised by an inherent stability so that any shift in the stress vector tends to be counterbalanced by the system. For example, magnetic microbeads can be coated with ligands that interact with cell surface receptor complexes (integrins) that link the ECM with the internal cytoskeleton (Wang et al., 1993). By applying magnetic fields across cells coated with such micro-beads, Wang and colleagues demonstrated that bead twisting was limited, suggesting that the cytoskeletal/ECM contact points to which the beads bound could counteract changes in stress applied to them. The idea is that such mechanical foci at the ECM/cytoskeleton interface sense changes in physical stress imposed on them and transduce such changes into secondary signals to influence both cytoskeletal structure and gene expression.

A characteristic of research in this area has been the utilisation of novel and imaginative approaches to manipulate and measure cellular growth and deformation. For example, Tan and colleagues recently developed a microfabricated post-array detector system in which spots of specific ECM material were brought into contact with growing cells to create defined mechanical landscapes (Tan *et al.*, 2003). As cell growth occurred over the surface of the spots, the individual elements of the array were deflected and the degree of deflection could be measured. By integrating the degree of spot deflection over time and space, estimates of traction forces at the subcellular level were made and, in conjunction with specific staining techniques, correlated with the behaviour of contact points between the ECM and the plasma membrane. This novel strategy allowed the simultaneous manipulation of the ECM, estimation of the traction forces in the system and observation of cell behaviour as a consequence of the manipulations performed.

A key element underpinning the advances in the animal field in this area has been the characterisation of the membrane-spanning molecular machinery that connects the ECM and the cytoskeleton. However, before considering the evidence in plants for such ECM/cytoskeletal adhesion points, let us first examine the evidence that physical force is involved at all in plant intercellular signalling.

First, it is clear that plant form is responsive to the mechanical stresses imposed on the organism. In the wild, the growth of plants grown in an environment with a constant wind direction is different from that of plants grown under still conditions. Experimental manipulations indicate that the plane of cell division in callus cultures can be manipulated by controlling the vector of external force applied (Yeoman & Brown, 1971; Linthilac & Vesecky, 1984) and that the growth form of regenerating protoplasts is influenced by the force applied on them (Wymer *et al.*, 1996). Moreover, molecule evidence has revealed that plants may be exquisitely sensitive to mechanical stimulation, with signal transduction pathways being induced by even the slightest transient touch (Braam & Davies, 1990; Knight *et al.*, 1991, 1992). Thus, it is apparent that plant cells possess the machinery to respond rapidly to external changes in pressure. The question is: do plant cells use this machinery to coordinate responses during the normal developmental program of growth and differentiation?

The observation that plants show responses to changes in physical stress, coupled with the view of plant tissue as a material through which force can easily and rapidly be transmitted, has led to the proposition that patterns of physical stress play a causal role in the coordination of morphogenic events. Essentially, because of the contiguous nature of the cell wall, a change in tension around one group of cells will inevitably influence the tensile forces in neighbouring tissue (Fig. 4.3). If cells can sense these changes and provide an output in terms of altered growth, then a simple, direct and rapid system would be generated for the control of form (Trewavas & Knight, 1994). These ideas are most closely associated with a progression of ideas from Green and colleagues (Green, 1992, 1994, 1999), and have been formulated into a number of theories based on minimal energy configurations and the mechanical process of buckling. In these models, plant tissue (in particular the outer epidermal layer) is thought of as a growing composite material whose physical attributes can be altered and whose growth is spatially constricted by set boundaries. In such a system, modelling methods demonstrate that growth may continue in a planar fashion but accompanied by a build-up of tensional forces within the material.

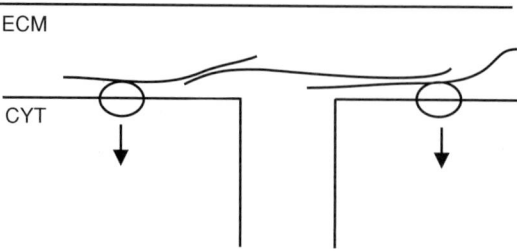

Figure 4.3 Mechanical signalling within the ECM. Components of the ECM form physical bridges between adjacent cells. Because of their rigidity, altered physical stress in one region of the ECM will be transmitted to the adjacent cells via these bridges. If these ECM components are connected to the cytosol (CYT) of the respective cells, the possibility exists of a direct mechanical interaction of neighbouring cells that might initiate intracellular signal transduction pathways.

Eventually, depending on the attributes of the material and the growth rates and boundaries set, the material may buckle to take up an energetically more stable configuration. Since the process of force redistribution within the material will occur very rapidly, the buckling process can lead to the simultaneous generation of a pattern of spatially distributed humps apparently growing out of an initially planar surface. These patterns of hump formation are reminiscent of the process of organ initiation observed in, for example, many flowers. In these models, the physical stresses within the tissue (primarily the cell wall) act as a type of signal. The pattern generated is dependent on the cellular growth characteristics of the tissue (local or global) and the extensibility of the cell wall (local or global).

Although the mathematical modelling approaches and arguments are convincing, experimental data to support the idea that physical forces in plant tissue act over a distance to coordinate morphogenesis are limited. For example, it can be shown that constraining the growth of an apex by physical means leads to alterations in morphogenesis that are consistent with the models put forward, but the question remains open as to whether these manipulations mimic the processes that occur during normal development (Hernandez & Green, 1993). One prediction of the biophysical models is that local alteration of cell wall extensibility should influence morphogenesis. Experiments in which the local activity of a cell wall protein (expansin) was manipulated on the apical meristem indicated that local alteration of cell wall characteristics could induce morphogenesis and, indeed, leaf formation (Fleming et al., 1997; Pien et al., 2001). However, although ectopic leaf formation was associated with loss (or delay) of the expected leaf morphogenesis on the opposite side of the meristem, it is unclear if altered patterns of mechanical stress across the meristem were instrumental in this phenotype. At present, it seems that although local discontinuities in cell wall tension allow the local coordinated outgrowth of tissue during organogenesis, a causal role in patterning needs to be substantiated.

To summarise, outgrowth of a tissue at a particular area requires an imbalance of forces to allow such outgrowth to occur. Cessation of growth must require the establishment of a new equilibrium of forces within the tissue. Molecules involved in regulation cell wall extensibility are likely to play a key role in this process. Since the wall of any cell contributes to and is part of an extracellular matrix that unites a number of cells throughout a tissue, morphogenesis in one region has the potential to alter the pattern of stress in a region of tissue at a distance. Depending on the biophysical character of the material, the variability of this character in space and time, and the variability of the compressive forces generated by growth, complex but entirely predictable changes in morphogenesis could result, as indicated by various modelling approaches. The question remains, however, whether these theoretical considerations reflect an endogenous mechanism. At the local cellular level, evidence in animal tissue strongly supports biophysical components of a signalling mechanism. In plants, the experimental evidence remains sporadic.

The plant cell wall is a complex composite material, and modelling of the physical characteristics of such a material is non-trivial. Moreover, the physical forces predicted to play a role in linking morphogenic events across a tissue are likely to be transduced via a number of cell walls associated with different cell layers. Again, predicting, measuring and manipulating the physical characteristics of such a complex material at cellular spatial resolution is not trivial. A key aspect of research in this area in animal systems has been the development and application of novel microtechniques to manipulate and measure the physical environment of cells in culture. As yet, such techniques have not been applied to plant systems and, until this happens, the role of biophysical parameters in intercellular communication in plants is likely to remain debatable.

4.4.1 Connections between the cell wall and the cytosol as a conduit for intercellular signalling

As mentioned in the previous section, advances in animal research with respect to the significance of the ECM in intercellular signalling have been dependent on the characterisation of the molecular bridges connecting the ECM and the cytoskeleton (Hynes, 1987). A key component of these bridges are the integrin receptors that (on the extracellular face) can interact with ECM proteins (such as fibronectin, vitronectin and laminin) whereas on the cytosolic face integrins can be anchored to various cytoskeletal-associated elements (such as vinculin and talin). These sites of adhesion play a key role in the perception and transduction of information. Although it is clear that plant cells must also contain some type of molecular bridge between the ECM and the internal components of the cell, the identity of these bridges remains essentially uncharacterised. Proteins that are immunologically cross-reactive with antibodies raised against vitronectins are present in plant tissue extracts, but the genes encoding such proteins have not yet been identified (Sanders *et al.*, 1991; Wagner *et al.*, 1992; Zhu *et al.*, 1993). More recently, interest in this area has focused on a novel family of WAKs.

WAKs consist of a cysteine-rich extracellular domain that has sequence similarity to epidermal growth factor repeats, suggesting that they can interact with an extracellular ligand. This extracellular domain is linked across the plasma membrane to a cytosolic domain with similarity to serine/threonine kinases implicated in signal transduction. Biochemical analysis indicates that the extracellular domain is closely associated or linked with pectin material in the cell wall and that at least some WAKs can interact with glycine-rich proteins in the cell wall, as well as AGPs (He *et al.*, 1996; Gens *et al.*, 2000; Wagner & Kohorn, 2001). Thus, WAKs have the structure and location to act as transducers of information from the ECM to the cytosol (Anderson *et al.*, 2001). Initial data indicated that WAKs might play a role in plant response to pathogen attack (He *et al.*, 1998), but recent results suggest a more basic involvement of WAKs in the regulation of plant growth. For example, use of an inducible antisense strategy to suppress WAK expression indicated that impaired WAK function led to a dramatic suppression of plant growth (Lally *et al.*, 2001; Wagner & Kohorn, 2001). In particular, although germination appeared to proceed normally, subsequent elongation of hypocotyls and root and expansion of leaves was blocked. How WAKs might impinge on cell growth processes is still unclear, but one interpretation of these experiments is that WAKs interact with extracellular ligands that play a role in stimulating plant growth. Reduction of WAK gene expression would limit the potential binding sites for these ligands, thus leading to the phenotypes observed. That this ligand might be a component of the ECM is intriguing. Mutation of a gene encoding a chimeric leucine-rich repeat/extensin protein has been shown to lead to abnormal root hair growth, indicating that hydroxyproline-rich extracellular proteins can be involved in the maintenance of appropriate cell shape (Baumberger *et al.*, 2001). Whether specific glycine-rich proteins might fulfil a similar function via interaction with WAKs is open to speculation. A fuller characterisation of the extracellular ligand(s) for WAKs promises to provide an important insight into how components of the cell wall might transmit information to the cell.

To summarise, a major prerequisite for physical signalling within or from the cell wall is that there must be a mechanism by which information in the ECM is transmitted across the plasma membrane to induce a signalling cascade within the cell. Although a number of receptor-like molecules have begun to be identified in plant systems (some of which are described in other chapters of this book), a major gap in our knowledge lies in our ignorance of the nature of the bridge between the ECM, the plasma membrane and components of the cytoskeleton. Advances in plant cell biology, coupled with the identification of gene sequences encoding proteins that might play a role in this connection, should lead to significant advances in this area in the near future. Such advances are essential and overdue.

4.5 Conclusions

From this summary of possible mechanisms of intercellular communication in plants, it is clear that in addition to the classical and well-investigated and accepted systems described elsewhere in this book, a number of other mechanisms have been postulated and investigated. In particular, the idea that the cell wall can be both a

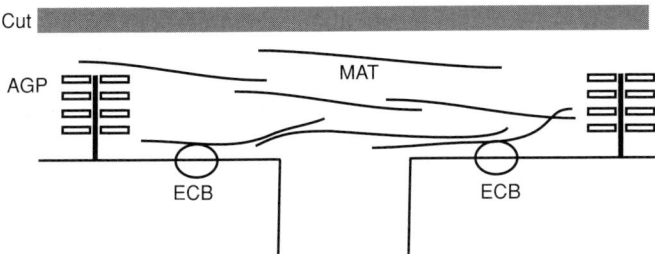

Figure 4.4 Composite model to demonstrate the potential variety of roles played by the ECM in signalling. Components of the matrix (MAT) may be a source of chemical signals, as well as acting as physical connections between neighbouring cells. Such physical connections could transduce force between spatially separated ECM/cytosolic bridges (ECB). AGPs might also act as a source of chemical signals. The flux of signals within the ECM might be controlled by the architecture of the matrix as well as by the structure of the cuticle (cut).

source of signals as well as a means of signal perception of modification is highly attractive (Fig. 4.4). The large body of evidence from the animal field clearly shows how the ECM of cells can influence growth and differentiation of groups of cells within the matrix. In plants, despite often intensive and widespread investigation, the topic is still somewhat unclear. The most promising lines of evidence at present lead from the recent findings on AGPs and their potential interaction and processing by chitinase-like enzymes. Significant progress in this area can be expected. With respect to oligosaccharides, although a role in defence signalling is accepted, a role in endogenous developmental programs and growth processes remains debatable. Finally, non-chemical-based signalling systems (i.e. biophysical) remain on the fringes but provide a number of challenging ideas as to the interpretation of biological events.

Note added in proof. The recent finding that the proteoglycan xylogen mediates vascular differentiation provides direct evidence of the importance of the plant ECM in differentiation (Motose *et al.*, 2004).

Acknowledgements

The author was supported during part of this work by a START Fellowship from the Swiss National Science Foundation.

References

Aldington, S. & Fry, S.C. (1993) Oligosaccharins. *Adv. Bot. Res.*, **19**, 1–101.
Anderson, C.M., Wagner, T.A., Perret, M., He, Z.-H., He, D. & Kohorn, B.D. (2001) WAKS: cell wall-associated kinases linking the cytoplasm to the extracellular matrix. *Plant Mol. Biol.*, **47**, 197–206.

Artavanis-Tsakonas, S., Rand, M.D. & Lake, R.J. (1999) Notch signalling: cell fate control and signal integration in development. *Science*, **284**, 770–776.

Baumberger, N., Ringli, C. & Keller, B. (2001) The leucine-rich repeat/extensin cell wall protein LRX1 is required for root hair morphogenesis in *Arabidopsis thaliana*. *Genes Dev.*, **15**, 1128–1139.

Becraft, P.W., Stinard, P.S., & McCarty, D.R. (1996) CRINKLY4: a TNFR-like receptor kinase involved in maize epidermal differentiation. *Science*, **273**, 1406–1409.

Berger, D. & Altmann, T. (2000) A subtilisin-like protease involved in the regulation of stomatal density and distribution in *Arabidopsis thaliana*. *Genes Dev.*, **14**, 1119–1131.

Berger, F., Taylor, A. & Brownlee, C. (1994) Cell fate determination by the cell wall in early *Fucus* development. *Science*, **263**, 1421–1423.

Beveridge, C.A., Symons, G.M. & Turnbull, C.G.N. (2000) Auxin inhibition of decapitation-induced branching is dependent on graft transmissible signals regulated by genes *rms1* and *rms2*. *Plant Physiol.*, **123**, 689–697.

Braam, J. & Davies, R.W. (1990) Rain-, wind-, and touch-induced expression of calmodulin and calmodulin related genes in *Arabidopsis*. *Cell*, **60**, 357–364.

Branca, C., DeLorenzo, G. & Cervone, F. (1988) Competitive inhibition of the auxin-induced elongation by oligogalacturonides in pea stem segments. *Physiol. Plant.*, **72**, 499–504.

Brownlee, C. & Berger, F. (1995) Extracellular matrix and pattern in plant embryos: on the lookout for developmental information. *Trends Genet.*, **11**, 344–348.

De Jong, A.J., Cordewener, J., LoSchiavo, F., Terzi, M., Vandekerckhove, J., Van Kammen, A. & De Vries, S. (1992) A carrot somatic embryo mutant is rescued by chitinase. *Plant Cell*, **4**, 425–433.

De Jong, A.J., Heidstra, R., Spaink, H.P., Hartog, M.V., Meijer, E.A., Hendriks, T., Lo Schiavo, F., Terzi, M., Bisseling, T., Van Kammen, A. & De Vries, S. (1993) Rhizobium lipooligosaccharides rescue a carrot somatic embryo mutant. *Plant Cell*, **5**, 615–620.

Dhugga, K.S., Barreiro, R., Whitten, B., Stecca, K., Hazebroek, J., Randhawa, G.S., Dolan, M., Kinney, A.J., Tomes, D., Nichols, S. & Anderson, P. (2004) Guar seed b-mannan synthase is a member of the cellulose synthase super gene family. *Science*, **303**, 363–366.

Dolan, L., Linstead, P. & Roberts, K. (1995) An AGP epitope distinguishes a central metaxylem initial from other vascular initials in the *Arabidopsis* root. *Protoplasma*, **189**, 149–155.

Du, H., Simpson, R.J., Clarke, A.E. & Bacic, A. (1996) Molecular characterization of a stigma-specific gene encoding an arabinogalactan protein (AGP) from *Nicotiana alata*. *Plant J.*, **9**, 313–323.

Eberhard, S., Doubrova, N., Marfa, V., Mohnen, D., Southwick, A., Darvill, A. & Albersheim, P. (1989) Pectic cell wall fragments regulate tobacco thin-cell-layer explant morphogenesis. *Plant Cell*, **1**, 747–755.

Finkelstein, R.R., Gampala, S.S.L. & Rock, C.D. (2002) Abscisic acid signaling in seeds and seedlings. *Plant Cell*, **14**, S15–S45.

Fleming, A.J., Mandel, T., McQueen-Mason, S. & Kuhlemeier, C. (1997) Induction of leaf primordia by the cell wall protein expansin. *Science*, **276**, 1415–1418.

Fletcher, J.C., Brand, U., Running, M.P., Simon, R. & Meyerowitz, E.M. (1999) Communication of cell fate decisions by CLAVATA3 in *Arabidopsis* shoot meristems. *Science*, **283**, 1911–1914.

Fortini, M.E. (2001) Notch and presenelin: a proteolytic mechanism emerges. *Curr. Opin. Cell Biol.*, **13**, 627–634.

Fukada, H. & Komamine, A. (1980) Establishment of an experimental system for the tracheary element differentiation from single cells isolated from the mesophyll of *Zinnia elegans*. *Plant Physiol.*, **52**, 57–60.

Gens, J.S., Fujiki, M. & Pickard, B.G. (2000) Arabinogalactan protein and wall associated kinase in plasmalemmal reticulum with specialised vesicles. *Protoplasma*, **212**, 115–134.

Gifford, M.L., Dean, S. & Ingram, G.C. (2003) The *Arabidopsis ACR4* gene plays a role in cell layer organization during ovule integument and sepal margin development. *Development*, **130**, 4249–4258.

Gray, J.E., Holyroyd, G.H., Van der Lee, F.M., Bahrami, A.R., Sijmons, P.C., Woodward, F.I., Schuch, W. & Hetherington, A.M. (2002) The *HIC* signalling pathway links CO_2 perception to stomatal development. *Nature*, **408**, 713–716.

Green, P.B. (1992) Pattern formation in shoots: a likely role for minimal energy configurations of the tunica. *Int. J. Plant Sci.*, **153**, S59–S75.
Green, P.B. (1994) Connecting gene and hormone action to form, pattern and organogenesis: biophysical transductions. *J. Expt. Bot.*, **45**, 1775–1778.
Green, P.B. (1999) Expression of pattern in plant: combining molecular and calculus-based biophysical paradigms. *Am. J. Bot.*, **86**, 1059–1076.
He, Z.H., Fujiki, M. & Kohorn, B.D. (1996) A cell wall-associated, treceptor-like protein kinase. *J. Biol. Chem.*, **271**, 19789–19793.
He, Z.H., He, D. & Kohorn, B.D. (1998) Requirement for the induced expression of a cell wall associated receptor kinase for survival during pathogen response. *Plant J.*, **14**, 55–63.
Hernandez. L.F. & Green, P.B. (1993) Transductions for the expression of structural pattern: analysis in sunflower. *Plant Cell*, **5**, 1725–1738.
Hoson, T. & Masuda, Y. (1991) Effect of xyloglucan nonasaccharide on cell elongation induced by 2,4-dichlorophenoxyacetic acid and indole-3-acetic acid. *Plant Cell Physiol.*, **32**, 777–782.
Hutchison, C.E. & Kieber, J.L. (2002) Cytokinin signaling in *Arabidopsis*. *Plant Cell*, **14**, S47–S59.
Hynes, R.O. (1987) Integrins: a family of cell surface receptors. *Cell*, **48**, 549–554.
Ingber, D.E. (2003) Mechanosensation through integrins: cells act locally but think globally. *Proc. Natl. Acad. Sci. U.S.A.*, **100**, 1472–1474.
Kessler, S., Seiki, S. & Sinha, N. (2002) Xcl1 causes delayed oblique periclinal cell divisions in developing maize leaves, leading to cellular differentiation by lineage instead of position. *Development*, **129**, 1859–1869.
Knight, M.R., Campbell, A.K., Smith, S.M. & Trewavas, A.J. (1991) Transgenic plant aequorin reports the effects of touch and cold shock and fungal elicitors on cytosolic calcium. *Nature*, **352**, 524–526.
Knight, M.R., Smith, S.M. & Trewavas, A.J. (1992) Wind-induced plant motion immediately increases cytosolic calcium. *Proc. Natl. Acad. Sci. U.S.A.*, **89**, 4967–4972.
Knox, J.P., Linstead, P.J., Peart, J., Cooper, C. & Roberts, K. (1991) Developmentally regulated epitopes of cell surface arabinogalactan proteins and their relation to root tissue pattern formation. *Plant J.*, **1**, 317–326.
Kolattukudy, P.E. (1980) Biopolyester membranes of plants: cutin and suberin. *Science*, **208**, 990–1000.
Lally, D., Ingmire, P., Tong, H.-Y. & He, Z.-H. (2001) Antisense expression of cell wall-associated protein kinase, WAK4, inhibits cell elongation and alters morphology. *Plant Cell*, **13**, 1317–1332.
Lake, J.A., Quick, W.P., Beerling, D.J. & Woodward, F.I. (2001) Signals from mature to new leaves. *Nature*, **411**, 154–155.
Lerouge, P., Roche, P. Faucher, C., Maillet, F., Truchet, G., Promé, J.C. & Dénarié, J. (1990) Symbiotic host-specificity of *Rhizobium meliloti* is determined by a sulphated and acylated glucosamine oligosaccharide signal. *Nature*, **344**, 781–784.
Lerouxel, O., Choo, T.S., Seveno, M., Usadel, B., Faye, L., Lerouge, P. & Pauly, M. (2002) Rapid structural phenotyping of plant cell wall mutants by enzymatic oligosaccharide fingerprinting. *Plant Physiol.*, **130**, 1754–1763.
Li, Y.Q., Faleri, C., Geitmann, A., Zhang, H.Q. & Cresti, M. (1995) Immunogold localization of arabinogalactan proteins, unesterified and esterified pectins in pollen grains and pollen tubes of *Nicotiana tabacum* L. *Protoplasma*, **189**, 251–261.
Linthilac, P.M. & Vesecky, T.B. (1984) Stress-induced alignment of division plane in plant tissues grown in vitro. *Nature*, **307**, 363–364.
Majewska-Sawska, A. & Nothnagel, E.A. (2000) The multiple roles of arabinogalactan proteins in plant development. *Plant Physiol.*, **122**, 3–9.
Mauch, F., Mauch-Mani, B. & Boller, T. (1988) Antifungal hydrolases in pea tissue. *Plant Physiol.*, **88**, 936–942.
McCabe, P.F., Valentine, T.A., Forsberg, L.S. & Pennell, R.I. (1997) Soluble signals from cells identified at the cell wall establish a developmental pathway in carrot. *Plant Cell*, **9**, 2225–2241.
McDougall, G.J. & Fry, S.A.C. (1988) Inhibition of auxin-stimulated growth of pea stem segments by a specific nonasaccharide of xyloglucan. *Planta*, **175**, 412–416.

McNeil, M., Darvill, A.G., Fry, S.C., Albersheim, P., & Dell, A. (1984) Structure and function of the primary cell walls of plants. *Ann. Rev. Biochem.*, **53**, 625–663.

Milioni, D., Sado, P.-E., Stacey, N.J., Roberts, K. & McCann, M. (2002) Early gene expression associated with the commitment and differentiation of a plant tracheary element is revealed by cDNA-amplified fragment length polymorphism analysis. *Plant Cell*, **14**, 2813–2824.

Moille, G., Robin, S., Lecompte, M., Pagant, S. & Höfte, H. (2003) Classification and identification of *Arabidopsis* cell wall mutants using Fourier transform infrared (FT-IR) microspectroscopy. *Plant J.*, **35**, 393–404.

Motose, H., Fukuda, H. & Sugiyama, M. (2001) An arabinogalactan protein(s) is a key component of a fraction that mediates local intercellular communication involved in tracheary element differentiation of *Zinnia* mesophyll cells. *Plant Cell Physiol.*, **42**, 129–137.

Motose, H., Sugiyama, M. & Fukuda, H. (2004) A proteoglycan mediates inductive interaction during plant vascular development. *Nature*, **429**, 873–878.

Oxley, D. & Bacic, A. (1999) Structure of the glycosylphosphatidylinositol anchor of an arabinogalactan protein from *Pyrus communis* suspension-cultured cells. *Proc. Natl. Acad. Sci. U.S.A.*, **96**, 14246–14251.

Park, M.H., Suzuki, Y., Chono, M., Knox, J.P. & Yamaguchi, I. (2003) CsAGP1, a gibberellin-responsive gene from cucumber hypocotyls, encodes a classical arabinogalactan protein and is involved in stem elongation. *Plant Physiol.*, **131**, 1450–1459.

Pennel, R.I. & Roberts, K. (1990) Sexual development in the pea is presaged by altered expression of arabinogalactan protein. *Nature*, **344**, 547–549.

Perrimon, N. & Bernfield, M. (2001) Cellular functions of proteoglycans – an overview. *Semin. Cell Dev. Biol.*, **3**, 1317–1326.

Pien, S., Wyrzykowska, J., McQueen-Mason, S., Smart, C. & Fleming A. (2001) Local expression of expansin induces the entire process of leaf development and modifies leaf shape. *Proc. Natl. Acad. Sci. U.S.A.*, **98**, 11812–11817.

Pruitt, R.E., Vielle-Calzada, J.P., Pionese, S.E., Grossniklaus, U. & Lolle, S.J. (2000) Fiddlehead, a gene required to suppress epidermal cell interactions in *Arabidopsis*, encodes a putative lipid biosynthetic enzyme. *Proc. Natl. Acad. Sci. U.S.A.*, **97**, 1311–1316.

Roberts, K. (1994) The plant extracellular matrix: in an expansive mood. *Curr. Opin. Cell Biol.*, **6**, 688–694.

Ryan, C.A. & Farmer, E.E. (1991) Oligosaccharide signals in plants: a current assessment. *Annu. Rev. Plant Physiol. Mol. Biol.*, **42**, 651–674.

Sanders, L.C., Wang, C.S., Walling, L.L. & Lord, E.M. (1991) A homolog of the substrate adhesion factor vitronectin occurs in four species of flowering plants. *Plant Cell*, **3**, 629–635.

Shi, H., Kim, Y.-S., Guo, Y., Stevenson, B. & Zhu, J.-K. (2003) The *Arabidopsis* SOS5 locus encodes a putative cell surface adhesion protein and is required for normal cell expansion. *Plant Cell*, **15**, 19–32.

Sieber, P., Schorderet, M., Ryser, U., Buchala, A., Kolattukudy, P., Metraux, J.-P. & Nawrath, C. (2000) Transgenic *Arabidopsis* plants expressing a fungal chitinase show alterations in the structure and properties of the cuticle and postgenital organ fusions. *Plant Cell*, **12**, 721–737.

Siegel, B.A. & Verbeke, J.A. (1989) Diffusible factors essential for epidermal cell redifferentiation in *Catheranthus roseus*. *Science*, **244**, 580–582.

Sorefan, K., Booker, J., Haurogne, K., Goussot, M., Bainbridge, K., Foo, E., Chatfield, S., Ward, S., Beveridge, C., Rameau, C. & Leyser, O. (2003) MAX4 and RMS1 are orthologous dioxygenase-like genes that regulate shoot branching in *Arabidopsis* and pea. *Genes Dev.*, **17**, 1469–1474.

Stacey, N.J., Roberts, K., Carpita, N.C., Wells, B. & McCann, M.C. (1995) Dynamic changes in cell surface molecules are very early events in the differentiation of mesophyll cells from *Zinnia elegans* into tracheary elements. *Plant J.*, **8**, 891–906.

Svetek, J., Yadav, M.P. & Nothnagel, E.A. (1999) Presence of a glycosylphosphatidlyinositol lipid anchor on rose arabinogalactan proteins. *J. Biol. Chem.*, **274**, 14724–14733.

Takeda, T., Furuta, Y., Awano, T., Mizuno, K., Mitsuishi, Y. & Hayashi, T. (2002) Suppression and acceleration of cell elongation by integration of xyloglucans in pea stem segments. *Proc. Natl. Acad. Sci. U.S.A.*, **99**, 9055–9060.

Tan, J.L., Tien, J., Pirone, D.M., Gray, D.S., Bhadriraju, K. & Chen, C.S. (2003) Cells lying on a bed of microneedles: an approach to isolate mechanical force. *Proc. Natl. Acad. Sci. U.S.A.*, **100**, 1484–1489.

Tanaka, H., Watanabe, M., Watanabe, D., Tanaka, T., Machida, C. & Machida, Y. (2002) ACR4, a putative receptor kinase gene of *Arabidopsis thaliana*, that is expressed in the outer cell layers of embryos and plants, is involved in proper embryogenesis. *Plant Cell Physiol.*, **43**, 419–428.

Tanaka, T., Tanaka, H., Machida, C., Watanabe, M. & Machida, Y. (2004) A new method for rapid visualization of defects on leaf cuticle reveals five intrinsic patterns of surface defects in *Arabidopsis*. *Plant J.*, **37**, 139–146.

Thompson, H.J.M. & Knox, J.P. (1998) Stage-specific responses of embryogenic carrot cell suspension cultures to arabinogalactan protein binding B-glucosyl Yariv reagent. *Planta*, **205**, 32–38.

Toonan, M.A.J., Schmidt, E.D.L., Van Kammen, A. & de Vries, S.C. (1997) Promotive and inhibitory effects of arabinogalactan proteins in *Daucus carota* somatic embryogenesis. *Planta*, **201**, 188–195.

Tran Thanh Van, K., Toubart, P., Cousson, A. Darvill, A.G., Gollin, D.J., Chelf, P. & Albersheim, P. (1985) Manipulation of the morphogenetic pathways of tobacco explants by oligosaccharins. *Nature*, **314**, 615–617.

Trewavas, A. & Knight, M. (1994) Mechanical signalling, calcium and plant form. *Plant Mol. Biol.*, **26**, 1329–1341.

Van Hengel, A.J. & Roberts, K. (2002) Fucosylated arabinogalactan proteins are required for full root cell elongation in *Arabidopsis*. *Plant J.* **32**, 105–113.

Van Hengel., A.J. & Roberts, K. (2003) AtAGP30, an arabinogalactan-protein in the cell walls of the primary root, plays a role in root regeneration and seed germination. *Plant J.*, **36**, 256–270.

Van Hengel, A.J., Guzzo, F., van Kammen, A. & de Vries, S.C. (1998) Expression pattern of the carrot EP3 endochitinase genes in suspension cultures and in developing seeds. *Plant Physiol.*, **117**, 45–53.

Van Hengel, A.J., Tadess, Z., Immerzeel, P., Schols, H., Van Kammen, A. & de Vries, S.C. (2001) *N*-Acetylglucosamine and glucosamine containing arabinogalactan proteins control somatic embryogenesis. *Plant Physiol.*, **125**, 1880–1890.

Verbeke, J.A. (1992) Fusion events during floral morphogenesis. *Annu. Rev. Plant Physiol. Plant Mol. Biol.*, **43**, 583–598.

Vincken, J.-P., Schols, H.A., Oomen, R.J.F.J., McCann, M., Ulskov, P., Voragen, A.G.J. & *Visser*, R.G.F. (2003) If homogalacturonan were a side chain of rhamnogalacturonan, I: Implications for cell wall architecture. *Plant Physiol.*,**132**, 1781–1789.

Wagner, T.A. & Kohorn, B.D. (2001) wall-associated kinases are expressed throughout plant development and are required for cell expansion. *Plant Cell*, **13**, 303–318.

Wagner, V.T., Brian, L. & Quatrano, R.S. (1992) Role of a vitronectin-like molecule in embryo adhesion of the brown alga *Fucus*. *Proc. Natl. Acad. Sci. U.S.A.*, **89**, 3644–3648.

Wang, K.L.-C., Li, H. & Ecker, J.R. (2002) Ethylene biosynthesis and signaling networks. *Plant Cell*, **14**, S131–S151.

Wang, N., Butler, J.P. & Ingber, D.E. (1993) Mechanotransduction across the cell surface and through the cytoskeleton. *Science*, **260**, 1124–1127.

Wellesen, K., Durst, F., Pinot, F., Benveniste, I., Nettesheim, K., Wisman, E., Steiner-lange, S., Saedler, H. & Yephremov, A. (2001) Functional analysis of the LACERETA gene of *Arabidopsis* provides evidence for different roles of fatty acid hydroxylation in development. *Proc. Natl. Acad. Sci. U.S.A.*, **98**, 9694–9699.

Wilson, R.H., Smith, A.C., Kacurakova, M., Saunders, P.K., Wellner, N. & Waldron, K.W. (2000) The mechanical properties and molecular dynamics of plant cell polysaccharides studied by Fourier-transform infrared spectroscopy. *Plant Physiol.*, **124**, 397–405.

Wymer, C.L., Wymer, S.A., Cosgrove, D.J. & Cyr, R.J. (1996) Plant cell growth responds to external forces and the response requires intact microtubules. *Plant Physiol.*, **110**, 425–430.

Yang, H., Matsubayashi, Y., Nakamura, K. & Sakagami, Y. (1999) *Oryza sativa* PSK gene encodes a precursor of phytosulfokine, a sulfated peptide growth factor found in plants. *Proc. Natl. Acad. Sci. U.S.A.*, **96**, 13560–13565.

Yeoman, M.M. & Brown, R. (1971) Effects of mechanical stress on the plane of cell division in developing callus cultures. *Ann. Bot.*, **35**, 1101–1112.

York, W.S., Darvill, A.G. & Albersheim, P. (1984) Inhibition of 2,4-dichlorophenoxyacetic acid-stimulated elongation of pea stem segments by a xyloglucan oligosaccharide. *Plant Physiol.*, **75**, 295–297.

Youl, J.J., Bacic, A. & Oxley, D. (1998) Arabinogalactan proteins from *Nicotiana alata* and *Pyrus communis* contain glycosylphospatidlyinositol membrane anchors. *Proc. Natl. Acad. Sci. U.S.A.*, **95**, 7921–7926.

Zablackis, E., York, W.S., Hantus, P.H., Reiter, W.D., Chapple, C.C., Albersheim, P. & Darvill, A. (1996) Substitution of L-fucose by L-galactose in cell walls of *Arabidopsis* mur1. *Science*, **272**, 1808–1810.

Zhu, J.-K., Shi, U., Wyatt, S.E., Bressan, R.A., Hasegawa, P.M. & Carpita, N.C. (1993) Enrichment of vitronectin- and fibronectin-like proteins in NaCl adapted plant cells and evidence for their involvement in plasma membrane-cell wall adhesion. *Plant J.*, **3**, 637–646.

5 Plasmodesmata – gateways for intercellular communication in plants

Trudi Gillespie and Karl J. Oparka

5.1 Introduction

The aim of this chapter is to provide an overview of the structure and function of plasmodesmata, highlighting their major role in controlling the flux of solutes, signals and proteins that pass from cell to cell in the growing plant. Several major pieces of research have been conducted on plasmodesmata, and as our knowledge and understanding of the spatial and temporal changes that occur within plasmodesmata grows, numerous theories are being developed to explain how plasmodesmata function in intercellular communication. Some of this work will be brought together in this chapter, with an emphasis on those critical pieces of work that have made a significant contribution to our current understanding of plasmodesmata. Inevitably, for a chapter of this type, it will be impossible to do justice to the complete field, and the reader will be directed to key texts for additional information.

5.1.1 Plasmodesmata – key components of the symplast

One feature of plant cells that makes them unique from other eukaryotic cells is the cell wall. To be able to respond to environmental stimuli and coordinate differentiation and growth, neighbouring plant cells (and often distant cells) must communicate (Hashimoto & Inze, 2003). Communication and nutrient transfer between higher plant cells can occur by two parallel aqueous pathways. The aqueous phase of the apoplast lies outside the plasma membrane and consists of the cell walls and conducting cells of the xylem (Fisher, 2000). Molecules such as hormones and secreted protein ligands are known to carry information between cells apoplastically, and some transcription factors may travel in the apoplast to subsequently regulate gene expression (Hashimoto & Inze, 2003). This apoplastic pathway is best characterised by CLAVATA-mediated signalling, which has been shown to regulate the stem cell population in the shoot apical meristem (SAM) of *Arabidopsis* (Doerner, 2003). However, as Ding (1998) comments, higher plants have evolved a unique intercellular organelle, the plasmodesma, that permits direct cell-to-cell communication of molecules through a cytoplasmic channel traversing the wall (for discussion see Fisher, 2000). Recent research has shown that traditional views of the role of the plant cell wall as simply providing mechanical strength and acting as a barrier to pathogens need to be reconsidered. Plant cells are now thought to be capable of monitoring changes in wall composition, and using this information to activate signalling cascades within the cell (Hashimoto & Inze, 2003). In a similar way, plasmodesmata, which were once thought to be static structures embedded within the cell wall, are being revealed as dynamic structures that have a fundamental role in controlling symplastic protein and RNA trafficking.

5.1.2 Plasmodesmata: simple description, complex function

Various descriptions of plasmodesmata have been given over the last century (for review see Roberts & Oparka, 2003). The historical concept of plasmodesmata as static, membrane-lined channels that facilitate the cell-to-cell movement of low-molecular-weight solutes has dominated much of the debate on the structure and function of plasmodesmata (Lucas *et al.*, 1993; Blackman & Overall, 2001). However, recent ultrastructural, physiological, biochemical and molecular studies on plasmodesmata have transformed this opinion, and plasmodesmata are now considered to be highly flexible and diverse structures that can exert considerable control on the flux of molecules that pass through them (Lucas *et al.*, 1993; Blackman & Overall, 2001; Ehlers & Kollmann, 2001; Roberts & Oparka, 2003). As Lee *et al.* (2000) comment, the sophistication of cell-to-cell communication through plasmodesmata may yet rival that of nuclear transport.

5.1.3 Discovery of plasmodesmata

The term *plasmodesmata* (singular, plasmodesma) was first used by Strasburger, in 1901 (Carr, 1976). However, it was in 1879 that Eduard Tangl first observed the intercellular striations between cells in the cotyledons of *Strychnos nuxvomica*, which he interpreted to be protoplasmic contacts (Roberts & Oparka, 2003, citing Carr, 1976). In the same year Albert Pfeffer commented that '...it is quite likely that the connections could be utilised in the transport of substances and even, in particular cases, principally or solely for this purpose' (cited by Carr, 1976). Tangl's work formed the basis of the concept that intercellular communication between living plant cells is necessary not only for the survival of individual cells, but for the plant itself. This notion was significant as it contested the then current view that plant cells functioned as autonomous units (Carr, 1976). It was not until 1930 that Münch took this concept further and used the term *symplasm* to describe the cytoplasmic continuity that exists between plant cells throughout the whole plant. Since the work of Tangl, many key papers have increased our understanding of plasmodesmata, and provided new insights into their structure and function.

5.2 Structure

5.2.1 The general ultrastructure of plasmodesmata

Structural models of plasmodesmata have generally been based on data from transmission electron microscopy. Since the first published model of plasmodesmata in 1968 (Robards, 1968a,b) various structural models have been proposed, together with considerable debate as to the effect of different fixation and staining procedures on their appearance (Gunning & Robards, 1976; Robards & Lucas, 1990; Beebe & Turgeon, 1991; Tilney *et al.*, 1991; Ding *et al.*, 1992b; Botha *et al.*, 1993; Turner *et al.*, 1994; White *et al.*, 1994; Overall & Blackman, 1996; Ding, 1997, 1999; Waigmann *et al.*, 1997; Radford *et al.*, 1998; Overall, 1999). On the basis of ultrastructure alone, plasmodesmata are generally classified into two basic types: simple and branched. Simple plasmodesmata consist of a single channel traversing the cell wall, whereas branched plasmodesmata have a more complex structure with two or more channels on either side of the middle lamella, often joined by a central cavity (Roberts & Oparka, 2003). Many structural models depict simple plasmodesmata only, usually assuming that simple and branched plasmodesmata have a common basic architecture (Blackman & Overall, 2001; Ehlers & Kollmann, 2001). As van Bel *et al.* (1999) comment, this may give the impression that higher plants possess only one type of plasmodesma, forcing interpretations of function into a single structural model. The basic structure of simple and branched plasmodesmata is shown in Fig. 5.1. When discussing plasmodesmata ultrastructure it is important to note that a wide range of plasmodesmal morphologies and substructural variations have been found both between plant species and within the tissues of the same plant (Robinson-Beers & Evert, 1991; Waigmann *et al.*, 1997). Furthermore, electron micrographs provide a two-dimensional image of an essentially dynamic structure (Robards & Lucus, 1990; Botha & Cross, 1999; van Bel *et al.*, 1999). However, since the first observation of plasmodesmata in the electron microscope (Buvat, 1957), images of plasmodesmata have revealed a remarkable structural consistency (Overall, 1999).

A longitudinal section through a plasmodesma reveals a plasma-membrane-lined cylindrical channel that transverses the cell wall (Fig. 5.2; Blackman & Overall, 2001); the plasma membrane defines the symplastic boundary of the plasmodesma and is continuous between adjacent cells (Overall & Blackman, 1996). Grabski *et al.* (1993) suggested that the plasma membrane in plasmodesmata may be modified as it fails to allow diffusion of lipids between neighbouring cells. In the centre of the channel lies a strand of modified cortical endoplasmic reticulum, the desmotubule (Robards & Lucas, 1990; Ding *et al.*, 1992b; Lucas & Wolf, 1993; Epel, 1994). The modified endoplasmic reticulum of the desmotubule is continuous with the endoplasmic reticulum in the adjoining cells, forming an endomembrane continuum throughout the plant (Robards & Lucus, 1990; Denecke, 2001). This continuity has been shown by 3,3′-dihexyl-oxacarocyanine iodide ($DiOC_6$) staining after plasmolysis (Oparka *et al.*, 1994), and lipids may apparently diffuse along the endoplasmic reticulum membranes between adjacent cells (Grabski *et al.*, 1993). Although the endoplasmic reticulum is continuous between cells, it is the cytoplasmic sleeve (the space between the desmotubule and plasma membrane) through which the bulk of cell-to-cell communication is thought to occur (Ding, 1997).

At both ends of the plasmodesma channel, just inside the cell wall, is the neck region (see Fig. 5.2). This area is often constricted at either or both ends, where the plasma membrane comes into tight association with the entrance of the cytoplasmic

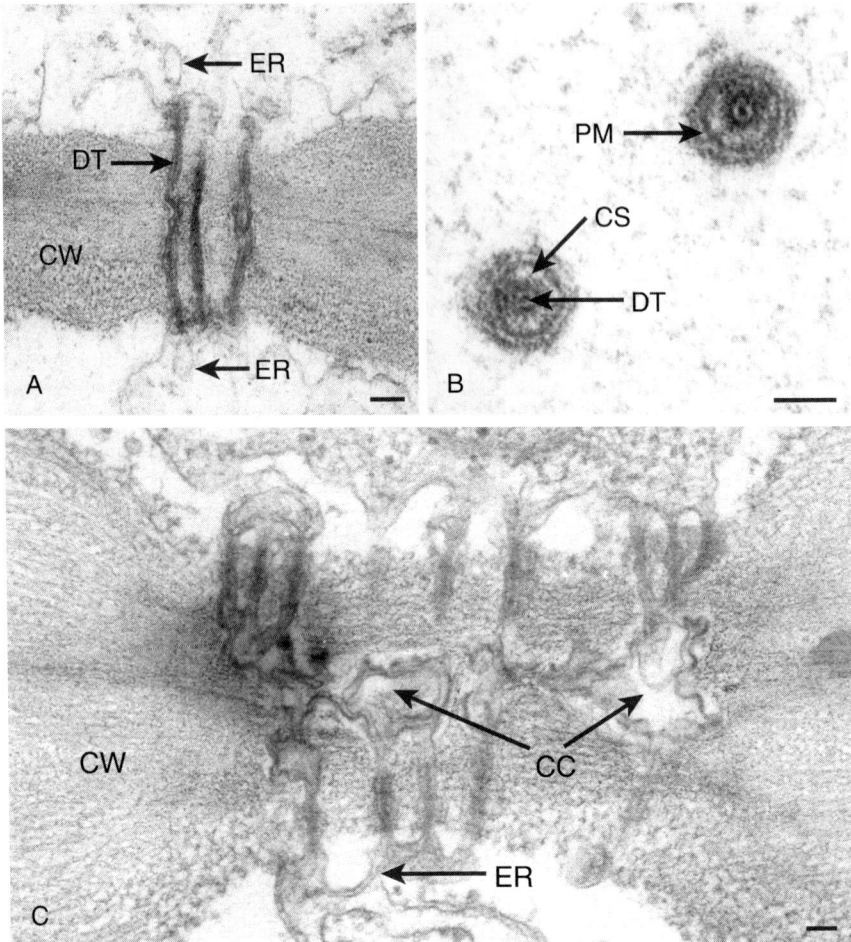

Figure 5.1 (A) Longitudinal section through simple plasmodesmata between adjacent parenchyma cells of potato tuber. ER, endoplasmic reticulum; DT, desmotubule; CW, cell wall. Bar: 25 nm. (B) Transverse section through simple plasmodesmata between parenchyma cells of potato tuber, showing the desmotubule and the plasma membrane. The central electron dense dot is thought to be formed by the appressed hydrophilic phospholipid head groups of the inner leaflet of the desmotubule. The cytoplasmic sleeve appears to be partially occluded; this is thought to be due to gobular subunits surrounding the desmotubule. PM, plasma membrane; CS, cytoplasmic sleeve. Bar: 20 nm. (C) Longitudinal section through branched plasmodesmata between parenchyma cells of potato tuber. Multiple branches can be seen extending from median central cavities. CC, central cavity; CW, cell wall; ER, endoplasmic reticulum. Bar: 25 nm.

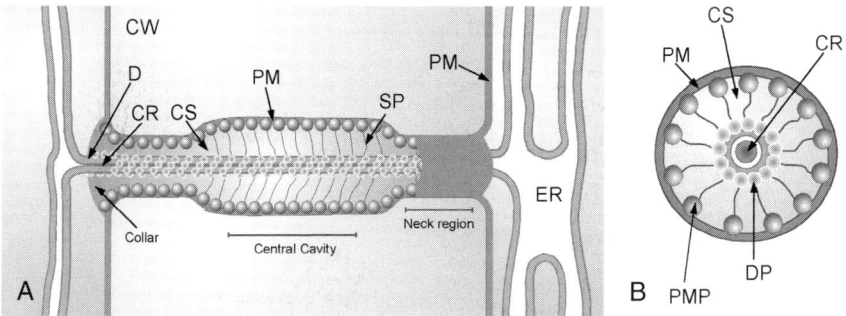

Figure 5.2 Schematic representation of a simple plasmodesmata in longitudinal (A) and transverse (B) sections. CW, cell wall; D, desmotubule; CR, central rod; CS, cytoplasmic sleeve; PM, plasma membrane; SP, spoke-like extensions; PMP, plasma-membrane-embedded proteins; DP, desmotubule-embedded proteins. Figure adapted from Roberts and Oparka (2003) and based on the model in Ding *et al.* (2003).

sleeve (Overall & Blackmann, 1996). In some cases this constriction is thought to be an artefact owing to physical wounding and glutaraldehyde fixation. Radford *et al.* (1998) showed that treatment with 2-deoxy-D-glucose (an inhibitor of callose synthesis) caused the neck of *Allium cepa* L. plasmodesmata to become funnel-shaped rather than constricted. When viewed in cross section, the cytoplasmic sleeve is seen to be partially occluded by the globular subunits surrounding the desmotubule (Ding *et al.*, 1992a; Overall & Blackmann, 1996). These subunits are thought to have a helical arrangement that divide the space in the cytoplasmic sleeve into a number of spiralling channels (Zee, 1969; Robards, 1976; Olesen, 1979; Overall *et al.*, 1982; Wolf *et al.*, 1989; Olesen & Robards, 1990; Robards & Lucas, 1990; Ding *et al.*, 1991; Robinson-Beers & Evert, 1991; Lucas *et al.*, 1993; Lucas & Wolf, 1993; Overall & Blackman, 1996; Waigman *et al.*, 1997). The functional diameter of these channels has been calculated to be between 3 and 4 nm (Terry & Robards, 1987; Fisher, 1999). An electron-dense ring of proteinaceous material in the wall surrounding the neck region has also been seen in electron micrographs (Olesen, 1979; Olesen & Robards, 1990; Beebe & Turgeon, 1991; Badelt *et al.*, 1994; Turner *et al.*, 1994). This wall collar, or 'sphincter', is most consistently visualised when tannic acid is included during fixation (Fisher, 2000). The collar can be closely associated with the plasma membrane, or it can be expanded out from the plasma membrane and connected to it via spokes or strings (Overall, 1999). The actual function of these structures in plasmodesmal regulation has not been verified; however, most authors believe them to be involved in modulating the outer dimensions of the plasmodesmal pore.

5.2.2 Primary and secondary; simple or branched

Plasmodesmata that develop between daughter cells during cell division are termed primary plasmodesmata (Jones, 1976). *Primary plasmodesmata* are formed across

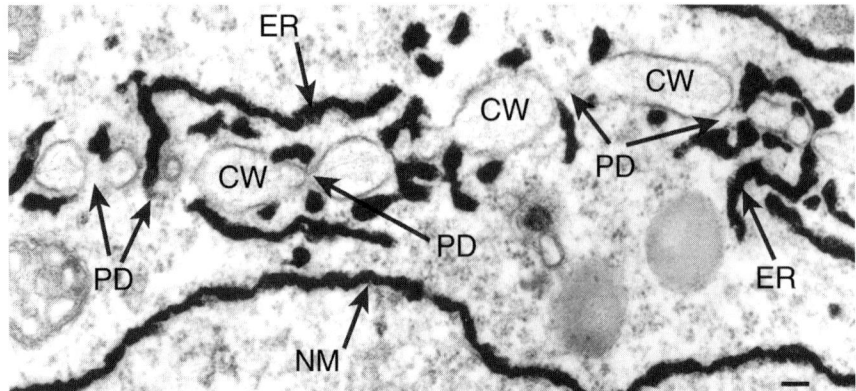

Figure 5.3 Cytokinesis in the root tip of *Zea mays* showing formation of primary plasmodesmata by endoplasmic reticulum entrapment in the developing cell plate. The tissue has been impregnated with zinc iodide/osmium tetroxide; this technique deposits osmium on the cisternal space of membrane-bound organelles such as the endoplasmic reticulum. Smooth domains of the endoplasmic reticulum crossing the growing cell plate indicate areas where primary plasmodesmata will form, constricted endoplasmic reticulum tubules (desmotubule) can also be seen between areas of the developing cell wall. ER, endoplasmic reticulum; CW, cell wall; PD, plasmodesmata; NM, nuclear membrane. Image courtesy of C. Hawes, Oxford Brookes University. Bar: 25 nm.

each developing phragmoplast, allowing cytoplasmic and endomembrane continuity to occur between daughter cells (see Fig. 5.3; Mezitt & Lucas, 1996). *Secondary plasmodesmata* are defined as those that form *de novo* across existing cell walls (Lucas *et al.*, 1993; Ehlers & Kollmann, 2001). Hepler (1982) provided the first detailed electron microscopy study of primary plasmodesmata formation during cell-wall deposition in roots of *Latuca sativa*. At the final stage of cytokinesis, a cell wall is formed to partition the two daughter cells. It is generally accepted that primary plasmodesmata are formed at sites where endoplasmic reticulum tubules become trapped within the fusing Golgi-derived vesicles of cell wall material forming the cell plate (Porter & Machado, 1960; Hepler, 1982; Staehelin & Hepler, 1996). The cytoplasmic strands become constricted during cell plate growth, and the endoplasmic reticulum tubules become transformed into the desmotubule, retaining endomembrane continuity between the two daughter cells (Ehlers & Kollmann, 2001). Overall (1999) has suggested that the dynamin-like molecule phragmoplastin, which has been identified in forming the cell plate (Gu & Verma, 1996), may induce the production of the tightly curled endoplasmic reticulum membrane tube as it passes through the plasmodesma.

Although plasmodesmata formed during cytokinesis are randomly distributed, they are typically grouped into pit fields in the fully elongated cell (Fisher, 2000). The number of primary plasmodesmata laid down in a given wall has been found to accurately predict the subsequent cell wall expansion that will take place (Overall, 1999). Little is known about how this process is regulated, although short-term treatment with a cellulose biosynthesis inhibitor led to the formation of thickened cell plates with increased callose and increased plasmodesmal density (Vaughn et al., 1996).

Secondary plasmodesmata are formed post-cytokinetically in existing cell walls (Jones, 1976; Lucas et al., 1993; Ehlers & Kollmann, 2001). Secondary plasmodesmata can form along any wall of the cell, allowing cells to increase their potential for molecular trafficking, and also to create connections between cells that are not cytokinetically related (van der Schoot et al., 1995; Ding & Lucas, 1996; Volk et al., 1996; Itaya et al., 1998; Oparka et al., 1999; van der Schoot & Rinne, 1999). Secondary plasmodesmata cannot be unambiguously distinguished from primary plasmodesmata on the basis of structural criteria. Therefore, most models of secondary plasmodesmata formation are based on studies of plasmodesmata formed in walls between cells that were initially separated (Ehlers & Kollman, 2001). Examples of such studies include the following: protoplast fusion in cell tissue culture (Monzer, 1990; Ehlers & Kollmann, 1996); graft unions (Kollmann & Glockmann, 1985, 1991); host–parasite connections (Dörr, 1987; Dawson et al., 1994); organs fused post-genitally (carpel-fusion walls) (van der Schoot et al., 1995); and the genetically distinct cells of plant chimaeras (Steinberg & Kollmann, 1994).

Various models of the mechanism of secondary plasmodesmata formation have been proposed (Jones, 1976; Juniper, 1977; Lucas & Gilbertson, 1994; Ding & Lucas, 1996). The most favoured model is described by Jones (1976), in which it is proposed that secondary plasmodesmata form by the endoplasmic reticulum becoming adnated to the plasma membrane, together with locally restricted enzymatic wall degradation. The wall thinning continues to a point where the plasma membrane and endoplasmic reticulum of either cell can penetrate the existing cell wall from one or both sides and subsequently fuse. This is followed by new wall deposition and the branching of endoplasmic reticulum (described by Ehlers & Kollman, 2001). To date, it is not known how the cell predicts the sites at which secondary plasmodesmata will form. Similarly, nothing is known concerning the process of synchronous wall thinning, or how the formation of each 'half plasmodesma' is coordinated between the adjacent cells (Kollmann & Glockmann, 1999). The formation of primary and secondary plasmodesmata has been reviewed in depth by Ehlers and Kollmann (2001).

Both primary and secondary plasmodesmata are initially simple in structure, but during tissue development may become complex, branched structures with a central cavity. Plasmodesmata in the newly formed cell wall may undergo post-cytokinetic structural modifications (Ehlers & Kollmann, 1996). During wall maturation, the secondary wall is deposited on the inner face of the primary wall (Fry, 2001), and further endoplasmic reticulum tubules, continuous with the desmotubule, may become trapped. This process can give rise to the formation of elongated simple plasmodesmata, or branched primary plasmodesmata (Kollmann & Glockmann, 1999). Branching of secondary plasmodesmata is also encompassed in the Jones model (Jones, 1976). As with primary plasmodesmata, endoplasmic reticulum tubules, continuous with the desmotubule, may become gradually embedded by Golgi-derived wall material as the cell wall is built up in the thinned areas where the secondary plasmodesmata are established (Ehlers & Kollmann, 2001). During tissue development, neighbouring primary plasmodesmata may develop into complex plasmodesmal morphotypes, with median branching planes and central cavities. For example, during the sink–source transition in leaves, simple plasmodesmata are converted to branched plasmodesmata that contain a central cavity aligned along the middle lamella of the cell wall (Oparka et al., 1999; Roberts et al., 2001). This transformation is thought to occur via an H-shaped intermediate that appears to form by the introduction of a new protoplasmic bridge between neighbouring pairs of simple plasmodesmata (Roberts & Oparka, 2003). At present, the genetic determinants of plasmodesmata formation are unknown.

5.2.3 *Plasmodesmal frequency and distribution: gain and loss*

Plasmodesmata are found between most living cells of higher plants, but are absent from key developmental interfaces, e.g. between sporophyte and gametophyte, between guard cells and epidermal cells, and between maternal tissues and embryos. In other tissues, plasmodesmal frequencies are within the range of $0.1–10$ mm^{-2} of cell wall, although exceptions can be found on either side of this range (Robards & Lucas, 1990). However, the relationship between plasmodesmal frequency and capacity for intercellular trafficking is not straightforward; plasmodesmata within the plant are not uniform and their transport capacity can vary considerably between cells and tissues (van Bel & Oparka, 1995). Thus, a paucity of plasmodesmata does not necessarily mean impaired transport (Fisher, 2000). As cells expand and differentiate, their fate determines the extent to which their cytoplasmic connectivity to other cells is maintained (Mezitt & Lucas, 1996). Various studies have found that plasmodesmal densities do not decline significantly as a result of cell elongation and wall expansion (Schnepf & Sych, 1983; Seagull, 1983). Such constancy in density can only be accounted for by the formation of secondary plasmodesmata. It is generally agreed that the formation of secondary plasmodesmata is an integral component of normal plant development (Jones, 1976; Robards & Lucas, 1990; Lucas et al., 1993). However, loss or restriction (occlusion) of plasmodesmata also appears to be common during differentiation. For example, in vascular tissue some cells greatly reduce the number of plasmodesmata in their adjoining cell walls (Gamalei, 1989). In plant species with a putative apoplastic loading mechanism, plasmodesmata in the vascular tissue are almost completely eliminated, with only a small number between the bundle sheath/phloem parenchyma cells and the sieve element–companion cell (SE-CC) complexes remaining (Oparka & Santa Cruz, 2000). During the sink–source transition in tobacco leaves, when there is a rapid phase of cell expansion in the spongy-mesophyll cells, primary plasmodesmata are thought to be lost, dramatically reducing the numbers of simple plasmodesmata (Roberts et al., 2001). At the same time, branched plasmodesmata are thought to be formed through the merging of any remaining simple plasmodesmata (Oparka et al., 1999). The immunocytological localisation of ubiquitin in those plasmodesmata destined for removal has led some authors to suggest that the removal process is accomplished, at least in part, by degradation of plasmodesmal proteins via the ubiquitin degradation pathway (Ehlers *et al.*, 1996; Fisher, 2000; Roberts & Oparka, 2003). Along with removal, plasmodesmata can also become occluded. For example, numerous plasmodesmata are found in the pits that connect immature xylem elements to mesophyll cells of *Sorbus torminalis*. However, during the final stages of programmed cell death, the pits become sealed off by deposition of cell wall material across both ends of the plasmodesmal pores (Lachaud & Maurousset, 1996). Callose and/or deposition of cell wall material has also been shown to block plasmodesmata at the bundle sheath–vascular parenchyma cell wall interface of a sucrose deficient (*sxd1*) mutant of maize that is unable to load sucrose into minor veins (Russin *et al.*, 1996; Botha *et al.*, 2000). The most extreme case of plasmodesmal 'downregulation' occurs around stomatal complexes (Oparka & Roberts, 2001). Guard cells are initially symplastically coupled to adjoining epidermal cells. However, as the guard cells mature new wall material is deposited across plasmodesmata, rendering them nonfunctional and symplastically isolating the mature guard cells from the surrounding epidermal cells (Wille & Lucas, 1984; Palevitz & Hepler, 1985). Lack of symplastic interconnection by plasmodesmata, such as that observed at the interface between maternal and embryonic tissue, suggests a compulsory apoplastic step in the transport pathway that can be precisely controlled (Ehlers *et al.*, 1999). In a developing embryo, symplastic discontinuity ensures autonomous development, restricting the entry of macromolecular complexes such as RNA or plant transcription factors that might influence the development of the genetically distinct embryo (Lucas, 1995; Ehlers & van Bel, 1999). The ability to add or lose plasmodesmata, both spatially and temporally, reveals the highly dynamic nature of intercellular communication in plants (Botha & Evert, 1988; Brown *et al.*, 1995).

5.2.4 *Plasmodesmal components*

Studies to elucidate the molecular structure of plasmodesmata have proven difficult because of the small size of plasmodesmata and their inaccessibility within the cell wall. Consequently, much of what we know comes from a combination of techniques such as immunological localisation and the random biochemical identification of novel proteins and polysaccharides. To date, relatively few plasmodesmal proteins have been positively identified (Roberts & Oparka, 2003).

One approach has been to use proteases and detergent treatments to remove components of plasmodesmata, although results from such studies have been contradictory. The desmotubule in the plasmodesmata of fern (*Onoclea sensibilis*) gametophytes was removed by digestion with papain, a protease, which also left the plasma membrane delineating the plasmodesmata swollen and irregular in profile. However, treatment with Triton X-100 did solubilise the plasma membrane limiting the plasmodesmata but the desmotubule was left intact (Tilney *et al.*, 1991). The authors of this study concluded that the desmotubule provides a stabilising cytoskeletal element for each plasmodesma. In maize root-tip wall fragments, Turner *et al.* (1994) found that light protease treatment (with trypsin or chymotrypsin) removed material from the exposed ends of the plasmodesmata, but there was little evidence of structural alteration in the plasmodesmal core. In contrast to Tilney et al. (1991), Turner et al. (1994) found that treatment with detergents removed both the plasma membrane and the desmotubule from within plasmodesmata, leaving the cell wall collar undisturbed. These authors concluded that the plasmodesmata collar is not proteinaceous, but proteins may bind it into the wall. Ritzenthaler et al. (2000) found that treatment with Pronase E and SDS could eliminate $DiOC_6$ staining of plasmodesmata in cell wall fragments of *Nicotiana clevelandii*; however, spots of callose were still identifiable. Treatment with chloroform–methanol, which should solubilise all membranous material, did not eliminate $DiOC_6$ staining of plasmodesmata. These results led Ritzenthaler et al. (2000) to conclude that $DiOC_6$ staining of plasmodesmata is unlikely to be due to the lipophilic nature of this stain but may be correlated to the presence of protein within the plasmodesma. Blackman and Overall (2001) have suggested that such proteins may be essential for maintaining the structure of the desmotubule.

Biochemical analysis of plasmodesmal proteins has proven to be problematical because of the difficulty in obtaining pure extracts that are free from wall contaminants (Epel, 1994; Epel et al., 1995, 1996). Filtration, nitrogen bomb disruption and wall-digesting enzymes have been used to improve plasmodesmata extraction and to identify several putative proteins enriched in these extracts (Kotlizky et al., 1992; Epel et al., 1996). Antibodies against a 41-kDa protein, isolated from a plasmodesmal extract, have been localised to plasmodesmata in maize mesocotyl cells (Epel et al., 1996). A putative plasmodesma-associated protein from maize mesocoytl, PAP26, was found to cross-react with antibodies raised against connexin from animal gap junctions (Yahalom et al., 1991). Two monoclonal antibodies (JIM76 and JIM64) raised against maize root plasmodesmata-associated proteins (Turner et al., 1994) were found to label plasmodesmata in trichomes and mesophyll of *N. clevelandii* (Waigmann et al., 1997). Blackman et al. (1998) have also identified four putative plasmodesmata-associated proteins from the alga, *Chara corallina*. Antibodies raised against one of these, a 45-kDa protein, showed localisation to plasmodesmata. However, to date, no genes encoding these proteins have been isolated or characterised.

Immunogold localisation and rhodamine phalloidin labelling have used to show that actin is a component along the entire length of plasmodesmata in higher plants (White et al., 1994), and also in the alga *C. corallina* (Blackman & Overall, 1998). Anitbodies raised against animal myosin have been localised to plasmodesmata in root tissue of *A. cepa*, *Zea mays* and *Hordeum vulgare* (Radford & White, 1998). Also, a mung bean myosin antibody was found to localise to plasmodesmata in *C. corallina* (Blackman & Overall, 1998). Myosin VIII, an unconventional myosin found only in plants, has also been localised to developing plasmodesmata in transverse cell walls of both meristematic and post-mitotically growing root cells (Reichelt et al., 1999). Tubulin has been found in protein extracts of walls containing plasmodesmata, but it has not been convincingly localised to plasmodesmata using immunogold procedures (Blackman & Overall, 1998). The carbohydrate callose, a b-1,3-glucan, is commonly associated with the extracellular neck region of the pore (for review see Stone & Clarke, 1992), and callose deposition at plasmodesmata has also been confirmed by localisation of antibodies raised against callose (Northcote *et al.*, 1989) and against a 65-kDa component of the callose synthase complex (Delmer *et al.*, 1993). Citovsky *et al.* (1993) showed that the activity of a cell-wall-associated protein kinase was correlated with the developmental maturation of plasmodesmata, and antibodies raised against an *Arabidopsis* calcium-dependent protein kinase were localised to the cell wall in a plasmodesmata-like distribution (Yahalom *et al.*, 1998). ATPase activity has been found at plasmodesmata in several plant species (Didehvar & Baker, 1986; Robards & Lucas, 1990; Chauhan *et al.*, 1991). In barley roots, ATPase activity is localised to the neck region of plasmodesmata, as are calcium-binding sites (Belitser *et al.*, 1982). The calcium-sequestering protein calreticulin, normally found in the lumen of the endoplasmic reticulum, has been found to be a component of the cortical endoplasmic reticulum elements associated with plasmodesmata (Baluška *et al.*, 1999, 2001), and centrin, a calcium-binding contractile protein, has also been localised to the neck region of plasmodesmata (Baluška *et al.*, 1999; Blackman *et al.*, 1999).

It is clear that cytoskeletal elements of the actin–myosin families are located in plasmodesmata. Overall (1999) has suggested that the helically arranged spiral of electron-lucent particles around the desmotubule is composed of actin microfilaments, and in the structural model proposed by Blackman and Overall (2001) putative myosin spokes radiate out from the actin, physically linking it to the plasma membrane. Myosins are a large superfamily of motor proteins that, in association with actin, are involved in intracellular motile processes (Reichelt *et al.*, 1999). It has been hypothesised that the actin filaments may form a static scaffold within plasmodesmata along which molecules move using a myosin-based motor (Roberts & Oparka, 2003).

Other cellular proteins/components that have been localised at or near plasmodesmata by cytochemical reactions include a -amylase (Gubler *et al.*, 1987), peroxidase (Schnepf & Sych, 1983), pectin methylesterase (Morvan *et al.* 1998), unesterified pectin (Casero & Knox, 1995), low-esterified pectin (Roy *et al.*, 1997) and 5′-nucleotidase (Nougaréde *et al.*, 1985). Additional cytochemical studies have shown plasmodesmata to be the location of high enzyme activity and strong reducing substances (see review by Olesen, 1979). However, as White *et al.* (1994) note, very little of this information has been successfully incorporated into a working model for plasmodesmal regulation.

The creation of mutants with defects in plasmodesmata structure and/or function, and the subsequent cloning of the genes concerned, is another approach that is being used to identify unique molecular components of plasmodesmata (Ding, 1998). However, both Ding (1998) and Zambryski (2004) note that the major obstacle to this approach is that alterations to plasmodesmata structure or function may be lethal or have drastic effects on plant development. To overcome this problem, Kim *et al.* (2002a) developed a heterozygous embryo–assay system in *Arabidopsis* to screen mutants that have an increased plasmodesmal size exclusion limit (SEL) at the torpedo stage of embryo development. This screen has so far identified two *ise* (*increased size exclusion*) mutants, although the genes for these are yet to be cloned and characterised. Some genes have been identified that affect plasmodesmata function but do not localise to plasmodesmata. For example, the *sxd1* mutation in maize (Russin *et al.*, 1996; Botha *et al.*, 2000) affects plasmodesmal development, but its primary role is in chloroplast-to-nucleus signalling (Mezitt Provencher *et al.*, 2001). Similarly, in transgenic plants expressing a yeast acid invertase gene, complex secondary plasmodesmata formation between mesophyll cells was found to be arrested during maturation (Ding *et al.*, 1993). These studies point to a complex interplay between basic metabolic processes and the formation of plasmodesmata.

In two recent studies, cDNA::GFP fusions have been used to probe for protein localisation and function *in planta*. Cutler *et al.* (2000) created a large number of *Arabidopsis* transgenic plants that expressed random GFP::cDNA fusions. Medina Escobar *et al.* (2003) took this high-throughput screening approach a stage further by using a Tobacco mosaic virus (TMV) vector to express libraries of random partial cDNAs fused to GFP. Over 20,000 infection foci expressing independent cDNA::GFP fusions were examined, and this screen isolated 12 putative plasmodesmal proteins. Further studies will reveal whether these proteins are structural components of plasmodesmata or whether they play a role in regulating transit through plasmodesmata (Ding *et al.*, 2003).

5.2.5 Passage through the cytoplasmic sleeve

The consensus in the literature is that diffusion of many small molecules such as sugars, metabolites, ions and amino acids are all thought to move by diffusion through the cytoplasmic sleeve of plasmodesmata (see Lucas *et al.*, 1993). Small fluorescent dyes that have been loaded passively (Duckett *et al.*, 1994; Roberts *et al.*, 1997) or microinjected (Goodwin, 1983; Erwee *et al.*, 1985; Madore & Lucas, 1986; Oparka *et al.*, 1991) utilise this pathway. Recent studies suggest that larger molecules, including GFP, may also move via the cytoplasmic sleeve (Roberts & Oparka, 2003). Plasmodesmata models derived from high-resolution electron micrographs show gaps in the range of 3 nm between the inner-plasmodesmal proteins that form the individual channels within the cytoplasmic sleeve (Ding *et al.*, 1991; Lucas *et al.*, 1993; Overall & Blackman, 1996; Fisher, 1999; Blackman & Overall, 2001), consistent with basal SEL of about 1 kDa (Tucker, 1982; Goodwin, 1983). However, a recalculation of data for diffusion through plasmodesmata (Terry & Robards, 1987) by Fisher (1999) showed that the basal SEL might be nearer to 4 nm, which significantly increases the potential SEL of the pore. As Roberts and Oparka (2003) note, an increase from 3 to 4 nm would mean a substantial increase in the Stokes radius for macromolecules that might diffuse through the cytoplasmic sleeve.

5.3 Macromolecular trafficking

There is now a considerable body of evidence to support the concept that plasmodesmata have the capacity to mediate cell-to-cell trafficking of endogenous macromolecules such as transcription factors, plant-defence-related proteins and RNA (Mezitt & Lucas, 1996; Ghoshroy *et al.*, 1997; McLean *et al.*, 1997; Haywood *et al.*, 2002). Intercellular protein trafficking through plasmodesmata is now thought to be an important means to regulate plant developmental processes, physiological functions and plant defence reactions (Ding, 1998).

Studies into the mechanisms by which plant viruses move from cell to cell within host tissue provided initial evidence that plasmodesmata mediate the trafficking of macromolecules (Deom *et al.*, 1992; Lucas & Gilbertston, 1994; Carrington *et al.*, 1996; Ding, 1997). Some viruses, such as Cowpea mosaic virus, move through plasmodesmata as intact virions, causing permanent modification to plasmodesmal structure (Hull, 1992; Storms *et al.*, 1995). Other viruses, typified by TMV, cause transient alterations to plasmodesmata in order to traffic the viral genome as a ribonucleoprotein complex (Deom *et al.*, 1992; Citovsky & Zambryski, 1993; McLean *et al.*, 1993; Lucas & Gilberston, 1994). Many viruses have been found to encode one or more non-structural proteins, termed movement proteins (MPs), that interact with plasmodesmata to facilitate viral spread between cells (reviewed in Wolf *et al.*, 1989; Atabekov & Taliansky, 1990; Robards & Lucas, 1990; Deom *et al.*, 1992; Lucas & Gilbertson, 1994; Carrington *et al.*, 1996; Gilbertson & Lucas, 1996; Nelson & van Bel, 1998). The most studied viral MP is the 30-kDa MP of TMV, which has a plasmodesmal targeting signal, increases the plasmodesmal SEL, mediates its own transport into neighbouring cells and potentiates the cell-to-cell movement of the virus in the form of complex with the viral genomic RNA (for recent reviews see Haywood *et al.*, 2002; Roberts & Oparka, 2003).

Based on the understanding gained from studying viral MPs, a considerable amount of evidence has accumulated to support the concept of cell-to-cell trafficking of plant gene products (Gilberston & Lucas, 1996; Ding, 1997; Ghoshroy *et al.*, 1997; Zambryski & Crawford, 2000; Haywood *et al.*, 2002). Endogenous proteins that move between plant cells have been termed non-cell autonomous proteins (NCAPs; Lee *et al.*, 2003). Although recent evidence has shown that some transcription factors may pass through meristematic cells by diffusion (Wu *et al.*, 2003), many of the NCAPs identified to date appear selectively to increase the SEL of plasmodesmata (recently reviewed by Oparka, 2004). The first NCAP identified was the transcription factor *KNOTTED* 1 (KN1; Jackson *et al.*, 1994). The KN1 protein was found to traffic between cells in the leaf and SAM by increasing plasmodesmal SEL, and was capable of trafficking its own RNA between tobacco mesophyll cells (Kim *et al.*, 2002b). Kragler *et al.* (1998) have since shown that KN1 requires the activity of a chaperone to unfold the protein, and a plasmodesmal receptor to allow selective movement from cell to cell. Studies performed on a number of other plant transcription factors involved in flower (*FLORICAULA* – Carpenter & Coen, 1990, 1995; *DEFICIENS* – Perbal *et al.*, 1996; *LEAFY* – Sessions *et al.*, 2000), leaf (*GNARLEY1* – Foster *et al.*, 1999) and root (*WEREWOLF* – Lee & Schiefelbein, 1999; *CAPRICE* – Dolan & Costa, 2001; *SHORT ROOT* – Nakajima *et al.*, 2001) development have shown that these endogenous proteins have the capacity to interact with and move through plasmodesmata (Haywood *et al.*, 2002). Work by

Nakajima *et al.* (2001) has highlighted the biological significance of the movement of a transcription factor beyond its site of synthesis. Experiments performed on the short-root (SHR) *Arabidopsis* mutant provided evidence that SHR protein acts as an NCAP, moving from its site of synthesis in the stele into adjacent cells of the endodermis to convey the positional information necessary for determining the fate of endodermal cells.

A number of endogenous proteins have also been found to traffic through the specialised plasmodesmata that connect sieve elements and companion cells, subsequently trafficking long distances within the phloem before being unloaded into sink tissues. Evidence that sieve element proteins may 'gate' mesophyll plasmodesmata has been demonstrated by microinjection experiments involving phloem proteins (PP2 – Balachandran *et al.*, 1997; RPP13-1 – Ishiwatari *et al.*, 1998; CmPP16 – Xoconostle-Cázares *et al.*, 1999; CmPP36 – Xoconostle-Cázares *et al.*, 2000). In addition to macromolecules, plant RNAs have also been found to traffic into and through the phloem pathway (see Chapter 3 for a full review). The CmPP16 protein identified from phloem sap of *Curcurbita maxima* allows the transport of both sense and antisense RNA in the phloem, and has been shown to move from cell to cell in tobacco mesophyll cells and bind its own RNA (Xoconostle-Cázares *et al.*, 1999). Grafting experiments have provided other examples of specific mRNA trafficking. For example, RNA transcripts encoding a mutated homeodomain protein from the mouse ears (*me*) mutant of tomato have been found to move from a rootstock via the phloem into the SAM of a grafted scion and to induce a change in leaf developmental patterns (Kim *et al.*, 2001). Post-transcriptional gene silencing signals, which result in the sequence-specific degradation of targeted mRNA, are also thought to move through plasmodesmata, and several studies have provided evidence that short RNA species can enter the phloem pathway and subsequently unload in sink tissues (Voinnet *et al.*, 1998; Hamilton & Baulcombe, 1999; for in depth review see Chapter 3). These studies illustrate the role of macromolecular and RNA trafficking in plant development, supporting the concept that plants routinely use endogenous proteins and mRNAs as intercellular signals that can traffic through plasmodesmata. Readers are referred to a series of comprehensive reviews (Ding, 1997, 1998; Lucas *et al.*, 2001; Ruiz-Medrano *et al.*, 2001; Ueki & Citovsky, 2001; Haywood *et al.*, 2002; Heinlein, 2002; Lindsey *et al.*, 2002; Wu *et al.*, 2002; Roberts & Oparka, 2003; Oparka, 2004) for further information on symplastic trafficking of proteins and RNA.

5.3.1 Passive transport and the basal SEL

Transport through plasmodesmata is described as passive/non-selective when molecules below the basal SEL traffic without inducing changes in plasmodesmata structure or modifying the basal SEL (Schulz, 1999). Initial studies of basal SEL gave rise to the consensus that only small molecules ranging between 850 and 900 Da move freely through plasmodesmata (Tucker, 1982; Goodwin, 1983; Erwee & Goodwin, 1985; Terry & Robards, 1987; Burnell, 1988). However, many exceptions were found to this rule. Kikuyama *et al.* (1982) were the first to report the passive movement of a 45-kDa protein labelled with FITC in the green alga *Nitella*. Studies

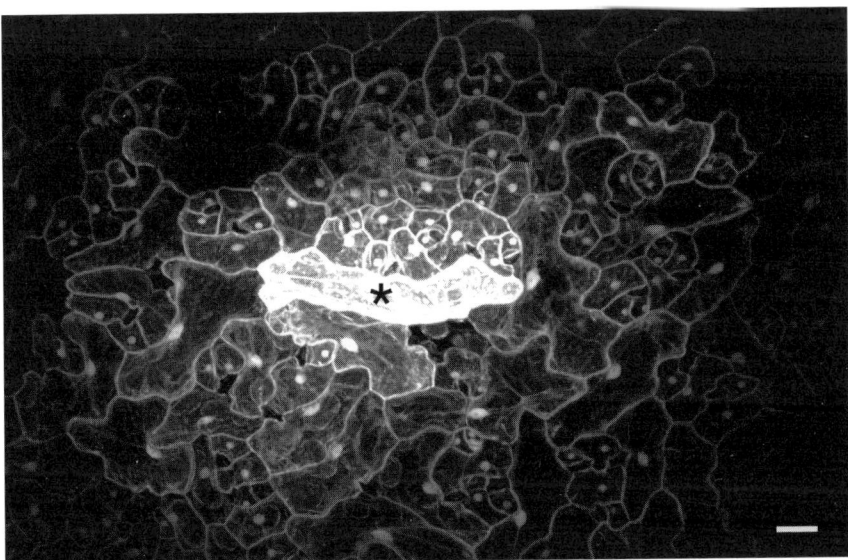

Figure 5.4 Non-selective movement of GFP in sink epidermal cells of *Arabidopsis*. Asterisk indicates the initial bombarded cell expressing GFP from the *CaMV* 35s promoter. Bar: 20 mm.

in other two-dimensional cell-file systems such as stamen hairs of *Setcreasea purpurea* (Yang et al., 1995) and leaf trichomes of tobacco (Waigmann & Zambryski, 1995) revealed that the SEL of plasmodesmata can be much greater than 1 kDa. Additionally, Wang and Fisher (1994) demonstrated that dextrans (M_r of 10 kDa) applied to the post-phloem crease of the wheat grain were capable of moving from cell to cell, and Kempers and van Bel (1997) showed that dextrans (M_r of 10 kDa) could move through plasmodesmata between sieve elements and companion cells in the phloem of *Vicia faba*. The 27-kDa GFP has also been shown to move passively in specific tissues (see Fig. 5.4; Imlau et al., 1999; Oparka et al., 1999). GFP has been shown to traffic through plasmodesmata from companion cells to sieve elements and migrate from the phloem into sink tissues such as the seed coat, root tips and sink-leaf mesophyll cells (Imlau et al., 1999). In sink-leaf tissue of tobacco, where plasmodesmata are predominately simple in structure, Oparka et al. (1999) found that GFP-fusion proteins with an M_r up to 50 kDa could move freely, revealing a high basal SEL for plasmodesmata in sink tissue. However, in mature source-leaf tissues of tobacco, where plasmodesmata are branched, movement of GFP was restricted, indicating that plasmodesmata in source tissue have a low SEL (Oparka et al., 1999). GFP can move passively between cells of *Arabidopsis* leaf and stem epidermis, but not in the epidermis of either tomato or cucumber (Itaya et al., 2000). Itaya et al. (2000) also found that GFP could move from the epidermis into the cortex of cucumber hypocotyls, but movement from the epidermis into the mesophyll in cotyledons was restricted. Studies using both fluorescent probes, such as dextrans, and GFP have shown that the basal SEL of plasmodesmata, and therefore passive trafficking of molecules through plasmodesmata can vary depending on plant species, between organs of the same plant, within different tissues of an organ and between cell types at different developmental stages (Goodwin & Cantrill, 1999; Oparka *et al.*, 1999). On the basis of these findings, some authors (Itaya *et al.*, 2000; Blackman & Overall, 2001; Zambryski, 2004) have questioned how specific metabolites, small proteins and peptides are retained within cells. Importantly, it is not the molecular mass of a protein that determines its permeability through a plasmodesma, but its Stokes radius (R_s). Terry and Robards (1987) emphasised that the mobility of a molecule through a plasmodesma is determined by the effective R_s, which is governed by the shape (width and depth) and the hydrodynamic drag of a molecule. Fisher and Cash-Clark (2000) point out that there is no unique relationship between M_r and molecular dimensions, for example, a 25-kDa dextran has the same R_s as a 51-kDa globular protein (Jørgensen & Møller, 1979). In the case of GFP, its compact beta-barrel structure (Ormö *et al.*, 1996), and physical dimensions (diameter 3 nm; length 4 nm; Phillips, 1997), rather than its molecular mass, may enable it to traffic passively through plasmodesmata.

5.3.2 *Selective transport and gating: modulation of the SEL*

Transport through a plasmodesma is described as selective when molecules larger than the basal SEL traffic between cells, mediated by conformational changes in plasmodesmata that permit an increase in the SEL. The term *gating* has generally been used to describe the specific interaction between a molecule and a plasmodesma that leads to dilation of the plasmodesmal pore. However, as Schulz (1999) remarks the term gating can be ambiguous, as passage of smaller solutes is also controlled by the width of the cytoplasmic sleeve. Thus, dilation/gating of a plasmodesma can also be non-selective. Not all authors use the same terminology to describe selective transport and plasmodesmal gating; for example, Zambryski and Crawford (2000) describe plasmodesmata as having different functional states, and comment that 'plasmodesmata can be assumed to be either closed or open'. They define a closed plasmodesma as not allowing any intercellular exchange, an open plasmodesma as allowing the movement of molecules below the basal SEL, and a dilated plasmodesma as allowing the movement of macromolecules larger than the basal SEL. Zambryski (2004) concludes that non-targeted protein movement may provide coordination between large groups of cells, whereas proteins with more limited movement allow the programming of more specific pathways.

5.3.3 *Physiological modulation of SEL*

Various abiotic and biotic factors have been found to alter the basal SEL of plasmodesmata (see Table 5.1). Turgor pressure differentials and elevated Ca^{2+} levels have been shown to decrease the SEL, although changes in pH may negate the effect of elevated calcium (Holdaway-Clarke *et al.*, 2001). Plasmodesmata are also sensitive to wounding, which induces rapid callose deposition at the neck of the pore. Decreases in ATP levels, anerobic stress, long-term osmotic stress,

Table 5.1 Biotic and abiotic factors that affect plasmodesmata (PD) size exclusion limit (SEL)

Area	Increase	Decrease	No effect	Reference
Turgor		Pressure differential of more than 200 kPa prevented dye movement in *Nicotiana clevelandii* trichome cells.		Oparka & Prior, 1992
		Pressure differences in the alga *Chara corallina* cause PD to seal off.		Ding & Tazawa, 1989; Reid & Overall, 1992
			Increasing turgor, by microinjection of oil droplets, in *Arabidopsis* root hairs has no effect on electrical coupling of cells.	Lew, 1996
Osmotic	Increasing mannitol concentrations caused constricted neck regions of root tip PD to widen, with increased phloem unloading.			Schulz, 1995
	SEL of deplasmolysed *Egeria densa* leaves increased for >24 h.	Hypertonic conditions stopped dye movement between cells of *Egeria densa*.		Erwee & Goodwin, 1984
	Increase in SEL in wheat roots after osmotic stress.	Callose deposited in PD of plasmolysed oat coleoptiles: transitory decrease as callose disappeared 4–6 h after cells were fully turgid.		Cleland et al., 1994
				Drake et al., 1978
		PD resistance high at cell borders with lasting turgor differences, e.g. SE-CC and surrounding cells.		Patrick, 1997; Kempers et al., 1998

Cont.

Table 5.1 (Continued)

Area	Increase	Decrease	No effect	Reference
Anaerobic stress	Reduced ATP caused SEL to increase from 1 kDa to between 5 and 10 kDa in stressed roots.			Cleland et al., 1994
Hypoxia			Membrane resistance (electrical coupling ratio), a measure of PD, resistance unaffected.	Zhang & Tyerman, 1997
Metabolic inhibitors	CCCP and probenicid increased symplastic movement from transport phloem in *Arabidopsis* roots.			Wright & Oparka, 1997
	Cytochalasin, which depolymerises actin, widened the neck of plasmodesmata in *Nephrolepis exaltata*.		Cytochalasin had no effect on plasmodesmal aperture in roots of *Hordeum vulgare* and *Azolla*.	White et al., 1994
	Cytochalasin increased SEL in epidermal cells of *Nicotiana tabacum* from 1 to 20 kDa.			Ding et al., 1996
	Actin-binding protein profilin increased SEL in *Nicotiana tabacum*.			Ding et al., 1996
		Phalloidin, which stabilises actin filaments, inhibited cell-to-cell movement.		Ding et al., 1996
		BDM, which can increase cytoplasmic Ca^{2+}, induced constriction in the neck region of *Allium cepa*, *Zea mays* and *Horedum vulgare* PD.		Radford & White, 1998

Secondary messengers		Myosin inhibitor BDM constricted PD in maize roots.	Samaj et al., 2000
		IP_2 and IP_3 inhibited cell-to-cell transport in *Setcreasea purpurea*, by releasing Ca^{2+} from internal stores into the cytosol via inositol triphosphate-sensitive channels.	Tucker, 1990; Tucker & Boss, 1996
Group II/ divalent cations		Elevated cytosolic Ca^{2+} decreased PD permeability in *Egeria densa* leaves.	Erwee & Goodwin, 1983
		Elevated Ca^{2+} restricts dye movement in soybean microcallus cells.	Baron-Epel et al., 1998
	No effect of Ca^{2+} on PD permeability in the green alga *Chara*.		Reid & Overall, 1992
		Free Ca^{2+} injected into cytosol transiently closed PD of maize cells increasing electrical resistance between cells.	Holdaway-Clarke et al., 2000
		Cytoplasmic increases in other group II ions, e.g. Mg^{2+} and Sr^{2+}, decrease PD permeability.	Erwee & Goodwin, 1983
		High level of cytosolic Ca^{2+} stopped dye movement in *Setcreasea purpurea* cells.	Tucker, 1990

Cont.

Table 5.1 (Continued)

Area	Increase	Decrease	No effect	Reference
pH/Ca^{2+}		Cold treatment, found to increase cytoplasmic-free Ca^{2+}, transiently closed PD.	Increase in Ca^{2+}, together with a decrease in pH, has no effect on the electrical conductivity of cells.	Holdaway-Clarke et al., 2000; Reid & Overall, 1992; Holdaway-Clarke et al., 2001
Ammonia		Application to the apoplasm of leaves and stems caused PD to close.		Fensom et al., 1990
Aromatic amino acids		Decrease in PD permeability.		Erwee & Goodwin, 1984
Azide	SEL of staminal trichomes of *Sercreasea* increased. Azide reduced ATP in wheat roots, mimicking anaerobis and causing PD to dilate.			Tucker, 1993; Cleland et al., 1994
Anaerobic conditions	Induced anaerobis by N$_2$ treatment dilates PD in detached wheat roots.			Cleland et al., 1994
Growth conditions	Glasshouse-grown plants show greater GFP movement in epidermal cells than do tissue culture plants.	Light pulses found to inhibit carboxyfluorescein transport in maize seedlings.		Crawford & Zambryski, 2000; Epel & Erlanger, 1991

deplasmolysis and some pharmcological treatments have been found to increase the SEL. Schulz (1999) hypothesised that the prime mechanism for control of non-selective trafficking is the rapid and transient conformational change in plasmodesmal proteins brought about by changes in calcium and/or ATP levels. Comparatively small and local changes (elevation) in Ca^{2+} concentration will cause a decrease in plasmodesmal SEL. However, longer term blockage caused by wounding may depend on a massive influx of calcium from the apoplast or the vacuole. Conformational changes in plasmodesmal proteins are rapidly reversible, whereas callose synthesis and breakdown is much slower. Conversely, low levels of ATP, that can be caused by anaerobic stress or induced by azide treatment, dilate plasmodesmata, allowing increased intercellular exchange (Schulz, 1999). However, there have been contradictory reports on the effect of turgor on plasmodesmal SEL. Roberts and Oparka (2003) propose that sudden pressure differentials between adjacent cells, as might occur during wounding, will cause isolation of plant cells from their neighbours, whereas an overall drop in tissue turgor, characteristic of water stress, will lead to an increase in SEL.

5.3.4 Fine regulation of plasmodesmal SEL – role of the cytoskeleton

Zambryski and Crawford (2000) have suggested that constriction and relaxation may occur along the entire length of the pore, correlated to the distribution of actin and myosin in plasmodesmata. Other authors (Radford & White, 1998; Reichelt *et al.*, 1999) have suggested that the helically arranged 'spokes' that connect the desmotubule to the plasma membrane (Overall & Blackman, 1996) may provide a possible contractile mechanism for controlling the aperture of the cytoplasmic sleeve. In mammalian cells, some myosins have a role in generating tension between adjacent membranes (Küssel-Andermann *et al.*, 2000). Myosin VIII has been implicated in plasmodesmata function (Baluška *et al.*, 2000, 2001) and has a characteristic motor-domain region and a C-terminal structure that includes a probable phosphorylation site for protein kinases, as well as four calmodulin-binding motifs (Reichelt & Kendrick-Jones, 2000). Myosin VIII has therefore been proposed to function as a calcium-regulated plasmodesmal motor protein that traffics macromolecular cargo along the actin filaments in plasmodesmata (Roberts & Oparka, 2003; Oparka, 2004).

Regulation of plasmodesmal aperture may also occur at the neck region of the plasmodesma. Blackman and Overall (2001) have suggested that centrin, a calcium-binding contractile protein found in the neck region, may have a role in regulating plasmodesmal aperture. Dephosphorylation of this protein, caused by an increase in cytosolic calcium, induces the centrin nanofilaments to rapidly contract, potentially closing the plasmodesmata (Martindale & Salisbury, 1990; Blackman *et al.*, 1999). The ATPase inhibitor 2,3-butanedione 2-monoxime (BDM), which is known to stimulate calcium release from cardiac sarcoplasmic reticulum (Phillips & Altschuld, 1996) and inhibit myosin by preventing microfilament depolymerisation, was found to contract the neck region of plasmodesmata in *A. cepa, Z. mays* and *H. vulgare*

(Radford & White, 1998; Samaj et al., 2000). The activity of ATPase, which is also found in the neck region together with calcium-binding sites, has been shown to decrease with the addition of a calcium-chelating agent (Belitser et al., 1982). Roberts and Oparka (2003) note that calreticulin, a calcium-sequestering protein shown to be a component of the cortical endoplasmic reticulum (Baluška et al., 1999, 2001), could potentially be involved in aperture regulation at the neck by buffering calcium levels.

5.3.5 'Coarse' regulation by callose

Callose (b-1,3-glucan) is known to be deposited at plasmodesmata during wounding (Hughes & Gunning, 1980). Mechanical wounding of leaf tissue stimulates a rapid deposition of callose at plasmodesmata (Currier, 1957; Schulz, 1999). However, the rate of callose deposition and subsequent degradation varies depending on the system studied (Overall & Blackman, 2001). Callose deposition at plasmodesmata is thought to be a ubiquitous plant response to pathogen spread (Roberts & Oparka, 2003), and has been observed in the late phase of the hypersensitive response to TMV infection (Susi, 2000). However, callose has also been detected in plasmodesmata outside the visible necrotic lesions of plants infected with PVX, suggesting that it may function as an early defence strategy against PVX (Allison & Shalla, 1974). Callose has also been reported to inhibit the long-distance trafficking of viruses (Leisner & Turgeon, 1993; Choi, 1999). Callose deposition has been found to block plasmodesmata during infection by the oomycete *Phytophthora sojae* (Enkerli et al., 1997), and fungal xylanase has been found to elicit a hypersensitive response in tobacco associated with callose deposition (Bailey et al., 1990). Interestingly, a low concentration of Brefeldin A, a fungal agent produced by *Penicillium brefeldianum*, which disrupts the endomembrane system in plants (Satiat-Jeunemaitre & Hawes, 1992a,b), triggers callose synthesis in onion epidermal cells (Kartusch et al., 2000). Callose deposition at plasmodesmata has also been shown in response to aluminium toxicity (Jones et al., 1998; Sivaguru et al., 2000).

Olesen and Robards (1990) and Lucas et al. (1993) have proposed a model that depicts the enzyme b-1,3-D-glucanase as an integral membrane protein that produces b-1,3-glucan between the plasma membrane and the cell wall. The precise regulation of b-1,3-glucan deposition is thought to regulate plasmodesmal SEL. Enhanced callose deposition at plasmodesmata has been reported in mutant plants where the activity of b-1,3-glucanase has been blocked, causing a reduction in SEL but not absolute closure of plasmodesmata (Beffa et al., 1996). Recently, three host proteins (TIPs) that interact with the 12-kDa protein of the triple gene block of PVX, a protein required for movement by PVX, have been identified (Fridborg et al., 2003). All three TIPs were found to interact with b-1,3-glucanase, suggesting that regulation of b-1,3-glucanase activity by viral MPs might be a means by which some viruses overcome host deposition of callose during infection (Iglesias & Meins, 2000).

In addition to wounding and pathogenesis, callose deposition is also implicated in plugging of the neck region plasmodesmata, for example, by the formation of neck 'sphincters' in cells of the SAM of birch buds entering endodormancy (Rinne et al., 2001). The proteinaceous sphincters were found to contain pronounced deposits of b-1,3-D-glucan and thought to prevent any form of symplastic communication between cells in endodormant tissue (Rinne & van der Schoot, 1998). These authors found that removal of callose from plasmodesmata during chilling coincided with the production of b-1,3-glucanase (Rinne et al., 2001). Blockage of plasmodesmata by callose in the SAM of poplar buds entering into endodormancy, together with elevated levels of cytosolic Ca^{2+}, has also been reported by Jian et al. (1997). Interestingly, Karlson et al. (2003) have recently reported the immunolocalisation of a 24-kDa dehydrin protein to plasmodesmata in cold acclimated xylem parenchyma cells of *Cornus sericea*. The pattern of labelling was identical to that observed with monoclonal antibodies specific for callose. The phloem of perennials has also been shown to become inactivated in the autumn by the deposition of b-1,3-glucan at the sieve-plate pores, which become unplugged in the spring (Esau, 1977) by the activation of b-1,3-glucanase (Krabel et al., 1993; Rinne et al., 2001). In the sucrose export deficient maize mutant (*sxd1*), phloem loading is also prevented by callose deposits that specifically block plasmodesmata at the interface between bundle sheath and vascular parenchyma cells (Botha et al., 2000). Hofius and Sonnewald (2003) have noted that the *sxd1* mutation in maize corresponds to a *VTE1*-tocopherol cyclase (vitamin E) *Arabidopsis* knockout mutant (Porfirova et al., 2002), and have therefore suggested a putative link between tocopherol cyclase and plasmodesmata function.

Caution may be required in interpreting some studies on callose deposition because the preparation of tissue for EM studies can itself induce callose by stimulating a wounding response (Radford et al., 1998). It is interesting to note that, although the effects of callose are generally thought to be coarser than other control mechanisms, the extent of callose synthesis and degradation is clearly highly regulated (Roberts & Oparks, 2003).

5.3.6 Phosphorylation, protein unfolding and chaperones

Plasmodesmal trafficking of protein and/or RNA is unique to plants (Ding et al., 2003). Haywood et al. (2002) remark that all viral MPs and NCAPs studied so far have been found to expose a motif(s) that can induce dilation of the SEL. This has led various authors to comment that selective movement of protein through plasmodesmata involves the interaction between this motif and cognate cellular factors (Haywood et al., 2002; Ding et al., 2003; Oparka, 2004). In a recent review of protein movement through plasmodesmata, Haywood et al. (2002) suggest that, in the simplest scenario, plasmodesmal dilation is necessary and sufficient to allow protein movement by diffusion into neighbouring cells (i.e. non-selective gating; see Schulz, 1999). However, Wu et al. (2003) recently provided evidence that the transcription factor LEAFY (Sessions et al., 2000) moves from the L1 layer of the *Arabidopsis* SAM into the L2 and L3 layers by diffusion, and therefore defined the movement of LEAFY as non-targeted (Crawford & Zambryski, 2001). Wu et al. (2003) concluded that the LEAFY protein sequence does not contain a specific movement or export signal. Unfortunately, these authors did not test to see if the basal SEL of plasmodesmata was increased in the presence of LEAFY, and therefore it is difficult to ascertain if the movement of the LEAFY protein is truly non-selective. Haywood *et al.* (2002) have compared cell-to-cell transport of proteins through plasmodesmata to import of protein into organelles, which is known to involve exposure of a targeting motif, binding to a translocation receptor complex, protein unfolding and/or structural modification to the translocation complex. Kragler *et al.* (2000) also remark that this process shares features in common with intracellular translocation mechanisms in which the transfer of macromolecules occurs in either a folded or unfolded state (Subramani, 1996; Kragler *et al.*, 1998; Schatz, 1998).

Few potential cellular factors that control regulation of selective plasmodesmal gating have been identified (Ding *et al.*, 2003). There has been some debate in the literature as to whether viral MPs may have been plant proteins that became incorporated into the viral genome to facilitate viral movement (Lucas & Gilbertson, 1994; Maule, 1994; Mezitt & Lucas, 1996). Viral MPs may use an endogenous plant pathway for intercellular trafficking of macromolecules (Jackson, 2001), and have provided clues about the factors that potentiate selective gating *in planta* (Ding *et al.*, 2003). The MPs of TMV, *Turnip vein-clearing virus* (TVCV) and *Cauliflower mosaic virus* (CaMV) have been shown to interact/bind with pectin methylesterase (PME) (Dorokhov *et al.*, 1999; Chen *et al.*, 2000). Chen *et al.* (2000) found that deletion of a PME-binding domain in TMV MP prevents cell-to-cell movement of TMV. PME is an enzyme that modifies pectin in plant cell walls and has been shown to have RNA-binding properties (Dorokhov *et al.*, 1999). PME has been shown to have a heterogeneous cellular distribution and has been localised to microdomains of the cell wall, the plasma membrane and the endoplasmic reticulum of flax hypercotol cortical cells (Morvan *et al.*, 1998). Oparka (2004) has suggested that MPs may interact with PME before its delivery to the cell wall, effectively hijacking the PME in order to 'piggy back' on the cells endogenous macromolecular transport pathway to plasmodesmata.

Citovsky *et al.* (1993) showed that the MP of TMV is phosphorylated *in vitro* at its C-terminal serine and threonine residues by a cell-wall-associated protein kinase. These authors concluded that phosphorylation of TMV MP may represent a mechanism for the host plant to sequester MP following its localisation to cell walls. *In vivo* studies of TMV MP phosphorylation showed that this process may negatively regulate the effect of MP on plasmodesmal permeability (Waigmann *et al.*, 2000). Phosphorylation of other viral MPs has also been demonstrated (Sokolova *et al.*, 1997; Matsushita *et al.*, 2000, 2002). Waigman *et al.* (2000) also found that activity of this MP-binding kinase required Mg^{2+} but not Ca^{2+} cations; however, a calcium-dependent protein kinase has been localised to the cell wall in a plasmodesmata-like distribution in *Arabidopsis* (Yahalom *et al.*, 1998). Blackman and Overall (2001) suggest that localisation of ATPase activity to plasmodesmata indicates that post-translational modification of proteins by kinases and phosphatase plays a role in the control of cell-to-cell communication. Schulz (1999) also hypothesised that a unifying mechanism of non-selective and selective gating of plasmodesmata may be due to the phosphorylation and dephosphorylation of plasmodesmal proteins. In the case of selective gating, each MP would directly interact with plasmodesmal proteins, 'relaxing' the putative myosin spokes bound to the plasma membrane, and enabling modulation of the SEL (Schulz, 1999). However, it is not known whether all endogenous macromolecular proteins that are selectively trafficked through plasmodesmata require phosphorylation to mediate their passage. Oparka (2004) has proposed a hypothetical model for the role of phosphorylation and dephosphorylation in protein movement through plasmodesmata. In this model the trafficked protein is bound to a motor protein by a specific chaperone for delivery to plasmodesmata. At the pore, the complex interacts with the putative 'docking' protein. Recognition of a trafficking motif on the protein cargo activates a myosin-specific kinase that phosphorylates the C-terminus of the myosin motor, releasing it from the plasma membrane. The released myosin, together with its cargo, then moves along the actin filaments associated with the desmotubule via its motor domain. Oparka (2004) proposes that phosphorylation and dephosphorylation of the motor protein might regulate the detachment and attachment of the protein cargo from the plasma membrane lining the plasmodesmal pore.

Haywood *et al.* (2002) have also suggested that other forms of protein structural modification may be involved in controlling selective transport through plasmodesmata. They cite the example of the phloem protein CmPP36, which is able to induce only an increase in SEL and move from cell to cell in an N-terminally truncated form, indicating that the capacity of this NCAP to target to and/or transport through plasmodesmata is controlled by proteolytic processing (Xoconostle-Cázares *et al.*, 2000).

In certain cases, selective gating may be achieved by conformational changes in the trafficking protein or macromolecular complex. Kragler *et al.* (1998) demonstrated that partial unfolding of KN1 appears to be involved during its translocation through plasmodesmata. Yeast two-hybrid screens using the putative MP of *Tomato spotted wilt virus* (TSWV) as bait identified two DnaJ-like interacting proteins from *Nictoiana tabacum* and *Arabidopsis* (Soellick *et al.*, 2000) and another DnaJ-like protein from *Lycopersicon esculentum* (von Bargen *et al.*, 2001). DnaJ proteins belong to the Hsp40 subclass of heat-shock proteins, and are known to be involved in protein import into organelles and in the regulation of the chaperone, Hsp70 (Kelley, 1999). Both Jackson (2001) and Blackman and Overall (2001) have suggested that chaperone activity may be involved in cell-to-cell trafficking of MPs, and could have a role in partially unfolding proteins for translocation through the pore (Kragler *et al.*, 1998). Alternatively, it may function directly as a motor to facilitate translocation through the plasmodesmal pore (Alzhanova *et al.*, 2001). Aoki *et al.* (2002) detected Hsc70 (Hsp70-related) proteins in a plasmodesmal-enriched cell wall fraction and isolated and characterised two Hsp70 chaperones from *C. maxima* phloem sap. Using mutational analysis, Aoki *et al.* (2002) showed these proteins contain a motif that mediates their intercellular trafficking. Fusion of this motif to a human Hsp70 chaperone allowed the latter to move from cell to cell. Aoki *et al.* (2002) have postulated that these pumpkin Hsp70 chaperones may play a role in refolding structurally altered proteins so that they can enter the sieve elements from the companion cell.

Recently, NCAPP1 (non-cell autonomous pathway protein 1) was isolated from a plasmodesmal-enriched cell wall protein fraction using the CmPP16 trafficking protein as bait for candidate components of the plasmodesmal translocation machinery (Lee *et al.*, 2003). Microinjection studies of a truncated version of NCAPP1 indicated that NCAPP1 plays a role in mediating the cell-to-cell trafficking of selective endogenous proteins. Fluorescent protein tagging, coupled with immunogold localisation, confirmed its localisation to the peripheral endoplasmic reticulum, approximately 200 nm away from the pore itself. Lee *et al.* (2003) proposed that NCAPP1 is anchored to the endoplasmic reticulum membrane, with the bulk of the protein exposed in the cytoplasm where it interacts with proteins moving towards the plasmodesmal pore. Although NCAPP1 was found to be obligatory for the movement of CmPP16 and TMV MP, other macromolecular proteins, including KN1 and CMV MP, were able to traffic without NCAPP1, suggesting that NCPP1 does not play a role in the trafficking of these other NCAPs (Jackson & Kim, 2003).

5.3.7 *The emerging picture of plasmodesmata*

Current research on plasmodesmata has indicated that macromolecular trafficking between plant cells may be a commonplace event. Plasmodesmata can no longer be viewed as simple channels through the plant cell wall that facilitate the movement of nutrients. Instead, they have a much more complex role in controlling intercellular communication and the flux of information passing through the symplasm. In the future, it will be necessary to examine the ways in which proteins and ribonucleoprotein complexes interact with components of the plasmodesmal pore. To do this will require a systematic isolation and characterisation of the genes that encode novel plasmodesmal components, and an understanding of the mechanism by which these integral proteins interact with the various 'cargos' that are being shuttled between plant cells. The genetics of plasmodesmata is in its infancy and there is a pressing need to identify the gene products that regulate the formation and development of plasmodesmata. Manipulation of plasmodesmal development is an exciting target for the future and should lead to a more comprehensive picture of how cells communicate with one another, as well as determining the internal and external forces that impinge on this process.

Acknowledgements

The authors are supported by grants awarded by the Scottish Executive Environment and Rural Affairs Department (SEERAD) and the Gatsby Foundation. The authors thank Prof. Chris Hawes for providing Fig. 5.3.

References

Allison, A.V. & Shalla, T.A. (1974) The ultrastructure of local lesions induced by potato virus X: a sequence of cytological events in the course of infection. *Phytopathology*, **64**, 784–793.

Alzhanova, D., Napuli, A.J., Creamer, R. & Dolja, V. (2001) Cell-to-cell movement and assembly of a plant closterovirus: roles for the capsid proteins and Hsp70 homolog. *EMBO J.*, **20**, 6997–7007.

Aoki, K., Kragler, F., Xoconostle-Cázares, B. & Lucas, W.J. (2002) A subclass of plant heat shock cognate 70 chaperones carries a motif that facilitates trafficking through plasmodesmata. *Proc. Natl. Acad. Sci. U.S.A.*, **99**, 16342–16347.

Atabekov, J.G. & Taliansky, M.E. (1990) Expression of a plant virus-coded transport by different viral genomes. *Adv. Virus Res.*, **38**, 201–248.

Badelt, K., White, R.G., Overall, R.L. & Vesk, M. (1994) Ultrastructural specialisations of the cell wall sleeve around plasmodesmata. *Am. J. Bot.*, **81**, 1422–1427.

Bailey, B.A., Dean, J.F.D. & Anderson, J.D. (1990) An ethylene biosynthesis-inducing endoxylanase elicits electrolyte leakage and necrosis in *Nicotiana tabacum* cv. Xanthi leaves. *Plant Physiol.*, **94**, 1849–1854.

Balachandran, S., Xiang, Y., Schobert, C., Thompson, G.A. & Lucas, W.J. (1997) Phloem sap proteins from *Cucurbita maxima* and *Ricinus communis* have the capacity to traffic cell to cell through plasmodesmata. *Proc. Natl. Acad. Sci. U.S.A.*, **94**, 14150–14155.

Baluška, F., Barlow, P.W. & Volkmann, D. (2000) Actin and myosin VIII in developing root apex cells. In: *A Dynamic Framework for Multiple Plant Cell Functions* (ed. C.J. Staiger), pp. 457–476. Kluwer, Dordrecht, The Netherlands.

Baluška, F., Cvrckova, F., Kendrick-Jones, J. & Volkmann, D. (2001) Sink plasmodesmata as gateways for phloem unloading. Myosin VIII and calreticulin as molecular determinants of sink strength? *Plant Physiol.*, **126**, 39–46.

Baluška, F., Samaj, J., Napier, R. & Volkmann, D. (1999) Maize calreticulin localizes preferentially to plasmodesmata in root apex. *Plant J.*, **19**, 481–488.

Baron-Epel, O., Hernandez, D., Jiang, L. W., Meiners, S. & Schindler, M. (1998) Dynamic continuity of cytoplasmic and membrane compartments between plant cells. *J. Cell Biol.*, **106**, 715–721.

Beebe, D.U. & Turgeon, R. (1991) Current perspectives on plasmodesmata: structure and function. *Physiol. Plant.*, **83**, 194–199.

Beffa, R.S., Hofer, R.M., Thomas, M. & Meins, J. (1996) Decreased susceptibility to viral disease of b-1,3-glucanase-deficient plants generated by antisense transformation. *Plant Cell*, **8**, 1001–1011.

Belitser, N.V., Zaalishvili, G.V. & Sytnianskaja, N.P. (1982) Ca^{2+}-binding sites and Ca^{2+}-ATPase activity in barley root tip cells. *Protoplasma*, **111**, 63–78.

Blackman, L.M. & Overall, R.L. (1998) Immunolocalisation of the cytoskeleton to plasmodesmata of *Chara corallina*. *Plant J.*, **14**, 733–741.

Blackman, L.M., Gunning, B.E.S. & Overall, R.L. (1998) A 45 kDa protein isolated from the nodal walls of *Chara corallina* is localised to plasmodesmata. *Plant J.*, **15**, 401–411.

Blackman, L.M., Harper, J.D.I. & Overall, R.L. (1999) Localization of a centrin-like protein to higher plant plasmodesmata. *Eur. J. Cell Biol.*, **78**, 297–304.

Blackman, L.M. & Overall, R.L. (2001) Structure and function of plasmodesmata. *Aust. J. Plant Physiol.*, **28**, 709–727.

Botha, C.E.J. & Cross, R.H.M. (1999) Plasmodesmal imaging: towards understanding structure. In: *Plasmodesmata: Structure, Function, Role in Cell Communication* (eds A.J.E. van Bel & W.J.P. van Kesteren), pp. 27–36. Springer-Verlag, Berlin.

Botha, C.E.J. & Evert, R.F. (1988) Plasmodesmatal distribution and frequency in vascular bundles and contiguous tissues of the leaf of *Themeda triandra*. *Planta*, **173**, 433–441.

Botha, C.E.J., Cross, R.H.M., van Bel, A.J.E. & Peter, C.I. (2000) Phloem loading in the sucrose-export-defective (*SXD-1*) mutant maize is limited by callose deposition at plasmodesmata in bundle sheath–vascular parenchyma interface. *Protoplasma*, **214**, 65–72.

Botha, C.E.J., Hartley, B.J., & Cross, R.H.M. (1993) The ultrastructure and computer-enhanced digital image analysis of plasmodesmata at the Kranz mesophyll–bundle sheath interface of *Themeda triandra* var. *imberbis* (Retz) A. Camus in conventionally-fixed leaf blades. *Ann. Bot.*, **72**, 255–261.

Brown, S.M., Oparka, K.J., Sprent, J.I., & Walsh, K.B. (1995) Symplastic transport in soybean root nodules. *Soil Biol. Biochem.*, **27**, 387–399.

Burnell, J.N. (1988) An enzymic method for measuring the molecular weight exclusion limit of plasmodesmata of bundle sheath cells of C_4 plants. *J. Exp. Bot.*, **39**, 1575–1580.

Buvat, R. (1957) L'infrastructure des plasmodesmes et la continuité des cytoplasmes. *C. R. Acad. Sci. Paris*, **245**, 198–201.

Carpenter, R. & Coen, E.S. (1990) Floral homeotic mutations produced by transposon-mutagenesis in *Antirrhinum majus*. *Gene Dev.*, **4**, 1483–1493.

Carpenter, R. & Coen, E.S. (1995) Transposon induced chimaeras show that floricaula, a meristem identity gene acts non-autonomously between cell layers. *Development*, **121**, 19–26.

Carr, D.J. (1976) Historical perspectives on plasmodesmata. In: *Intercellular Communication in Plants: Studies on Plasmodesmata* (eds B.E.S. Gunning & A.W. Robards), pp. 291–295. Springer-Verlag, Berlin.

Carrington, J.C., Kasschau, K.D., Mahajan, S.K. & Schaad, M.C. (1996) Cell-to-cell and long-distance transport of viruses in plants. *Plant Cell*, **8**, 1669–1681.

Casero, P.J. & Knox, J.P. (1995) The monoclonal antibody JIM5 indicates patterns of pectin deposition in relation to pit fields at the plasma-membrane-face of tomato pericarp cell walls. *Protoplasma*, **188**, 133–137.

Chauhan, E., Cowan, D.S. & Hall, J.L. (1991) Cytochemical localisation of plasma membrane ATPase activity in plant cells. A comparison of lead and cerium-based methods. *Protoplasma*, **165**, 27–36.

Chen, M.H., Sheng, J., Hind, G., Handa, A.K. & Citovsky, V. (2000) Interaction between the tobacco mosaic virus movement protein and host cell pectin methylesterases is required for viral cell-to-cell movement. *EMBO J.*, **19**, 913–920.

Choi, C.W. (1999) Modified plasmodesmata in *Sorghum* (*Sorghum bicolor* L. Moench) leaf tissues infected by maize dwarf mosaic virus. *J. Plant Biol.*, **42**, 63–70.

Citovsky, V. & Zambryski, P. (1993) Transport of nucleic acids through membrane channels: snaking through small holes. *Annu. Rev. Microbiol.*, **47**, 167–197.

Citovsky, V., McLean, B.G., Zupan, J.R. & Zambryski, P. (1993) Phosphorylation of tobacco mosaic virus cell-to-cell movement protein by a developmentally regulated plant cell wall-associated protein kinase. *Gene Dev.*, **7**, 904–910.

Cleland, R.E., Fujiwara, T. & Lucas, W.J. (1994) Plasmodesmal-mediated cell-to-cell transport in wheat roots is modulated by anaerobic stress. *Protoplasma*, **178**, 81–85.

Crawford, K.M. & Zambryski, P.C. (2000) Subcellular localization determines the availability of non-targeted proteins to plasmodesmatal transport. *Curr. Biol.*, **10**, 1032–1040.

Crawford, K.M. & Zambryski, P.C. (2001) Non-targeted and targeted protein movement through plasmodesmata in leaves in different developmental and physiological states. *Plant Physiol.*, **125**, 1802–1812.

Currier, H. (1957) Callose substance in plant cells. *Am. J. Bot.*, **44**, 478–488.

Cutler, S.R., Ehrhardt, D.W., Griffitts, J.S. & Somerville, C.R. (2000) Random GFP::cDNA fusions enable visualization of subcellular structures in cells of *Arabidopsis* at a high frequency. *Proc. Nat. Acad. Sci. U.S.A.*, **97**, 3718–3723.

Dawson, J.H., Musselman, L.J., Wolswinkel, P. & Dörr, I. (1994) Biology and control of *Cuscuta*. *Rev. Weed Sci.*, **6**, 265–317.

Delmer, D.P., Volokita, M., Solomon, M., Fritz, U. & Delphendahl, W. (1993) A monoclonal antibody recognises a 65 kDa higher plant membrane polypeptide which undergoes cation dependent association with callose synthase in vitro and colocalizes with sites of high callose deposition in vivo. *Protoplasma*, **76**, 33–42.

Denecke, J. (2001) Plant endodplasmic reticulum. In: *Nature Encyclopedia of Life Sciences.* Nature Publishing Group, London.
Deom, C.M., Lapidot, M. & Beachy, R.N. (1992) Plant virus movement proteins. *Cell*, **69**, 221–224.
Didehvar, F. & Baker, D.A. (1986) Localization of ATPase in sink tissues of *Ricinus. Ann. Bot.*, **57**, 823–828.
Ding, B. (1997) Cell-to-cell transport of macromolecules through plasmodesmata: a novel signalling pathway in plants. *Trends Cell Biol.*, **7**, 5–9.
Ding, B. (1998) Intercellular protein trafficking through plasmodesmata. *Plant Mol. Biol.*, **38**, 279–310.
Ding, B. (1999) Tissue preparation and substructure of plasmodesmata. In: *Plasmodesmata: Structure, Function, Role in Communication* (eds A.J.E. van Bel & W.J.P. van Kesteren), pp. 37–49. Springer-Verlag, Berlin.
Ding, B. & Lucas, W.J. (1996) Secondary plasmodesmata: biogenesis, special functions and evolution. In: *Membranes: Specialised Functions in Plants* (eds M. Smallwood, J.P. Knox & D.J. Bowels), pp. 489–506. BIOS Scientific, Oxford, UK.
Ding, B., Haudenshield, J.S., Hull, R.J., Wolf, S., Beachy, R.N. & Lucas, W.J. (1992a) Secondary plasmodesmata are specific sites of localization of the tobacco mosaic virus movement protein in transgenic tobacco plants. *Plant Cell*, **4**, 915–928.
Ding, B., Haudenshield, J.S., Willmitzer, L. & Lucas, W.J. (1993) Correlation between arrested secondary plasmodesmal development and onset of accelerated leaf senescence in yeast acid invertase transgenic tobacco plants. *Plant J.*, **4**, 179–189.
Ding, B., Itaya, A. & Qi, Y. (2003) Symplasmic protein and RNA traffic: regulatory points and regulatory factors. *Curr. Opin. Cell Biol.*, **6**, 1–7.
Ding, B., Kwon, M.O. & Warnberg, L. (1996) Evidence that actin filaments are involved in controlling the permeability of plasmodesmata in tobacco mesophyll. *Plant J.*, **10**, 157–164.
Ding, B., Turgeon, R. & Parthasarathy, M.V. (1991) Plasmodesmatal substructure in cryofixed developing tobacco leaf tissue. In: *Recent Advances in Phloem Transport and Assimilate Partitioning* (eds J.L. Bonnemain, S. Delrot, J. Dainty & W.J. Lucas), pp. 317–323. Ouest Editions, Nantes.
Ding, B., Turgeon, R. & Parthasarathy, M.V. (1992b) Substructure of freeze-substituted plasmodesmata. *Protoplasma*, **169**, 28–41.
Ding, D.Q. & Tazawa, M. (1989) Influence of cytoplasmic streaming and turgor pressure gradient on the transnodal transport of rubidium and electrical conductance in *Chara corallina. Plant Cell Physiol.*, **30**, 739–748.
Doerner, P. (2003) Plant meristems: a merry-go-round of signals. *Curr. Biol.*, **13**, R368–R374.
Dolan, L. & Costa, S. (2001) Evolution and genetics of root hair stripes in the root epidermis. *J. Exp. Bot.*, **52S**, 413–417.
Dorokhov, Y.L., Mäkinen, K., Frolova, O.Y., Merits, A., Saarinen, J., Kalkkinen, N., Atabekov, J.G. & Saarma, M. (1999) A novel function for a ubiquitous plant enzyme pectin methylesterase: the host-cell receptor for the tobacco mosaic virus movement protein. *FEBS Lett.*, **461**, 223–228.
Dörr, I. (1987) The haustorium of *Cuscuta* – new structural results. In: *Parasitic Flowering Plants, Proceedings on 4th International Symposium on Parasitic Flowering Plants* (eds H.C. Weber & W. Forstreuter), pp. 163–170. Philipps-Universität, Marburg, Germany.
Drake, G.A., Carr, D.J. & Anderson, W.P. (1978) Plasmolysis, plasmodesmata, and the electrical coupling of oat coleoptyle cells. *J. Exp. Bot.*, **29**, 1205–1214.
Duckett, C.M., Oparka, K.J., Prior, D.A.M., Dolan, L. & Roberts, K. (1994) Dye-coupling in the root epidermis of *Arabidopsis* is progressively reduced during development. *Development*, **120**, 3247–3255.
Ehlers, K. & Kollmann, R. (1996) Formation of branched plasmodesmata in regenerating *Solanum nigrum* protoplasts. *Planta*, **199**, 126–138.
Ehlers, K. & Kollmann, R. (2001) Primary and secondary plasmodesmata: structure, origin and functioning. *Protoplasma*, **216**, 1–30.

Ehlers, K. & van Bel, A.J.E. (1999) The physiological and developmental consequences of plasmodesmal connectivity. In: *Plasmodesmata: Structure, Function, Role in Cell Communication* (eds A.J.E. van Bel & W.J.P. van Kesteren), pp. 243–260. Springer-Verlag, Berlin.

Ehlers, K., Binding, H. & Kollmann, R. (1999) The formation of symplastic domains by plugging of plasmodesmata: a general event in plant morphogenesis? *Protoplasma*, **209**, 181–192.

Ehlers, K., Schulz, M. & Kollmann, R. (1996) Subcellular localization of ubiquitin in plant protoplasts and the function of ubiquitin in selective degradation of outer-wall plasmodesmata in regenerating protoplasts. *Planta*, **199**, 139–151.

Enkerli, K., Hahn, M.G.& Mims, C.W. (1997) Immunogold localization of callose and other plant cell wall components in soybean roots infected with the oomycete *Phytophthora sojae*. *Can. J. Bot.*, **75**, 1509–1517.

Epel, B.L. (1994) Plasmodesmata: composition, structure and trafficking. *Plant Mol. Biol.*, **26**, 1343–1356.

Epel, B.L. & Erlanger, M.A. (1991) Light regulates symplastic communication in etiolated corn seedlings. *Physiol. Plant.*, **83**, 149–153.

Epel, B.L., Kuchuck, B., Kotlizky, G., Shurtz, S., Erlanger, M. & Yahalom, A. (1995) Isolation and characterization of plasmodesmata. *Method Cell Biol.*, **50**, 237–253.

Epel, B.L., van Lent, J.W.M., Cohen, L., Kotlizky, G., Katz, A. & Yahalom, A. (1996) A 41 kDa protein isolated from maize mesocotyl cell walls immunolocalises to plasmodesmata. *Protoplasma*, **191**, 70–78.

Erwee, M.G. & Goodwin, P.B. (1983) Characterisation of the *Egeria densa* Planch. leaf symplast. Inhibition of the intercellular movement of fluorescent probes by group II ions. *Planta*, **158**, 320–328.

Erwee, M.G. & Goodwin, P.B. (1984) Characterization of the *Egeria densa* leaf symplast: response to plasmolysis, deplasmolysis and to aromatic amino acids. *Protoplasma*, **122**, 162–168.

Erwee, M.G. & Goodwin, P.B. (1985). Symplast domains in extrastelar tissues of *Egeria densa* Planch. *Planta*, **163**, 9–19.

Erwee, M.G., Goodwin, P.B. & van Bel, A.J.E. (1985) Cell–cell communication in the leaves of *Commelina cyanea* and other plants. *Plant Cell Environ.*, **8**, 173–178.

Esau, K. (1977) *Anatomy of Seed Plants*, 2nd edn. John Wiley and Sons Inc., New York.

Fensom, D.S., Thompson, R.G. & Caldwell, C.D. (1990) Ammonia gas temporarily interrupts translocation of ^{11}C photosynthate in sunflower. *J. Exp. Bot.*, **41**, 11–14.

Fisher, D.B. (1999) The estimated pore diameter for plasmodesmal channels in the *Abutilon* nectary trichome should be about 4 nm, rather than 3 nm. *Planta*, **208**, 299–300.

Fisher, D.B. (2000) Long-distance transport. In: *Biochemistry and Molecular Biology of Plants* (eds B. Buchanan, W. Gruissem & R. Jones), pp. 730–784. American Society of Plant Physiologists, Rockville, MD.

Fisher, D.B. & Cash-Clark, C.E. (2000) Sieve tube unloading and post-phloem transport of fluorescent tracers and proteins injected into sieve tubes via severed aphid stylets. *Plant Physiol.*, **123**, 125–137.

Foster, T., Veit, B. & Hake, S. (1999) Mosaic analysis of the dominant mutant, *Gnarley1-R*, reveals distinct lateral and transverse signalling pathways during maize leaf development. *Development*, **126**, 305–313.

Fridborg, I., Grainger, J., Page, A., Coleman, M., Findlay, K. & Angell, S. (2003) TIP: a novel host factor linking callose degradation with the cell-to-cell movement of potato virus X. *Mol. Plant Microbe Interact.*, **16**, 132–140.

Fry, S.C. (2001) Plant cell walls. In: *Nature Encyclopedia of Life Sciences*. Nature Publishing Group, London.

Gamalei, Y.V. (1989) Structure and function of leaf minor veins in trees and herbs. *Trees*, **3**, 96–110.

Ghoshroy, S., Lartey, R., Sheng, J. & Citovsky, V. (1997) Transport of proteins and nucleic acids through plasmodesmata. *Annu. Rev. Plant Phys.*, **48**, 27–50.

Gilbertson, R.L. & Lucas, W.J. (1996) How do viruses traffic on the vascular highway? *Trends Plant Sci.*, **1**, 260–268.

Goodwin, P.B. (1983) Molecular size limit for movement in the symplast of the *Elodea* leaf. *Planta*, **157**, 124–130.
Goodwin, P.B. & Cantrill, L. (1999) Use and limitations of fluorochromes for plasmodesmal research. In: *Plasmodesmata: Structure, Function, Role in Cell Communication* (eds A.J.E. van Bel & W.J.P. van Kesteren), pp. 67–84. Springer-Verlag, Berlin.
Grabski, S., de Feijter, A.W. & Schindler, M. (1993) Endoplasmic reticulum forms a dynamic continuum for lipid diffusion between contiguous soybean root cells. *Plant Cell*, **5**, 25–38.
Gu, X. & Verma, D.P.S. (1996) Phragmoplastin, a dynamin-like protein associated with cell plate formation in plants. *EMBO J.*, **15**, 695–704.
Gubler, F., Ashford, A.E. & Jacobsen, J.V. (1987) The release of a-amylase through gibberellin-treated barley aleurone cell walls. *Planta*, **172**, 155–161.
Gunning, B.E.S. & Robards, A.W. (1976) *Intercellular Communication in Plants: Studies on Plasmodesmata*. Springer-Verlag, Berlin.
Hamilton, A.J. & Baulcombe, D.C. (1999) A species of small antisense RNA in posttranscriptional gene silencing in plants. *Science*, **286**, 950–952.
Hashimoto, T. & Inze, D. (2003) Cell biology: how unique is the plant cell? *Curr. Opin. Plant Biol.*, **6**, 1–3.
Haywood, V., Kragler, F. & Lucas, W.J. (2002) Plasmodesmata: pathways for protein and ribonucleoprotein signaling. *Plant Cell*, (Suppl.), S303–S325.
Heinlein, M. (2002) Plasmodesmata: dynamic regulation and role in macromolecular cell-to-cell signalling. *Curr. Opin. Plant Biol.*, **5**, 543–552.
Hepler, P.K. (1982) Endoplasmic reticulum in the formation of the cell plate and plasmodesmata. *Protoplasma*, **111**, 121–133.
Hofius, D. & Sonnewald, U. (2003) Vitamin E biosynthesis: biochemistry meets cell biology. *Trends Plant Sci.*, **8**, 6–8.
Holdaway-Clarke, T.L., Walker, N.A., Hepler, P.K. & Overall, R.L. (2000) Physiological elevations in cytoplasmic free calcium by cold or ion injection result in transient closure of higher plant plasmodesmata. *Planta*, **210**, 329–335.
Holdaway-Clarke, T.L., Walker, N.A., Reid, R.J., Hepler, P.K. & Overall, R.L. (2001) Cytoplasmic acidification with butyric acid does not alter the ionic conductivity of plasmodesmata. *Protoplasma*, **215**, 184–190.
Hughes, J.E. & Gunning, B.E.S. (1980) Glutaraldehyde-induced deposition of callose. *Can. J. Bot.*, **58**, 250–258.
Hull, R. (1992) Down the tube. Tubules extending from protoplasts infected with cowpea mosaic virus may reveal how some viruses move from cell to cell in infected plants. *Curr. Biol.*, **2**, 224–226.
Iglesias, V.A. & Meins, F., Jr (2000) Movement of plant viruses is delayed in a a-1,3-glucanase-deficient mutant showing a reduced plasmodesmatal size exclusion limit and enhanced callose deposition. *Plant J.*, **21**, 157–166.
Imlau, A., Truernit, E. & Sauer, N. (1999) Cell-to-cell and long-distance trafficking of the green fluorescent protein in the phloem and symplastic unloading of the protein into sink tissues. *Plant Cell*, **11**, 309–322.
Ishiwatari, Y., Fujiwara, T., McFarland, K.C., Nemoto, K., Hayashi, H., Chino, M. & Lucas, W.J. (1998) Rice phloem thioredoxin h has the capacity to mediate its own cell-to-cell transport through plasmodesmata. *Planta*, **205**, 12–22.
Itaya, A., Liang, G., Woo, Y.M., Nelson, R.S. & Ding, B. (2000) Nonspecific intercellular protein trafficking probed by green-fluorescent protein in plants. *Protoplasma*, **213**, 165–175.
Itaya, A., Woo, Y.M., Masuta, C., Bao, Y., Nelson, R.S. & Ding, B. (1998) Developmental regulation of intercellular protein trafficking through plasmodesmata in tobacco leaf epidermis. *Plant Physiol.*, **118**, 373–385.
Jackson, D. (2001) The long and the short of it: signalling development through plasmodesmata. *Plant Cell*, **13**, 2569–2572.

Jackson, D. & Kim, J.Y. (2003) Intercellular signalling: an elusive player steps forth. *Curr. Biol.*, **13**, R349–R350.

Jackson, D., Veit, B. & Hake, S. (1994) Expression of maize *KNOTTED1* related homeobox genes in the shoot apical meristem predicts patterns of morphogenesis in the vegetative shoot. *Development*, **120**, 405–413.

Jian, L.C., Li, P.H., Sun, L.H. & Chen, T.H.H. (1997) Alterations in ultrastructure and subcellular localisation of Ca^{2+} in poplar apical bud cells during the induction of dormancy. *J. Exp. Bot.*, **48**, 1195–1207.

Jones, D.L., Gilroy, S., Larson, P.B., Howell, S.H. & Kochian, L.V. (1998) Effect of aluminium on cytoplasmic Ca^{2+} homeostasis in root hairs of *Arabidopsis thaliana* (L.). *Planta*, **206**, 378–387.

Jones, M.G.K. (1976) The origin and development of plasmodesmata. In: *Intercellular Communication in Plants: Studies on Plasmodesmata* (eds B.E.S. Gunning & A.W. Robards), pp. 81–105. Springer-Verlag, Berlin.

Juniper, B.E. (1977) Some speculations of the possible roles of the plasmodesmata in the control of differentiation. *Theor. Biol.*, **66**, 583–592.

Jørgensen, K.E. & Møller, J.V. (1979) Use of flexible polymers as probes of glomerular pore size. *Am. J. Phys.*, **236**, F103–F111.

Karlson, D.T., Fujino, T., Kimura, S., Baba, K., Itoh, T. & Ashworth, E.N. (2003) Novel plasmodesmata association of dehydrin-like proteins in cold-acclimated red-osier dogwood (*Cornus sericea*). *Tree Physiol.*, **23**, 759–767.

Kartusch, R., Lichtscheidl, I.K. & Weidinger, M.L. (2000) Brefeldin A induces callose formation in onion inner epidermal cells. *Protoplasma*, **212**, 250–261.

Kelley, W.L. (1999) Molecular chaperones: how J domains turn on Hsp70s. *Curr. Biol.*, **9**, R305–R308.

Kempers, R. & van Bel, A.J.E. (1997) Symplasmic connections between sieve element and companion cell in the stem of *Vivia faba* L. have a molecular exclusion limit of at least 10 kDa. *Planta*, **201**, 195–201.

Kempers, R., Ammerlaan, A. & van Bel, A.J.E. (1998) Symplasmic constriction and ultrastructural features of sieve element/companion cell complex in the transport phloem of apoplasmically and symplasmically phloem-loading species. *Plant Physiol.*, **116**, 271–278.

Kikuyama, M., Hara, Y., Shimada, K., Yamamoto, K. & Hiramoto, Y. (1982) Intercellular transport of macromolecules in *Nitella*. *Plant Cell Physiol.*, **33**, 413–417.

Kim, I., Hempel, F.D., Sha, K., Pfluger, J. & Zambryski, P.C. (2002a). Identification of a developmental transition in plasmodesmatal function during embryogenesis in *Arabidopsis thaliana*. *Development*, **129**, 1261–1272.

Kim, J.Y., Yuan, Z., Cilia, M., Khalfan-Jagani, Z. & Jackson, D. (2002b) Intercellular trafficking of a *KNOTTED1* green fluorescent protein fusion in the leaf and shoot meristem of *Arabidopsis*. *Proc. Nat. Acad. Sci. U.S.A.*, **99**, 4103–4108.

Kim, M., Canio, W., Kessler, S. & Sinha, N. (2001) Developmental changes due to long-distance movement of a homeobox fusion transcript in tomato. *Science*, **293**, 287–289.

Kollmann, R. & Glockmann, C. (1985). Studies on graft unions, I: Plasmodesmata between cells of plants belonging to different unrelated taxa. *Protoplasma*, **124**, 224–235.

Kollmann, R. & Glockmann, C. (1991) Studies on graft unions, III: On the mechanism of secondary formation of plasmodesmata at the graft interface. *Protoplasma*, 165, 71–85.

Kollmann, R. & Glockmann, C. (1999) Multimorphology and nomenclature of plasmodesmata in higher plants. In: *Plasmodesmata: Structure, Function, Role in Cell Communication* (eds A.J.E. van Bel & W.J.P. van Kesteren), pp. 149–172. Springer-Verlag, Berlin.

Kotlizky, G., Shurtz, S., Yahalom, A., Malik, Z., Traub, O. & Epel, B. L. (1992) An improved procedure for the isolation of plasmodesmata embedded in clean maize cell walls. *Plant J.*, **2**, 623–630.

Krabel, D., Eshrich, W., Wirth, S. & Wolf, G. (1993) Callose- (1,3-b-d-glucanase) activity during spring reactivation in deciduous trees. *Plant Sci.*, **93**, 19–32.

Kragler, F., Monzer, J., Shash, K., Xoconostle-Cázares, B. & Lucas, W.J. (1998) Cell-to-cell transport of proteins: requirement of unfolding and characterisation of binding to a putative plasmodesmatal receptor. *Plant J.*, **15**, 367–381.

Kragler, F., Monzer, J., Xoconostle-Cázares, B. & Lucas, W. J. (2000) Peptide antagonists of the plasmodesmal macromolecular trafficking pathway. *EMBO J.*, **19**, 2856–2868.

Küssel-Andermann, P., El Amraoui, A., Safieddine, S., Nouaille, S., Perfettini, I., Lecuit, M., Cossart, P., Wolfrum, U. & Petit, C. (2000) Vezatin, a novel transmembrane protein, bridges myosin VIIA to the cadherin–catenin complex. *EMBO J.*, **19**, 6020–6029.

Lachaud, S. & Maurousset, L. (1996) Occurrence of plasmodesmata between differentiating vessels and other xylem cells in *Sorbus torminalis* L. Crantz and their fate during xylem maturation. *Protoplasma*, **191**, 220–226.

Lee, J.Y., Yoo, B.C., & Lucas, W.J. (2000) Parallels between nuclear-pore and plasmodesmal trafficking of information molecules. *Planta*, **210**, 177–187.

Lee, J.Y., Yoo, B.C., Rojas, M.R., Gomez-Ospina, N., Staehelin, L.A. & Lucas, W.J. (2003) Selective trafficking of non-cell-autonomous proteins mediated by NtNCAPP1. *Science*, **299**, 392–396.

Lee, M.M. & Schiefelbein, J.W. (1999) WEREWOLF, a MYB-related protein in *Arabidopsis*, is a position dependent regulator of epidermal cell patterning. *Cell*, **9**, 473–483.

Leisner, S.M. & Turgeon, R. (1993) Movement of virus and photoassimilate in the phloem: a comparative analysis. *Bioessays*, **15**, 741–748.

Lew, R.R. (1996) Pressure regulation of the electrical properties of growing *Arabidopsis thaliana* L. root hairs. *Plant Physiol.*, **112**, 1089–1100.

Lindsey, K., Casson, S. & Chilley, P. (2002) Peptides: new signalling molecules in plants. *Trends Plant Sci.*, **7**, 78–83.

Lucas, W.J. (1995) Plasmodesmata – intercellular channels for macromolecular transport in plants. *Curr. Opin. Cell Biol.*, **7**, 673–680.

Lucas, W.J. & Gilbertson, R.L. (1994) Plasmodesmata in relation to viral movement within leaf tissues. *Annu. Rev. Phytopathol.*, **32**, 387–411.

Lucas, W.J. & Wolf, S. (1993) Plasmodesmata: the intercellular organelles of green plants. *Trends Cell Biol.*, **3**, 308–315.

Lucas, W.J., Ding, B. & van der Schoot, C. (1993) Tansley Review No. 58. Plasmodesmata and the supracellular nature of plants. *New Phytol.*, **125**, 435–476.

Lucas, W.J., Yoo, B.C. & Kragler, F. (2001) RNA as a long-distance information macromolecule in plants. *Nat. Rev. Mol. Cell Biol.*, **2**, 849–857.

Madore, M.A. & Lucas, W.J. (1986) Characterization of the source leaf symplast by means of lucifer yellow CH. In: *Phloem Transport* (eds J. Cronshaw, W.J. Lucas & R.T. Giaquinta), pp. 129–133. Alan R. Liss, Inc., New York.

Martindale, V.E. & Salisbury, J.L. (1990) Phosphorylation of algal centrin is rapidly responsive to changes in the external milieu. *J. Cell Sci.*, **96**, 395–402.

Matsushita, Y., Hanazawa, K., Yoshioka, K., Oguchi, T., Kawakami, S., Watanabe, Y., Nishiguchi, M. & Nyunoya, H. (2000) *In vitro* phosphorylation of the movement protein of tomato mosaic tobamovirus by a cellular kinase. *J. Gen. Virol.*, **81**, 2095–2102.

Matsushita, Y., Yoshioka, K., Shigyo, T., Takahashi, H. & Nyunoya, H. (2002) Phosphorylation of the movement protein of *Cucumber mosaic virus* in transgenic tobacco plants. *Virus Genes*, **24**, 231–234.

Maule, A.J. (1994) Plant virus movement: *de novo* process or redeployed machinery. *Trends Microbiol.*, **2**, 305–306.

McLean, B.G., Hempel, F.D. & Zambryski, P.C. (1997) Plant intercellular communication *via* plasmodesmata. *Plant Cell*, **9**, 1043–1054.

McLean, B.G., Waigmann, E., Citovsky, V. & Zambryski, P. (1993) Cell-to-cell movement of plant viruses. *Trends Microbiol.*, **1**, 105–109.

Medina Escobar, N., Haupt, S., Thow, G., Boevink, P., Chapman, S. & Oparka, K.J. (2003) High-throughput viral expression of cDNA–green fluorescent protein fusions reveals novel subcellular addresses and identifies unique proteins that interact with plasmodesmata. *Plant Cell*, **15**, 1507–1523.

Mezitt, L.A. & Lucas, W.J. (1996) Plasmodesmal cell-to-cell transport of proteins and nucleic acids. *Plant Mol. Biol.*, **32**, 251–273.

Mezitt Provencher, L., Miao, L., Sinha, N. & Lucas, W.J. (2001) *Sucrose Export Defective1* encodes a novel protein implicated in chloroplast-to-nucleus signalling. *Plant Cell*, **13**, 1127–1141.

Monzer, J. (1990) Secondary formation of plasmodesmata in cultured cells. In: *Parallels in Cell-to-Cell Junctions in Plants and Animals* (eds A.W. Robards, W.J. Lucas, J.D. Pitts, H.J. Jongsma & D.C. Spray), pp. 185–197. Springer-Verlag, Berlin.

Morvan, O., Quentin, M., Jauneau, A., Mareck, A. & Morvan, C. (1998) Immunogold localization of pectin methylesterases in the cortical tissues of flax hypocotyl. *Protoplasma*, **202**, 175–184.

Nakajima, K., Sena, G., Nawy, T. & Benfey, P.N. (2001) Intercellular movement of the putative transcription factor SHR in root patterning. *Nature*, **413**, 307–311.

Nelson, R.S. & van Bel, A.J.E. (1998) The mystery of virus trafficking into, through and out of vascular tissue. *Prog. Bot.*, **59**, 476–533.

Northcote, D.H., Davey, R. & Lay, J. (1989) Use of antisera to localise callose, xylan and arabinogalactan in the cell-plate, primary and secondary walls of plant cells. *Planta*, **178**, 353–366.

Nougaréde, A., Landré, P., Rembur, J. & Hernandez, M.N. (1985) Are variations in the activities of 5′-nucleotidase and adenylate cyclase components in the release of inhibition in the pea cotyledonary bud? *Can. J. Bot.*, **63**, 309–323.

Olesen, P. (1979) The neck constriction in plasmodesmata. *Planta*, **144**, 349–358.

Olesen, P. & Robards, A.W. (1990) The neck region of plasmodesmata: general architecture and some functional aspects. In: *Parallels in Cell-to-Cell Junctions in Plants and Animals* (eds A.W. Robards, W.J. Lucas, J.D. Pitts, H.J. Jongsma & D.C. Spray), pp. 145–170. Springer-Verlag, Berlin.

Oparka, K.J. (2004) Getting the message across: how do plant cells exchange macromolecular complexes. *Trends Plant Sci.*, **9**, 33–40.

Oparka, K.J. & Prior, D.A.M. (1992) Direct evidence for pressure-generated closure of plasmodesmata. *Plant J.*, **2**, 741–750.

Oparka, K.J. & Roberts, A.G. (2001) Plasmodesmata. A not so open-and-shut case. *Plant Physiol.*, **125**, 123–126.

Oparka, K.J. & Santa Cruz, S. (2000). The great escape: phloem transport and unloading of macromolecules. *Annu. Rev. Plant Phys.*, **51**, 323–347.

Oparka, K.J., Murphy, R., Derrick, P., Prior, D.A.M. & Smith, J.A.C. (1991) Modification of the pressure-probe technique permits controlled intracellular microinjection of fluorescent probes. *J. Cell Sci.*, **98**, 539–544.

Oparka, K.J., Prior, D.A.M. & Crawford, J.W. (1994) Behaviour of plasma-membrane, cortical ER and plasmodesmata during plasmolysis of onion epidermal-cells. *Plant Cell Environ.*, **17**, 163–171.

Oparka, K.J., Roberts, A.G., Boevink, P., Santa Cruz, S., Roberts, I.M., Pradel, K.S., Imlau, A., Kotlizky, G., Sauer, N. & Epel, B. (1999) Simple, but not branched, plasmodesmata allow the nonspecific trafficking of proteins in developing tobacco leaves. *Cell*, **97**, 743–754.

Ormö, M., Cubitt, A.B., Kallio, K., Gross, L.A., Tsien, R.Y. & Remington, S.J. (1996) Crystal structure of the *Aequorea victoria* green fluorescent protein. *Science*, **273**, 1392–1395.

Overall, R.L. (1999) Substructure of plasmodesmata. In: *Plasmodesmata: Structure, Function, Role in Cell Communication* (eds A.J.E. van Bel & W.J.P. van Kesteren), pp. 129–130. Springer-Verlag, Berlin.

Overall, R.L. & Blackman, L.M. (1996) A model of the macromolecular structure of plasmodesmata. *Trends Plant Sci.*, **1**, 307–311.

Overall, R.L., Wolfe, J. & Gunning, B.E.S. (1982) Intercellular communication in *Azolla* roots, I: Ultrastructure of plasmodesmata. *Protoplasma*, **111**, 134–150.

Palevitz, B.A. & Hepler, P.K. (1985) Changes in dye coupling of stomatal cells of *Allium* and *Commelina* demonstrated by microinjection of Lucifer yellow. *Planta*, **164**, 473–479.

Patrick, J.W. (1997) Pholem unloading: sieve element unloading and post-phloem transport. *Annu. Rev. Plant Phys.*, **48**, 191–222.

Perbal, M.C., Haughn, G., Saedler, H. & Schwarz-Sommer, Z. (1996) Non-cell-autonomous function of the *Antirrhinum* floral homeotic proteins *DEFICIENS* and *GLOBOSA* is exerted by their polar cell-to-cell trafficking. *Development*, **122**, 3433–3441.

Phillips, G.N., Jr. (1997) Structure and dynamics of green fluorescent protein. *Curr. Opin. Struct. Biol.*, **7**, 821–827.
Phillips, R.M. & Altschuld, R.A. (1996) 2,3-Butanedione 2-monoxime (BDM) induces calcium release from canine cardiac sarcoplasmic reticulum. *Biochem. Biophys. Res. Commun.*, **229**, 154–157.
Porfirova, S., Bergmüller, E., Tropf, S., Lemke, R. & Dörmann, P. (2002) Isolation of an *Arabidopsis* mutant lacking vitamin E and identification of a cyclase essential for all tocopherol biosynthesis. *Proc. Nat. Acad. Sci. U.S.A.*, **99**, 12495–12500.
Porter, K.R. & Machado, R.D. (1960) Studies on the endoplasmic reticulum, IV: Its form and distribution during mitosis in cells of onion root tip. *J. Biophys. Biochem. Cytol.*, **7**, 167–180.
Radford, J.E. & White, R.G. (1998) Localization of a myosin-like protein to plasmodesmata. *Plant J.*, **14**, 743–750.
Radford, J.E., Vesk, M. & Overall, R.L. (1998) Callose deposition at plasmodesmata. *Protoplasma*, **201**, 30–37.
Reichelt, S. & Kendrick-Jones, J. (2000) Myosins. In: *Actin: A Dynamic Framework for Multiple Plant Cell Functions* (eds C.J. Staiger, F. Baluska, D. Volkmann & P. Barlow), pp. 29–44. Kluwer Academic Publishers, Dordrecht, The Netherlands.
Reichelt, S., Knight, A.E., Hodge, T.P., Baluska, F., Samaj, J., Volkmann, D. & Kendrick-Jones, J. (1999) Characterization of the unconventional myosin VIII in plant cells and its localization at the post-cytokinetic cell wall. *Plant J.*, **19**, 555–567.
Reid, R.J. & Overall, R.L. (1992) Intercellular communication in *Chara*: factors affecting transnodal electrical resistance and solute fluxes. *Plant Cell Environ.*, **15**, 507–517.
Rinne, P.L.H. & van der Schoot, C. (1998) Symplasmic fields in the tunica of the shoot apical meristem coordinate morphogenetic events. *Development*, **125**, 1477–1485.
Rinne, P.L.H., Kaikuranta, P.M. & van der Schoot, C. (2001) The shoot apical meristem restores its symplastic organisation during chilling-induced release from dormancy. *Plant J.*, **26**, 249–264.
Ritzenthaler, C., Finlay, K., Roberts, K. & Maule, A.J. (2000) Rapid detection of plasmodesmata in purified cell walls. *Protoplasma*, **211**, 165–171.
Robards, A.W. (1968a) A new interpretation of plasmodesmatal ultrastructure. *Planta*, **82**, 200–210.
Robards, A.W. (1968b) Desmotubule – a plasmodesmatal substructure. *Nature*, **218**, 784.
Robards, A.W. (1976) Plasmodesmata in higher plants. In: *Intercellular Communication in Plants: Studies on Plasmodesmata* (eds B.E.S. Gunning & A.W. Robards), pp. 15–57. Springer-Verlag, Berlin.
Robards, A.W. & Lucas, W.J. (1990) Plasmodesmata. *Annu. Rev. Plant Physiol.*, **41**, 369–419.
Roberts, A.G. & Oparka, K.J. (2003) Plasmodesmata and the control of symplasmic transport. *Plant Cell Environ.*, **26**, 103–124.
Roberts, A.G., Santa Cruz, S., Roberts, I.M., Prior, D.A.M., Turgeon, R. & Oparka, K.J. (1997) Phloem unloading in sink leaves of *Nicotiana benthamiana*: comparison of a fluorescent solute with a fluorescent virus. *Plant Cell*, **9**, 1381–1396.
Roberts, I.M., Boevink, P., Roberts, A.G., Sauer, N., Reichel, C. & Oparka, K.J. (2001) Dynamic changes in the frequency and architecture of plasmodesmata during the sink–source transition in tobacco leaves. *Protoplasma*, **218**, 31–44.
Robinson-Beers, K. & Evert, R.F. (1991) Fine structure of plasmodesmata in mature leaves of sugarcane. *Planta*, **184**, 307–318.
Roy, S., Watada, A.E. & Wergin, W.P. (1997) Characterization of the cell wall microdomain surrounding plasmodesmata in apple fruit. *Plant Physiol.*, **114**, 539–547.
Ruiz-Medrano, R., Xoconostle-Cázares, B. & Lucas, W.J. (2001) The phloem as a conduit for inter-organ communication. *Curr. Opin. Plant Biol.*, **4**, 202–209.
Russin, W.A., Evert, R.F., van der Veer, P.J., Sharkey, T.D. & Briggs, S.P. (1996) Modification of a specific class of plasmodesmata and loss of sucrose export ability in the *sucrose export defective1* maize mutant. *Plant Cell*, **8**, 645–658.

Samaj, J., Peters, M., Volkmann, D. & Baluška, F. (2000) Effects of myosin ATPase inhibitor 2,3-butanedione 2 monoxime on distributions of myosin, F-actin, microtubules, and cortical endoplasmic reticulum in maize root apices. *Plant Cell Physiol.*, **41**, 571–582.

Satiat-Jeunemaitre, B. & Hawes, C. (1992a) Redistribution of a Golgi glycoproteinin plant cells treated with brefeldin A. *J. Cell Sci.*, **103**, 1153–1166.

Satiat-Jeunemaitre, B. & Hawes, C. (1992b) Reversible dissociation of the plant Golgi apparatus by Brefeldin A. *Biol. Cell*, **74**, 325–328.

Schatz, G. (1998) Protein transport–the doors to organelles. *Nature*, **395**, 439–440.

Schnepf, E. & Sych, A. (1983) Distribution of plasmodesmata in developing *Sphagnum* leaflets. *Protoplasma*, **116**, 51–56.

Schulz, A. (1995) Plasmodesmal widening accompanies the short-term increase in symplasmic phloem unloading in pea root tips under osmotic stress. *Protoplasma*, **188**, 22–37.

Schulz, A. (1999) Physiological control of plasmodesmal gating. In: *Plasmodesmata: Structure, Function, Role in Cell Communication* (eds A.J.E. van Bel and W.J.P. van Kesteren), pp. 173–204. Springer-Verlag, Berlin.

Seagull, R.W. (1983) Differences in the frequency and disposition of plasmodesmata resulting from root cell elongation. *Planta*, **159**, 497–504.

Sessions, A., Yanofsky, M.F. & Weigel, D. (2000) Cell–cell signalling and movement by the floral transcription factors LEAFY and APETALA1. *Science*, **289**, 779–781.

Sivaguru, M., Fujiwara, T., Samaj, J., Baluska, F., Yang, Z., Osawa, H., Maeda, T., Mori, T., Volkmann, D. & Matsumoto, H. (2000) Aluminium-induced $1 \rightarrow 3$-a-D-glucan inhibits cell-to-cell trafficking of molecules through plasmodesmata. A new mechanism of aluminium toxicity in plants. *Plant Physiol.*, **124**, 991–1005.

Soellick, T.R., Uhrig, J.F., Bucher, G.L., Kellmann, J.W. & Schreier, P.H. (2000) The movement protein NSm of tomato spotted wilt tospovirus (TSWV): RNA binding, interaction with the TSWV N protein, and identification of interacting plant proteins. *Proc. Natl. Acad. Sci. U.S.A.*, **97**, 2373–2378.

Sokolova, M., Prüfer, D., Tacke, E. & Rohde, W. (1997) The potato leafroll virus 17K movement protein is phosphorylated by a membrane-associated protein kinase from potato with biochemical features related to protein kinase C. *FEBS Lett.*, **400**, 201–205.

Staehelin, L.A. & Hepler, P.K. (1996) Cytokinesis in higher plants. *Cell*, **84**, 821–824.

Steinberg, G. & Kollmann, R. (1994) A quantitative analysis of the interspecific plasmodesmata in the non-division walls of the plant chimera *Laburnocytisus adamii* (Poit.) Schneid. *Planta*, **192**, 75–83.

Stone, B.A. & Clarke, A.E. (1992) *Chemistry and Biochemistry of (1-3)-b-Glucans*. La Trobe University Press, Melbourne, Australia.

Storms, M.M.H., Kormelink, R., Peters, D., van Lent, J.W.M. & Goldbach, R.W. (1995) The nonstructural NSm protein of tomato spotted wilt virus induces tubular structures in plant and insect cells. *Virology*, **214**, 485–493.

Subramani, S. (1996) Protein transport into peroxisomes. *J. Biol. Chem.*, **271**, 27943–27948.

Susi, P. (2000) Dye-coupling in tobacco mesophyll cells surrounding growing tobacco mosaic tobamovirus-induced local lesions. *J. Phytopathol.*, **148**, 379–382.

Terry, B.R. & Robards, A.W. (1987) Hydrodynamic radius alone governs the mobility of molecules through plasmodesmata. *Planta*, **171**, 145–157.

Tilney, L.G., Cooke, T.J., Connelly, P.S. & Tilney, M.S. (1991) The structure of plasmodesmata as revealed by plasmolysis, detergent extraction, and protease digestion. *J. Cell Biol.*, **112**, 739–747.

Tucker, E.B. (1982) Translocation in the staminal hairs of *Setcreasea purpurea*, I: A study of cell ultrastructure and cell-to-cell passage of molecular probes. *Protoplasma*, **113**, 193–201.

Tucker, E.B. (1990) Calcium-loaded 1,2-bis(2-aminophenoxy)ethane-N,N,N',N'-tetra acetic acid blocks cell-to-cell diffusion of carboxyfluorescein in staminal hairs of *Setcreasea purpurea*. *Planta*, **182**, 34–38.

Tucker, E.B. (1993) Azide treatment enhances cell-to-cell diffusion in staminal hairs of *Setcreasea purpurea*. *Protoplasma*, **174**, 45–49.

Tucker, E.B. & Boss, W.F. (1996) Mastoparan induced intracellular Ca^{2+} fluxes may regulate cell-to-cell communication in plants. *Plant Physiol.*, **111**, 459–467.
Turner, A., Wells, B. & Roberts, K. (1994) Plasmodesmata of maize root tips: structure and composition. *J. Cell Sci.*, **107**, 3351–3361.
Ueki, S. & Citovsky, V. (2001) RNA commutes to work: regulation of plant gene expression by systemically transported RNA molecules. *Bioessays*, **23**, 1087–1090.
van Bel, A.J.E. & Oparka, K.J. (1995) On the validity of plasmodesmograms. *Bot. Acta*, **108**, 174–182.
van Bel, A.J.E., Günther, S. & van Kesteren, W.J.P. (1999) Plasmodesmata, a maze of questions. In: *Plasmodesmata: Structure, Function, Role in Cell Communication* (eds A.J.E. van Bel & W.J.P. van Kesteren), pp. 1–26. Springer-Verlag, Berlin.
van der Schoot, C. & Rinne, P.L.H. (1999) The symplasmic organisation of the shoot apical meristem. In: *Plasmodesmata: Structure, Function, Role in Cell Communication* (eds A.J.E. van Bel & W.J.P. van Kesteren), pp. 225–242. Springer-Verlag, Berlin.
van der Schoot, C., Dietrich, M.A., Storms, M., Verbeke, J.A. & Lucas, W.J. (1995) Establishment of a cell-to-cell communication pathway between separate carpels during gynoecium development. *Planta*, **195**, 450–455.
Vaughn, K.C., Hoffman, J.C., Hahn, M.G. & Stahelin, L.A. (1996) The herbicide dichlobenil disrupts cell plate formation: immunogold characterisation. *Protoplasma*, **194**, 117–132.
Voinnet, O., Vain, P., Angell, S. & Baulcombe, D.C. (1998) Systemic spread of sequence-specific transgene RNA degradation in plants is initiated by localized introduction of ectopic promoterless DNA. *Cell*, **95**, 177–187.
Volk, G.M., Turgeon, R. & Beebe, D.U. (1996) Secondary plasmodesmata formation in the minor-vein phloem of *Cucumis melo* L. and *Curcubita pepo* L. *Planta*, **199**, 425–432.
von Bargen, S., Salchert, K., Paape, M., Piechulla, B. & Kellmann, J.W. (2001) Interactions between the tomato spotted wilt virus movement protein and plant proteins showing homologies to myosin, kinesin and DnaJ-like chaperones. *Plant Physiol. Biochem.*, **39**, 1083–1093.
Waigmann, E. & Zambryski, P. (1995) Tobacco mosaic virus movement protein-mediated protein transport between trichome cells. *Plant Cell*, **7**, 2069–2079.
Waigmann, E., Chen, M.H., Bachmaier, R., Ghoshroy, S. & Citovsky, V. (2000) Regulation of plasmodesmal transport by phosphorylation of tobacco mosaic virus cell-to-cell movement protein. *EMBO J.*, **19**, 4875–4844.
Waigmann, E., Turner, A., Peart, J., Roberts, K. & Zambryski, P. (1997) Ultrastructural analysis of leaf trichome plasmodesmata reveals major differences from mesophyll plasmodesmata. *Planta*, **203**, 75–84.
Wang, N. & Fisher, D.B. (1994) The use of fluorescent tracers to characterize the post-phloem transport pathway in maternal tissues of developing wheat grains. *Plant Physiol.*, **104**, 17–27.
White, R.G., Badelt, K., Overall, R.L. & Vesk, M. (1994) Actin associated with plasmodesmata. *Protoplasma*, **180**, 169–184.
Wille, A.C. & Lucas, W.J. (1984) Ultrastructural and histochemical studies on guard cells. *Planta*, **160**, 129–142.
Wolf, S., Deom, C.M., Beachy, R.N. & Lucas, W.J. (1989) Movement protein of tobacco mosaic virus modifies plasmodesmatal size exclusion limit. *Science*, **246**, 377–379.
Wright, K.M. & Oparka, K.J. (1997) Metabolic inhibitors induce symplastic movement of solutes from the transport phloem of *Arabidopsis* roots. *J. Exp. Bot.*, **48**, 1807–1814.
Wu, X., Dinneny, J.R., Crawford, K.M., Rhee, Y., Citovsky, V., Zambryski, P. & Weigel, D. (2003) Modes of intercellular transcription factor movement in *Arabidopsis* apex. *Development*, **130**, 3735–3745.
Wu, X., Weigel, D. & Wigge, P.A. (2002) Signalling in plants by intercellular RNA protein movement. *Gene Dev.*, **16**, 151–158.
Xoconostle-Cázares, B., Ruiz-Medrano, R. & Lucas, W.J. (2000) Proteolytic processing of CmPP36, a protein from the cytochrome b_5 reductase family, is required for entry into the phloem translocation pathway. *Plant J.*, **24**, 735–747.

Xoconostle-Cázares, B., Xiang, Y., Ruiz-Medrano, R., Wang, H.I., Monzer, J., Yoo, B.C., McFarland, K.C., Franceschi, V.R. & Lucas, W.J. (1999) Plant paralog to viral movement protein that potentiates transport of mRNA into the phloem. *Science*, **283**, 94–98.

Yahalom, A., Lando, R., Katz, A. & Epel, B.L. (1998) A calcium-dependent protein kinase is associated with maize mesocotyl plasmodesmata. *J. Plant Physiol.*, **153**, 354–362.

Yahalom, A., Warmbrodt, R.D., Laird, D.W., Traub, O., Revel, J.P., Willecke, K. & Epel, B.L. (1991) Maize mesocotyl plasmodesmata proteins cross-react with connexin gap junction protein antibodies. *Plant Cell*, **3**, 407–417.

Yang, S.J., Li, M.Y. & Zhang, X.Y. (1995) Changes in plasmodesmatal permeability of the stamen hairs of *Setcresea purpurea* during development. *Acta Phytophysiol. Sin.*, **21**, 355–362.

Zambryski, P. (2004) Cell-to-cell transport of proteins and fluorescent tracers via plasmodesmata during plant development. *J. Cell Biol.*, **162**, 165–168.

Zambryski, P. & Crawford, K. (2000) Plasmodesmata: gatekeepers for cell-to-cell transport of developmental signals in plants. *Annu. Rev. Cell Dev. Biol.*, **16**, 393–421.

Zee, S.-Y. (1969) The fine structure of differentiating sieve elements of *Vica faba*. *Aust. J. Bot.*, **17**, 441–456.

Zhang, W.H. & Tyerman, S.D. (1997) Effect of low oxygen concentration on the electrical properties of cortical cells of wheat roots. *J. Plant Physiol.*, **150**, 567–572.

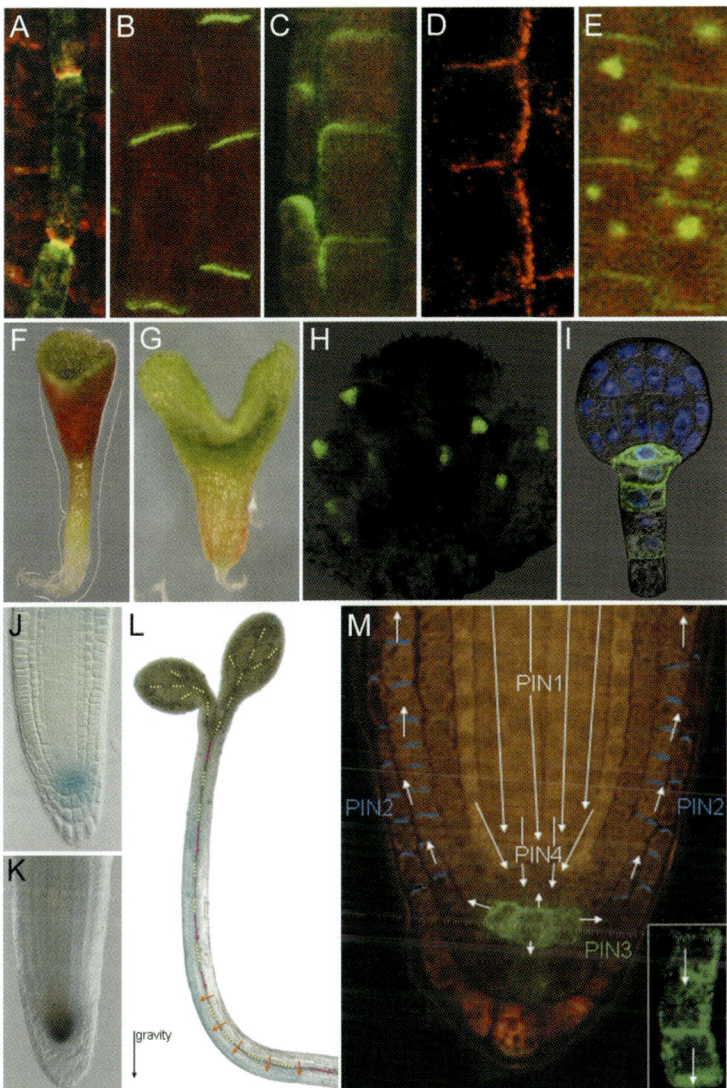

Plate 1.1 (A) Apical localization of AUX1 protein (green) and basal localization of PIN1 (red) in root protophloem cells. (B–D) Basal localization of PIN1 in stele (B), apical localization of PIN2 in epidermis and lateral root cap (C) and lateral localization of PIN3 in pericycle (D) cells of *Arabidopsis* root. (E) Internalization of PIN1 protein into aggregations of endosomes upon Brefeldin A treatment. (F, G) Seedling phenotypes of *pin1,3,4,7* quadruple (F) and *gnom* (G) mutant with rudimentary root and improperly specified cotyledons. Reproduced from Friml *et al*. (2003), with permission. (H) Developing flower with increased *DR5::GFP* signal at tips of developing primordia of floral organs. Reproduced from Benková *et al*. (2003), with permission. (I) *DR5rev::GFP* globular embryo showing increased auxin response (in green) at the basal pole. Nuclei depicted in blue. (J) Increased auxin response visualized by *DR5::GUS* expression in central root meristem. Reproduced from Benková *et al*. (2003), with permission. (K) Increased auxin levels in the central root meristem visualized by anti-IAA immunolocalization. Reproduced from Benková *et al*. (2003), with permission. (L) Auxin response and transport in gravistimulated *Arabidopsis* hypocotyl. Nonpolar (dashed line) and polar (red arrows) auxin transport systems. During gravitropic or phototropic bending, increased auxin response (*DR5::GUS*, displayed as blue staining) visualizes higher auxin levels at the more elongated, outer side of bending shoot. Reproduced from Friml (2003), with permission. (Continued)

Plate 1.1 (Continued) (M) Immunolocalization of PIN3 protein in the *Arabidopsis* root apex with probable routes of polar auxin transport (white arrows). PIN3 (in green) is localized symmetrically in columella cells and apparently mediates auxin distribution in lateral direction to all sides of root cap. After gravistimulation, PIN3 rapidly relocates to the bottom side of columella cells (inset), where it apparently mediates auxin flux to the lower side of root for gravitropic bending. Reproduced from Friml (2003), with permissions.

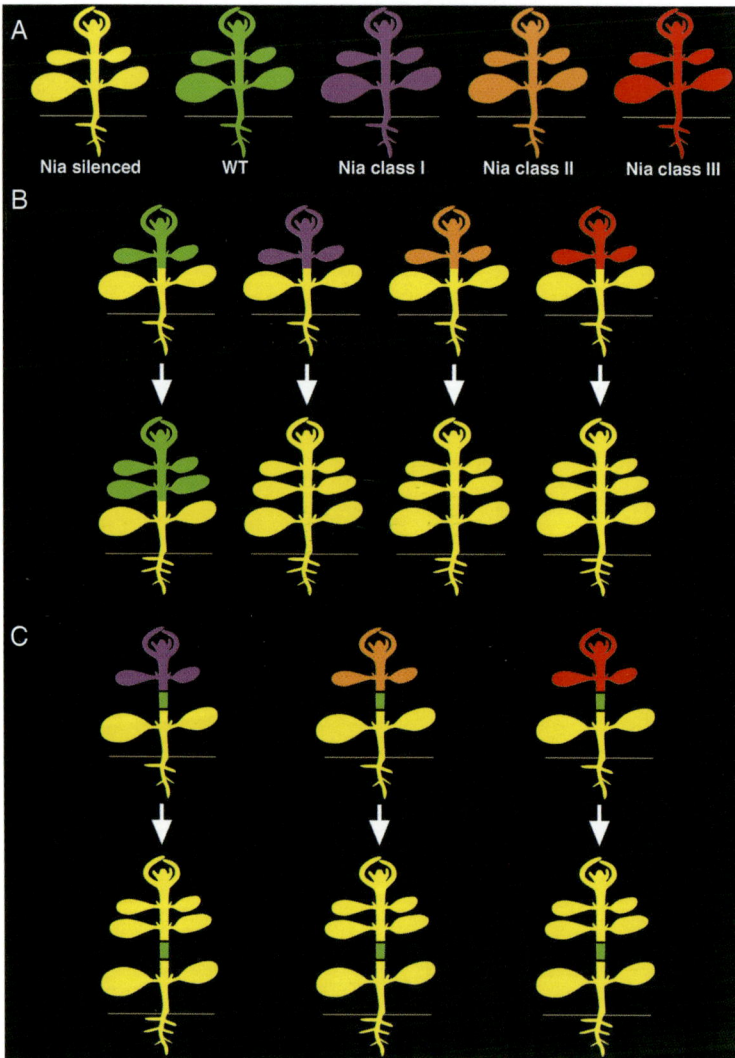

Plate 3.1 Graft transmission of nitrate reductase (*Nia*) co-suppression. (A) Schematic of the tobacco genotypes used. (B) Graft transmission of *Nia* silencing occurs in Class I, II and III, but not in wild-type (wt) plants. (C) Silencing can be transmitted through a section of up to 20 cm of wt plant.

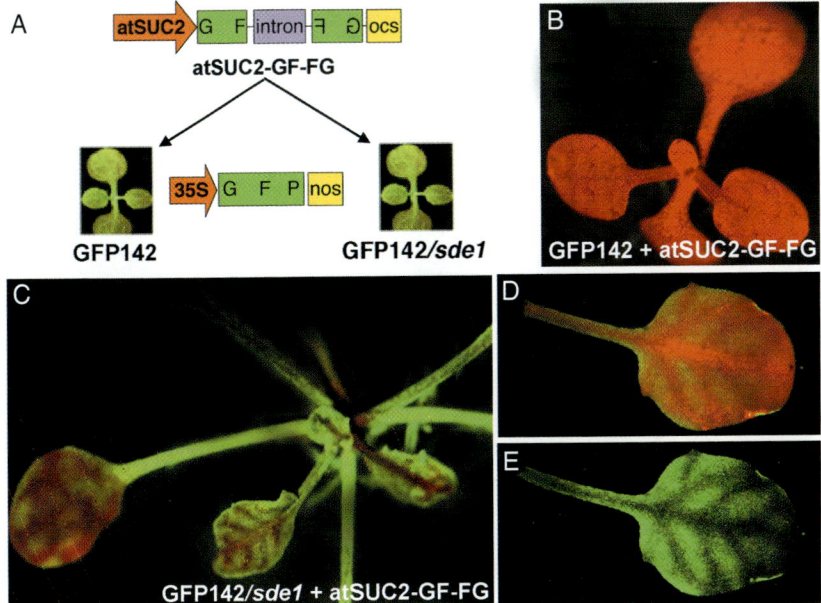

Plate 3.2 A system to study non-cell autonomous RNA silencing in *Arabidopsis*. (A) A pan-handled transgene construct designed to produce GFP dsRNA was mobilized under the control of the phloem-specific AtSUC2 promoter. The resulting construct was then used to transform *Arabidopsis* plants expressing constitutively a GFP transgene. These plants carried either a wt (GFP142) or a mutated copy (GFP142/*sde1*) of the *SDE1* gene that encodes a putative RNA-dependent RNA polymerase. (B) Uniform silenced phenotype of the wt transformants. Note that the tissues fluoresce in red under UV light, owing to chlorophyll fluorescence. (C–E) Vein-centred silencing of GFP in the *sde1* transformants. GFP silencing expands 10–15 cells outside the vasculature, indicating the existence of an SDE1-independent short-distance silencing movement process in *Arabidopsis*. The leaf in (E) was imaged with a band-pass filter removing chlorophyll fluorescence. Adapted from Himber *et al.* (2003).

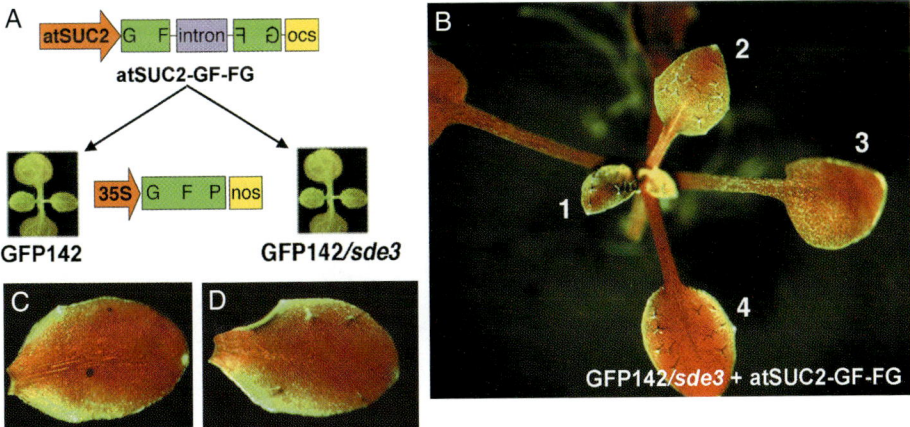

Plate 3.3 The extent of cell-to-cell silencing movement is also influenced by the SDE3 putative RNA helicase in *Arabidopsis*. (A) Setup of the experiments. (B–D) The extent of GFP silencing from the vasculature is higher in *sde3* than in *sde1* mutant plants, indicating a less stringent requirement for SDE3 in the movement process. Adapted from Himber *et al.* (2003).

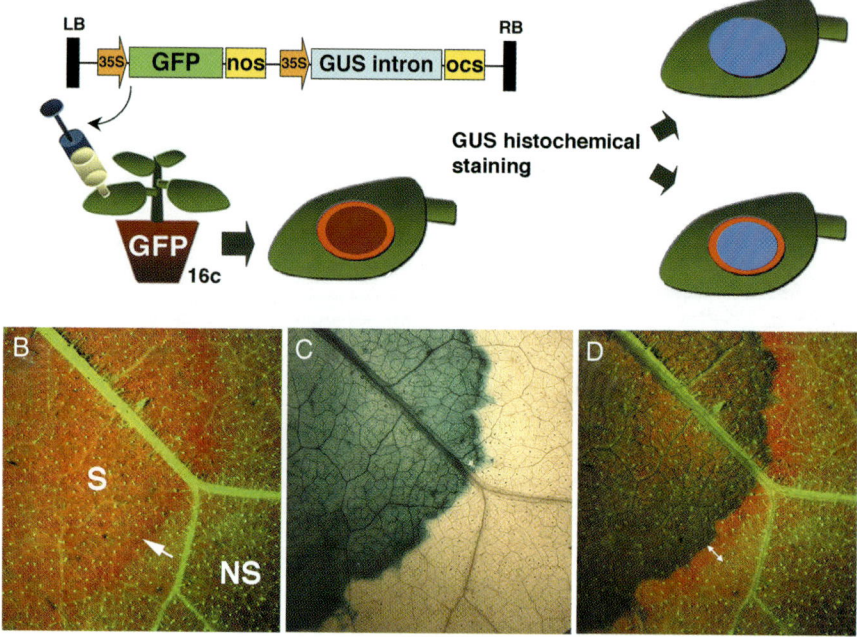

Plate 3.4 Short-range movement of GFP silencing in mature tobacco leaves. (A) Principle of the experiment. (B) The red border (arrow) triggered by the construct in (A) at 10 days post-infiltration. (C) GUS histochemical staining of the leaf in (B). (D) Overlay of the images in (B) and (C). S, silenced; NS, non-silenced tissue Adapted from Himber et al. (2003).

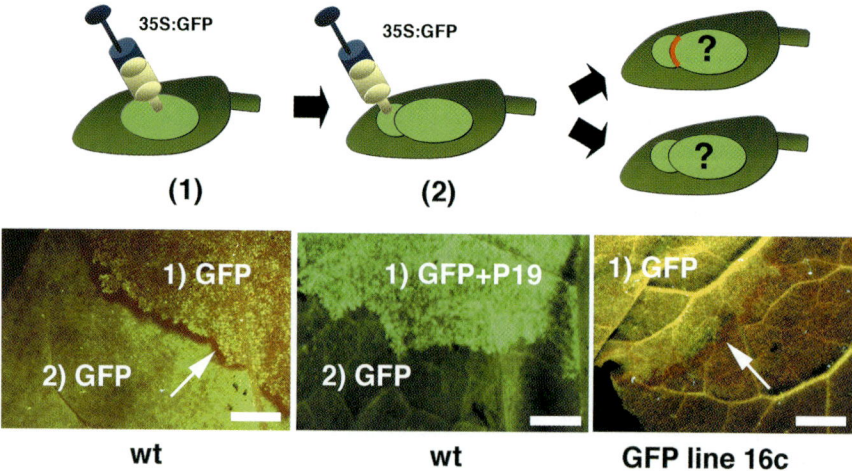

Plate 3.5 Short-range silencing movement in non-transgenic tobacco. The two sequential infiltrations with 35S:GFP are performed at a 5-day interval (numbered 1 and 2, respectively). In the middle panel, the first infiltration was performed with the P19 suppressor protein, hence the very high green fluorescence due to absence of GFP silencing. Arrows: short-range movement of silencing. Bar: 5 mm. Adapted from Himber et al. (2003).

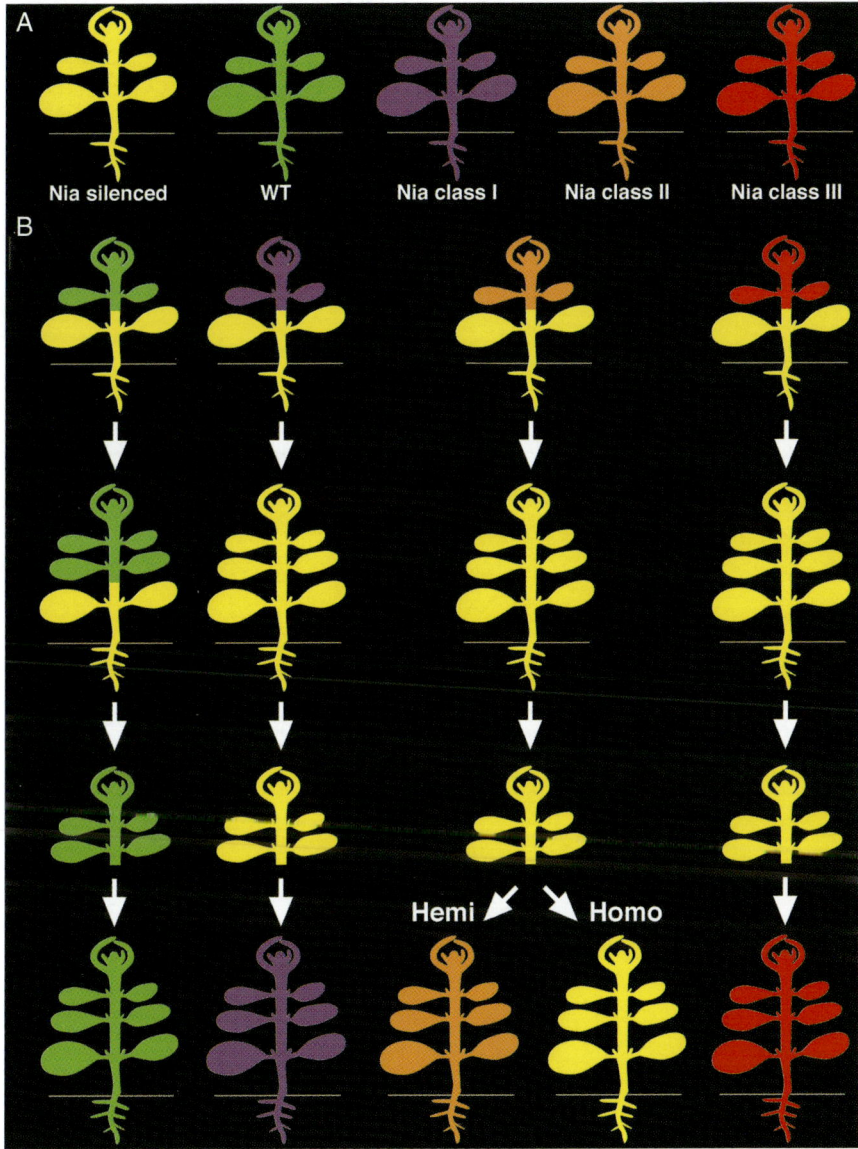

Plate 3.6 Maintenance of *Nia* co-suppression. (A) Schematic of the tobacco genotypes used. (B) Silencing is maintained only in homozygous but not in hemizygous Class-II plants after removal of the rootstocks.

Plate 3.7 Short-range silencing of the *SULPHUR* mRNA in *Arabidopsis*. The experimental setup is the same as in Figure 3.2, except that the pan-handled construct has homology to the endogenous *SULPHUR* mRNA, involved in pigment production. Silencing is manifested by development of chlorosis.

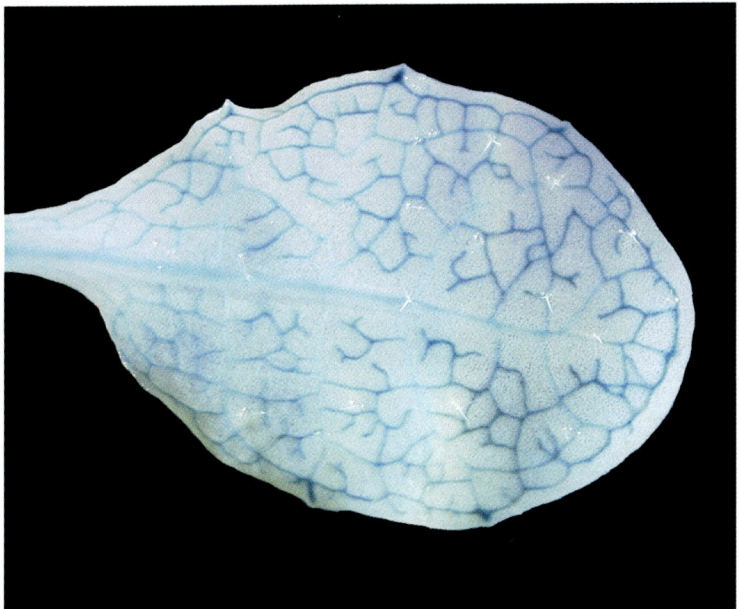

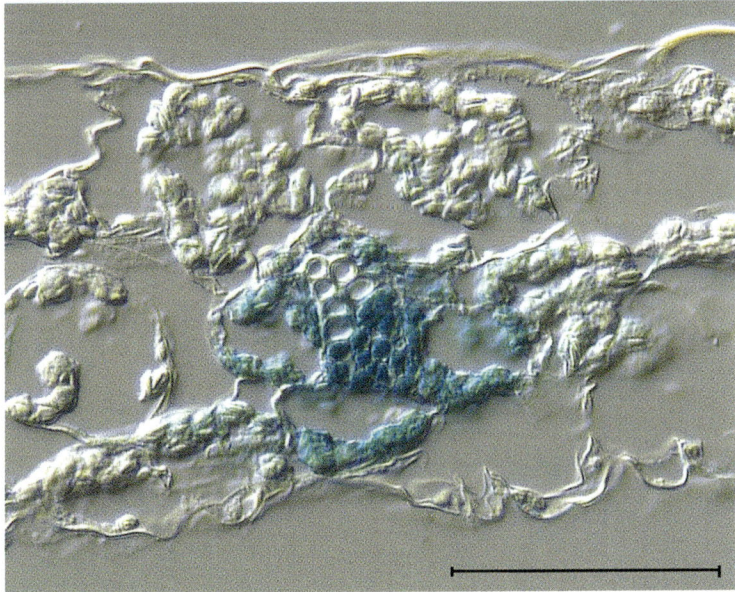

Plate 7.1 *CONSTANS* is expressed in the vascular tissue and regulates a long-distance signal that induces flowering at the meristem. Staining of a transgenic *Arabidopsis* plant carrying a fusion of the *CO* promoter to the *GUS* marker gene illustrates the pattern of expression of *CO*. Top: staining of a whole leaf detects *CO:GUS* expression in the vascular tissue. Bottom: A cross section of the leaf detects *CO:GUS* expression in the phloem. Bar: 50 μM. The pattern of expression of *CO* is discussed further in Takada and Goto (2003) and An *et al.* (2004). Images Courtesy of Coral Vincent, Max Planck Institute, Köln.

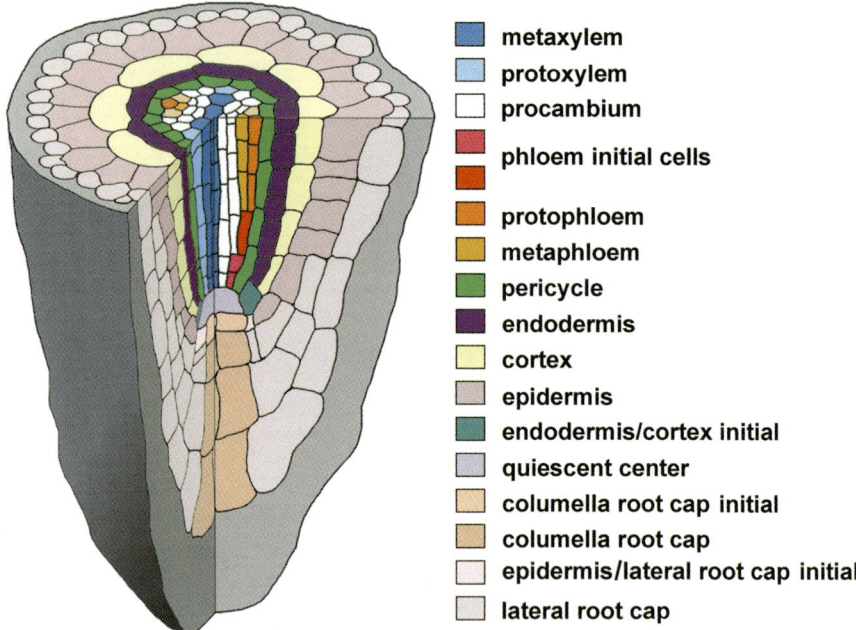

Plate 8.1 Cell lineages in the *Arabidopsis* root. Schematic: grey colours refer to cell lineages only, the differentiation state is unknown.

6 Lessons from the vegetative shoot apex
John F. Golz

6.1 Introduction

Post-embryonic growth and development of plants is largely determined by the activity of meristems, structures that contain self-renewing populations of undifferentiated cells. During embryogenesis, the primary meristems of the plant are formed at opposite poles of the embryo. The shoot apical meristem (Fig. 6.1A), which forms all above-ground organs, such as leaves, bracts and the tissue of the stem, is located at the apical pole; whereas the root apical meristem, which is the source of all subterranean organs, forms at the basal pole. Several different types of meristems also form during post-embryonic shoot growth. The most common are the bud-forming axillary meristems that arise in the axils of leaves. These buds may either continue to develop into organ-bearing branches or remain dormant indefinitely until growth is triggered. Floral meristems, which are specialised axillary meristems, arise from the axils of bracts following the transition to flowering. They produce a limited number of floral organs before ceasing activity. Other meristems, which do not form lateral organs, are also active in the shoot. The intercalary meristem drives shoot elongation whereas the lateral meristem causes an increase in the width of the shoot axis. In most dicots and some monocots, the activity of the lateral meristem is usually associated with secondary growth such as the formation of wood and bark. This chapter focuses on the vegetative shoot apical meristem and the role that signalling plays in its function. Descriptions of the root apical meristem and floral meristem can be found in Chapters 7 and 8 of this volume.

Studies over the last decade have shown that cell signalling plays important roles in regulating the size of the stem cell population at the summit of the shoot apical meristem, determining the position of organ formation and patterning the emerging organs. Not surprisingly the types of signals involved in these processes are varied and involve some of the molecules and pathways that have been mentioned in previous chapters of this book. How these and other pathways are integrated into a functioning shoot apical meristem is the focus of this chapter.

6.2 Structure of the angiosperm shoot apical meristem

In order to place signalling within the context of meristem function, the structure of the shoot apex is briefly considered.

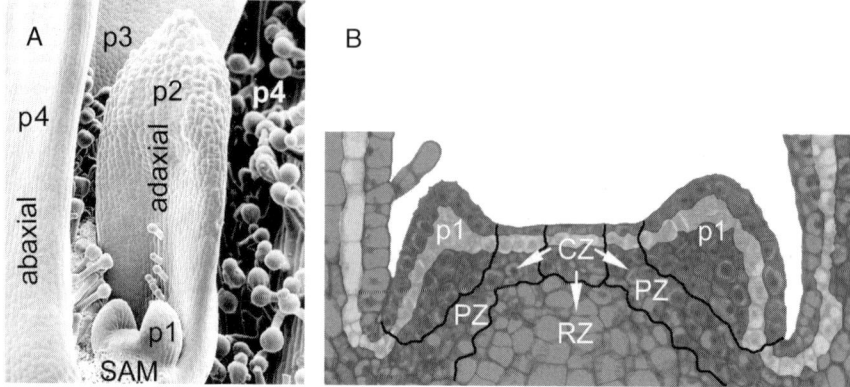

Figure 6.1 Shoot apical meristem structure. (A) View of an *Antirrhinum* shoot apical meristem (SAM) showing progressively older leaves (p1–p4). Two opposing leaves arise at each node and are displaced by 90° from the previous node, an arrangement called descussate phyllotaxis. In this image one of the p2 and p4 organs has been removed to allow a view of the meristem. (B) A cross section through the centre of an *Antirrhinum* apex shown in (A), highlighting the morphologically distinct zones and layers of the SAM. The epidermal L1 layer (upper layer of dark cells) and the sub-epidermal L2 layer (layer of light cells) consists of anticlinally dividing cells that are collectively termed the tunica. Underlying the tunica is the corpus, which is derived from the L3 layer and is composed of cells dividing in any orientation (lower layer of dark cells). Black lines separate the meristem into functionally distinct zones, although often the boundary between these zones is not well defined. Cells in the peripheral zone (PZ) either form lateral organs or the outer tissue of the stem. The inner stem tissue is derived from the rib zone (RZ). Within the central zone (CZ), pluripotent stem cells continually divide, producing cells that move into the PZ and RZ, as well as maintaining the stem cell population.

6.2.1 Zones of the meristem

The arrangement of cells within the shoot apical meristem of flowering plants has traditionally been described in one of two ways (Steeves & Sussex, 1989; Lyndon, 1998). The first divides the shoot apical meristem into three zones on the basis of cytoplasmic densities and cell division rates, although the boundaries between these zones are often unclear (for example see Laufs *et al.*, 1998b). Cells at the summit of the shoot apical meristem form the central zone and are distinguished by having a large, vacuolated appearance and a lower rate of cell division than that of surrounding tissue. Based on cell lineage studies in several species there are approximately six to nine pluripotent stem cells within the central zone that continually divide, producing cells that both maintain the integrity of the central zone and replace cells that have been consumed in organ formation (Stewart & Dermen, 1970; Furner & Pumfrey, 1992; Irish & Sussex, 1992). Flanking the central zone is the peripheral zone, which is composed of smaller, more rapidly dividing cells. Arising from this zone are lateral organs, such as leaves, and the outer tissue of the stem. The rib zone lies beneath the central zone and generates the central (pith) tissue of the stem (Fig. 6.1B).

6.2.2 Layers of the meristem

The arrangement of cells in distinct layers is another characteristic feature of the shoot apical meristem (Fig. 6.1B). The outer or *tunica* region of the meristem usually consists of two cell layers in most dicots, called L1 and L2, or a single layer in monocots (L1). Cells within these layers divide anticlinally (perpendicular to the surface), whereas cells in the underlying L3 layer divide in any orientation to form the *corpus*. The stereotypical pattern of cell divisions within each layer means that cells rarely move between layers and thus each layer is essentially clonally distinct. This is elegantly demonstrated by the occurrence of periclinal chimaeras, plants in which one or more layers of the shoot apical meristem are genetically distinct from adjacent layers (see Szymkowiak & Sussex, 1996). Owing to the stability of periclinal chimaeras, derivatives of each layer can be traced into the stem and lateral organs (Satina *et al.*, 1940; Dermen, 1953; Stewart & Burk, 1970). This shows that in most organs, the epidermis is derived from the L1 layer, the sub-epidermal tissue from the L2 layer and the vasculature and internal tissue of the organs and stem from the L3 layer. However this pattern may vary in some organs, for instance sub-epidermal tissue in the margins of petals is derived from the L1 layer.

6.2.3 Symplastic fields within the meristem

A further level of organisation within the apex, which might be relevant to meristem function, is the arrangement of plasmodesmatal connections that potentially allow communication between cells. Plasmodesmata interconnect symplastic fields that may include cells from the same or different lineages. They are also able to regulate the passage of potential signalling molecules, including developmentally important transcription factors or their RNA precursors (see Chapter 5; Haywood *et al.*, 2002) and may be involved in determining the polarity of movement. Studies of the birch and *Arabidopsis* meristems have shown that symplastic fields predict some aspects of meristem fate and are highly dynamic (Rinne & van der Schoot, 1998; Gisel *et al.*, 1999). Whether symplastic signalling is important in maintaining the function within a field or for specifying different functions remains to be determined.

6.3 Periclinal chimaeras reveal a role for signalling in plant development

If the function of cells within the meristem and lateral organs is determined by the inheritance of information from a progenitor cell, then cells with a common ancestry are expected to follow similar fates. Evidence that lineage does not control cell fate comes from the analysis of periclinal chimaeras in which a cell from one layer has been displaced into an adjacent layer following a periclinal division. Such events can occur within the meristem as well as in the developing leaf primordium, although the frequency of cell displacement varies between these tissues and even between species. Regardless of when or where the displacement occurs, all descendants of a displaced cell adopt fates identical to those of the surrounding cells in the receiving layer. This implies that cell fate is largely determined by positional rather than inherited information, and that cells must continually assess their position presumably by communicating with their neighbours.

Surgical experiments first conducted in the early part of the last century also showed the importance of cell–cell interactions in meristem development. In this case destroying part or all of central zone by surgical treatment, or more recently by laser ablation, causes cells in the adjacent peripheral zone to adopt central zone functions and re-establish a morphologically normal meristem (Steeves & Sussex, 1989; Sussex, 1989; Reinhardt et al., 2003a). This implies that cells of the central zone impart peripheral zone identity on adjacent cells, and that in the absence of this signal, stem cell identity is re-established in the peripheral zone (see later).

In addition to the role of signalling in determining cell fate, there is extensive evidence that it is also required to coordinate growth between the layers of the meristem and developing organ. This is particularly evident in leaf development, where analysis of periclinal chimaeras has shown that the contribution of cells from the L2 and L3 layers to the internal tissue of the blade is highly variable (Stewart & Dermen, 1975, 1979). This implies that growth of each layer is coordinated, such that a smaller contribution from one layer is matched by a correspondingly larger contribution from the other layer in any particular region of the leaf.

Further evidence for interlayer communication comes from the analysis of periclinal chimaeras that arise from a graft union between two related species with different patterns of growth. As the meristem of these plants have cell layers derived from different species, the contribution of each layer to the size and shape of an organ can be assessed. In some instances leaf and flower morphology is determined by the identity of the L2 layer, implying that signals from this layer regulate growth in adjacent layers (Jorgenson & Crane, 1927). However, other studies have shown that the L1 layer is more influential (Stewart et al., 1972). The consequences of signalling between the layers is therefore variable and is dependent on the type of organ and the particular developmental stage.

By analogy to animal development, there are various types of signalling systems that might regulate cell function within the meristem and developing organ. Signalling molecules, such as secreted proteins, RNAs and hormones, may be actively moved via membrane-bound transporters or through symplastic networks, forming a concentration gradient extending over many cells. Cells might respond according to the concentration of signal they experience, effectively converting a gradient into discontinuous domains of gene expression and cell identity. At the other extreme, signalling may involve interactions between adjacent cells, utilising a receptor/ligand system. This latter system seems to be important in plants, as hundreds of transmembrane receptors have been identified in the *Arabidopsis* genome (Shiu & Bleecker, 2003). Examples of these various types of signalling systems are considered in the following sections.

6.4 Signalling involved in meristem maintenance

The shoot apical meristem of most plants remains relatively constant in size unless responding to a changing environment or developmental signals. The stability of size reflects a balance between cell proliferation in the central and peripheral zones and the recruitment of cells into organ formation in the peripheral and rib zones. Meristems that have lost this balance occur naturally and are distinguished by generating broad flat stems and/or flowers that have many more organs than normal. The shoot apical meristem of these plants often grows as a band or ring rather than as a point, a property termed *fasciation*.

The genetic basis of meristem maintenance has been extensively studied in *Arabidopsis*, where mutant screens have identified plants with abnormal meristems. One class of mutant has meristems that get progressively larger as the plant grows, whereas the other class fails to form and/or maintain the shoot apical meristem. Much progress has been made in characterising these mutants and identifying genes involved in meristem function. In many instances these genes encode proteins that are involved in cell signalling, some of which are examined in detail in the following sections. Other genes whose protein products are either not known to be involved in signalling or have yet to be identified are listed in Table 6.1.

6.4.1 The CLAVATA mutants

Mutations in the three *Arabidopsis CLAVATA* loci cause a progressive enlargement of the shoot apical meristem during the growth of the plant. *clv* mutants also have more leaves and flowers than do wild-type plants and produce flowers with an increased number of organs (Leyser & Furner, 1992; Clark *et al.*, 1993, 1995; Kayes & Clark, 1998). The presence of additional carpels in the *clv* gynoecium gives it a club-shaped appearance, which led to these mutants being named '*clavatus*' from the Latin meaning club-like. The dramatic increase in the size of *clv* meristems results from an accumulation of cells within the central zone, a defect that may arise in several ways. One possibility is that the CLV pathway might regulate the rate of cell divisions within the central zone. The loss of CLV activity would therefore lead to an increase in the rate of cell proliferation within the central zone and an accumulation of cells. Alternatively, the CLV pathway might promote the transition of cells from the central zone into the peripheral zone. In this scenario, the increased accumulation of cells in the central zone of *clv* mutants is not associated with a change in the rate of cell division. Careful observations of both cell size and frequency of divisions in *clv* meristems have shown that there is a slight decrease in cell division rates within the central zone, supporting the second scenario (Laufs *et al.*, 1998b).

Genetic interactions between all three *clv* loci show that they function in the same genetic pathway. However, unlike the other *clv* mutants, *clv2* displays additional organ defects implying that its function is not limited to meristem maintenance (Kayes & Clark, 1998). In addition, genetic studies show that the activity of CLV1

Table 6.1 Genes involved in *Arabidopsis* meristem formation and maintenance

BELLRINGER (*BLR*)/*PENNYWISE* (*PNY*)	The *blr/pny* mutant phenotype is similar to the *brevipedicellus* mutant (see text) having both short internodes resulting in a stunted appearance and excessive development of axillary meristems (Byrne *et al.*, 2003; Smith & Hake, 2003). Although *blr/pny* mutants have normal SAMs, in different genetic backgrounds *BLR/PNY* is required for meristem maintenance. For instance *blr/pny* enhances the weak *stm* mutant phenotype and is required for meristem maintenance in an *as1 stm* mutant background (Bryne *et al.*, 2003). BLR/PNY encodes a *BELL1*-like homeodomain transcription factor physically interacts with the STM and KNAT1 proteins, resulting in high affinity binding to target sequences (Byrne *et al.*, 2003; Smith & Hake, 2003).
CUP-SHAPED COTYLEDON1 (*CUC1*) *CUP-SHAPED COTYLEDON2* (*CUC2*) *CUP-SHAPED COTYLEDON3* (*CUC3*)	*CUC1*, *CUC2* and *CUC3* encode plant-specific NAC domain proteins with extensive homology to the *NO APICAL MERISTEM* (*NAM*) gene of *Petunia* and the recently identified *CUPILLIFORMIS* (*CUP*) gene of *Antirrhinum* (Souer *et al.*, 1996; Aida *et al.*, 1997; Takada *et al.*, 2001; Vroemen *et al.*, 2003; Weir *et al.*, 2004). All three genes are required redundantly to promote the formation of the shoot apical meristem and the separation of the cotyledons during embryogenesis. In the developing embryo all three *CUC* genes are expressed in cells that will form the SAM and are then subsequently expressed at the boundary between organ primordia and the meristem during post-embryonic development. Based on the expression of these genes during embryogenesis and their combined mutant phenotype, it has been proposed that they either promote *STM* expression directly or provide signals that establish a programme of SAM development (Aida *et al.*, 1997, 1999; Long & Barton, 1998; Takada *et al.*, 2001; Vroemen *et al.*, 2003). Consistent with this is the formation of ectopic meristems on cotyledons of plants expressing *CUC1* ubiquitously (Takada *et al.*, 2001).
DORNROSCHEN (*DRN*)/*ENHANCER OF SHOOT REGENERATION1* (*ESR1*)	Identified as a gain-of-function mutation, *DRN/ESR1* is expressed in lateral organs and the meristem and encodes an AP2/EREB transcription factor. Misexpression in cultured roots leads to accelerated regeneration of shoots (Banno *et al.*, 2001), whereas increased *DRN/ESR1* levels in shoots causes an enlargement of the meristem and the successive radialisation of leaves, before the arrest of the meristem. DRN/ESR1 regulates *STM*, *CLV3* and *WUS* expression and functions to repress stem cell fate. The role of DRN/ESR1 in organ formation is unclear, but appears to be required non-cell autonomously (Kirch *et al.*, 2003).
FASCIATA1 (*FAS1*) *FASCIATA2* (*FAS2*)	Mutations in these genes severely disrupt cellular organisation of the SAM and RAM, leading to fasciation (Leyser & Furner, 1992). In the SAM, *WUS* expression expands both laterally and apically into all three layers. FAS1/2 encode subunits of the *Arabidopsis chromatin assembly factor-1* (CAF-1) and presumably functions in maintaining epigenetic states during cell division within meristems (Kaya *et al.*, 2001).
FOREVER YOUNG (*FEY*)	The *fey* mutant has an altered phyllotaxis, and produces radially symmetric organs at a high frequency. In addition cells within the vegetative meristem and leaf primordia are more vacuolated than normal. *FEY* encodes a nodulin-like protein, although how this protein functions in meristem and leaf development has yet to be determined (Callos *et al.*, 1994)

HAIRY MERISTEM (*HAM*)	A *Petunia* mutation that has a similar mutant phenotype to *wus*. *HAM* encodes a *GRAS*-like transcription factor that is predominantly expressed in the L3 layer of lateral organs and in the provascular tissue of the stem. *HAM* therefore promotes a non-cell autonomous signal that keeps meristem cells in an undifferentiated state. Absence of this signal prevents meristem cells from responding to *WUS* or *STM* expression (Stuurman et al., 2002)
MGOUN1 (*MGO1*) *MGOUN2* (*MGO2*)	Mutations in these two loci cause a reduction in the number of leaves and floral organs and an enlargement of the meristem. Within the meristem, the organisation of the layers is disrupted. Following the transition to flowering *mgo* meristems often fasciate, although the structure of these apices are different to the fasciated *clv* meristems, in that there appears to be many individual meristems grouped together within the large dome. The enlarged meristem may be a secondary affect arising from the reduction in organ primordia formation and the consequential increase in cells within the peripheral zone. The *MGO*s are therefore likely to function primarily in the peripheral zone where they promote organ formation (Laufs et al., 1998a).
PINHEAD (*PNH*)/*ZWILLE* (*ZWL*)	*PNH* is required for efficient SAM formation during embryogenesis and for its maintenance post-embryonically. Termination of the SAM is often associated with the formation of central radialised organ. *pnh* mutant also has fewer axillary meristems (McConnell & Barton, 1995). PNH belongs to a family of proteins that include the translation factor eIF2C and ARGONAUTE, and is expressed in the vasculature, the adaxial domain of organs and the throughout the SAM (Moussian et al., 1998; Lynn et al., 1999). One function of *PNH* is to maintain *STM* expression in the SAM.
STRUWWELPETER (*SWP*)	The *swp* mutant has fewer cells in all aerial organs and a SAM that gradually becomes disorganised. SWP is similar to proteins found in the Mediator complex and therefore may function in the recruitment of RNA polymerase II to promoters (Autran et al., 2002). The SAMs of *swp* mutants are frequently fasciated, a feature that is correlated with ectopic *WUS* expression. Interestingly, *STM* expression is also patchy within these meristems (Autran et al., 2002).
WIGGUM (*WIG*)/*ERA1*	*wig*/*era1* mutants are pleiotropic, displaying alterations in seed germination, flowering time, senescence, internode elongation, phyllotaxis as well as having meristem defects similar to the *clv* mutants (Running et al., 1998; Yalovsky et al., 2000). *WIG*/*ERA1* encodes an FTase b-subunit which is involved in farnesylation, a process that increases the hydrophobicity of target proteins, facilitating either membrane attachment or protein–protein interactions (Cutler et al., 1996; Ziegelhoffer et al., 2000).
ULTRAPETALA (*ULT*)	The inflorescence meristem of the *utl* mutant is larger than normal, and produces more floral meristems. *ult* flowers have an increased number of organs as well as more whorls of organs, indicating that the floral meristems are also bigger than normal. One function of *ULT* is to restrict *CLV1* expression to the central zone, as loss of *ULT* activity results in lateral expansion of *CLV1*. However, genetic analysis indicates that *ULT* has functions independent of *CLV1*. Thus the lateral expansion of *CLV1* in the *ult* mutant may simply reflect an increased proliferation of L3 cells. The precise function of *ULT* will become apparent only when the gene has been identified (Fletcher, 2001).

is sensitive to levels of CLV3, raising the possibility that products of both genes physically interact (Clark *et al.*, 1995).

6.4.2 The CLAVATA signalling pathway

All three *CLV* loci have been cloned and their molecular identity indicates they are components of a signalling pathway. *CLV1* encodes a receptor-like kinase (RLK) with an extracellular domain containing leucine-rich repeats (LRR), a transmembrane domain and an intracellular serine/threonine kinase domain (Clark *et al.*, 1997). The protein encoded by *CLV2* is related to CLV1, having both an LRR and transmembrane domains, but lacking a kinase domain (Jeong *et al.*, 1999). LRR is a motif that is commonly found in protein-binding domains, suggesting that these RLKs are likely to interact with protein ligands. Based on genetic interactions (see above), the likely ligand for CLV1 is the product of the *CLV3* locus, which encodes a small secreted protein of 96 amino acids (Fletcher *et al.*, 1999). As expected, two of the *CLV* genes are expressed specifically within the shoot apical and floral meristems (Clark *et al.*, 1997; Fletcher *et al.*, 1999), with *CLV3* accumulating predominantly in the L1 and L2 layers of the central zone and *CLV1* in the underlying L3 layer (Fig. 6.2A). *CLV2* is apparently expressed in the shoot apex as well as in other tissues of the plant, which is consistent with it having a broader role in development (Jeong *et al.*, 1999). The molecular identities of the *CLAVATA* loci imply that cell-to-cell signalling between the layers of the meristem is involved in regulating the size of the central zone. Elegant biochemical and genetic studies have provided insight into how this signalling pathway functions.

Purification of CLV1 from the *Arabidopsis* meristem shows that it is present in both 185- and 450-kDa protein complexes (Trotochaud *et al.*, 1999). The 450-kDa complex appears to be the active form, requiring the presence of both CLV2 and CLV3 for its formation (Jeong *et al.*, 1999; Trotochaud *et al.*, 1999). Absence of both forms of the complex in *clv2* mutants is consistent with CLV2 forming a heterodimer with CLV1, a complex that is predicted to be ~185 kDa in size. Surprisingly, CLV1 is associated with a larger 600-kDa complex in *clv2* mutants, indicating that CLV1 may also interact with other receptor-like proteins (Jeong *et al.*, 1999). The mild meristem defects of *clv2* mutants may result from these larger CLV1-containing complexes functioning, albeit partially, in CLV signalling. The finding that mutations in the LRR domain of CLV1 are semi-dominant, and condition a more severe meristem defect than loss-of-function *clv1* alleles, can also be explained if CLV1 forms complexes with other RLKs. According to this model, the severity of the missense mutations is caused by non-functional CLV1 proteins forming an inactive complex with these RLKs, effectively eliminating their function from the meristem (Dievart *et al.*, 2003).

Several lines of evidence indicate that CLV3 is the likely ligand for the putative CLV1/CLV2 receptor complex. Periclinal chimaeras that have a wild-type L1 layer and *clv3* mutant L2 and L3 layers form normal meristem, consistent with CLV3 functioning non-cell-autonomously (Fletcher *et al.*, 1999). Also, immunological studies have shown that CLV3 protein is secreted and accumulates in the extracellular space, where it might interact with the LRR domain of the CLV1/2 complex (Rojo et al., 2002). These findings support a model in which CLV3 is secreted from cells in the L1 and L2 layers and subsequently moves through the apoplast into the underlying L3 layer where it binds to the extracellular domains of the putative CLV1/2 receptor complex (stippled zone in Fig. 6.2A). According to this model, increasing the concentration of ligand, and therefore activity of the CLV pathway, should reduce the number of stem cells within the central zone, whereas less ligand

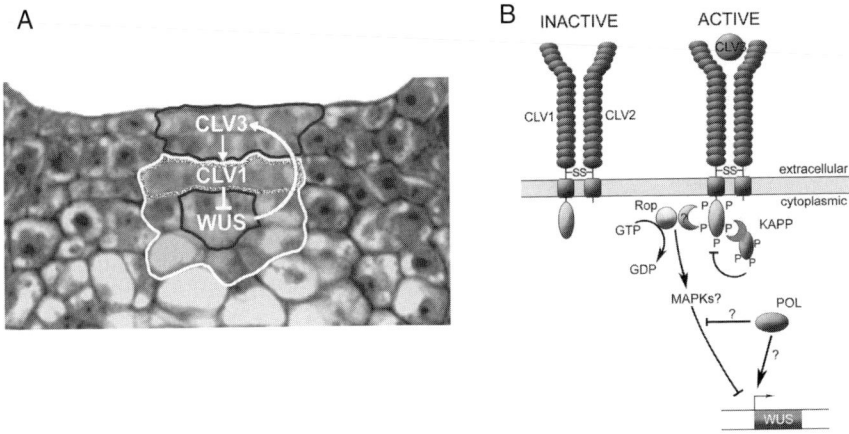

Figure 6.2 The *CLAVATA* signalling pathway. (A) Products of the three *CLAVATA* loci (*CLV1*, *CLV2* and *CLV3*) encode components of a signalling pathway that regulate the expression domain of the homeodomain transcription factor *WUSCHEL* (*WUS*). *CLV3* is expressed predominantly in the L1 and L2 cells of the central zone (black outline), whereas *CLV1* is expressed in L3 cells of the central and rib zones (white outline). *WUS* expression is confined to just a few central zone cells underlying the stem cells (black outline). These cells form an organising centre that promotes stem cell identity on the overlying layers of the central zone through a non-cell autonomous signal. This signal also promotes CLV3 expression, which subsequently activates the CLV1 signalling pathway (cells indicated with a stippled outline), leading to repression of *WUS* expression. This negative feedback loop limits the size of the stem cell population by regulating the size of the organising centre. (B) The molecular basis of the CLV–WUS feedback loop. The receptor-like kinase CLV1 and receptor-like protein CLV2 are likely to form an inactive 185-kDa heterodimer in the membrane of central zone cells via disulphide linkages (S) between conserved cysteine pairs flanking the LRR. CLV1 is also likely to form dimers with other receptor-like proteins (see text). Interactions between the CLV3 ligand and the external LRR domain of the heterodimer cause autophosphorylation of CLV1 and the subsequent recruitment of other intracellular proteins. The activated 450-kDa complex is associated with a Rho-like GTPase (Rop), which is likely to be involved in signal transduction. Rop may bind to the activated complex via a small linker protein (indicated with a question mark). Another component of the active complex is the protein phosphatase (KAPP) that is likely to modulate CLV1 activity. Currently the downstream components of the *CLV* signalling pathway are not known but it seems likely that a MAPK signalling cascade might be involved. The role of the phosphatase 2C POLTERGEIST (POL) in the CLV signalling pathway is ambiguous (see Table 6.2). It may act as a negative regulator of CLV signalling pathway, or as a positive regulator of *WUS*. Redrawn from Carlos and Fletcher (2003).

is predicted to increase the number of stem cells and the size of the central zone (see below). Although CLV3 is likely to be a ligand, there is currently no biochemical evidence to support a physical interaction between CLV3 and CLV1.

6.4.3 The wuschel mutant

The function of the CLV pathway in limiting the number of stem cells in the central zone is balanced by the cell-proliferating activity of *WUSCHEL* (*WUS*). *wus* mutants have meristems that terminate after the emergence of the first few leaves. However, adventitious meristems subsequently form and produce a few more leaves before terminating. Repeated rounds of initiation and termination cause *wus* plants to appear bushy or tousled (*wuschel* is German for 'tousled hair'; Laux *et al.*, 1996). *WUS* encodes a putative homeobox transcription factor that is expressed in a few cells of the L3 layer directly beneath the stem cells (Mayer *et al.*, 1998). It has been proposed that *WUS*-expressing cells define a region of the meristem called the organising centre, which promotes stem cell identity on overlying cell layers non-cell autonomously (Mayer *et al.*, 1998). Recent work supports this view, as induced *WUS* expression in patches of cells within the meristem causes stem cell identity and proliferation in adjacent non-*WUS*-expressing cells (Gallois *et al.*, 2002).

6.4.4 The CLAVATA–WUSCHEL regulatory loop

Genetic studies place *WUS* downstream of the CLV pathway and therefore a likely target of *CLV* regulation (Laux *et al.*, 1996; Schoof *et al.*, 2000). In the absence of CLV activity, the domain of *WUS* expands, suggesting that the larger meristem size of *clv* mutants might be caused by an increase of *WUS* activity (Brand *et al.*, 2000; Schoof *et al.*, 2000). This was tested directly by enlarging the domain of *WUS* expression using the *CLV1* promoter. Transgenic plants expressing this construct had a *clv*-like phenotype (Schoof *et al.*, 2000). Conversely, when the domain of *CLV3* expression was increased, the resulting *wus*-like phenotype correlated with a reduction of *WUS* expression (Brand *et al.*, 2000). This phenotype was dependent on CLV1 activity, suggesting that the function of the CLV signalling pathway is to restrict *WUS* expression and in doing so antagonise stem cell accumulation.

But what promotes CLV expression? Several lines of evidence point to a *WUS*-regulated signal promoting *CLV3* expression. Firstly, stem cell formation and *CLV3* expression can be induced ectopically when *WUS* is expressed in lateral organs (Schoof *et al.*, 2000; Brand *et al.*, 2002; Lenhard *et al.*, 2002). And secondly, when *CLV3* is ectopically expressed in the L3 region of the meristem, both *WUS* and endogenous *CLV3* expression (in L1 and L2 layers) is reduced or lost completely (Brand *et al.*, 2000). Thus, *WUS* acts non-cell autonomously to promote *CLV3* expression in the overlying cells of the central zone. However, the dependence of *CLV3* expression on WUS activity is not absolute, as *CLV3* is still expressed in the adventitious and axillary meristems of *wus* mutants (Brand *et al.*, 2002). Expression

Table 6.2 Other components of the *CLAVATA* pathway

Protein-phospatase KAPP

The KINASE ASSOCIATED PROTEIN PHOSPATASE (KAPP) is recruited to the active 450-kDa CLV1 receptor complex (Trotochaud *et al.*, 1999) where it binds to phosphorylated CLV1 (Williams *et al.*, 1997; Stone *et al.*, 1998). As increased levels of KAPP protein in transgenic plants trigger an enlargement of the shoot apical meristem, it is likely that KAPP functions as a negative regulator of CLV signalling by dephosphorylating CLV1 in the activated complex (William *et al.*, 1997).

ROP, a small GTPase

Another component of the 450-kDa CLV1 complex is a small GTPase belonging to the Rho subfamily, called ROP (Trotochaud *et al.*, 1999). It is thought that ROP either binds directly or via a smaller linker protein to the phosphorylated kinase domain of CLV1. Although the downstream components of the CLV signalling pathway are not known, it seems likely, by analogy with RLK associated with innate immunity, that a mitogen-activated protein kinase (MAPK) signalling cascade might be involved. Activation of the MAPK pathway may be mediated by ROP (Asai *et al.*, 2002).

SHEPHERD

SHEPHERD (*SHD*) mutants are fasciated, but also display reduced pollen tube growth and disorganisation of the root apical meristem (Ishiguro *et al.*, 2002). The *shd* mutant suppresses the effects of constitutive CLV signalling, showing that it functions in the CLV signalling pathway. *SHD* encodes a GRP94-like chaperone protein, which is likely to reside in the endoplasmic reticulum, where it may promote folding of all three CLV proteins or associated proteins (Ishiguro *et al.*, 2002).

POLTERGEIST

Mutations in *POLTERGEIST* (*POL*) suppress the *clv* mutant phenotype (Yu *et al.*, 2000), suggesting that *POL* is a negative regulator of the *CLV* pathway. However, genetic interactions with *wus* suggest *POL* works redundantly with *WUS* to promote stem cell identity (Yu *et al.*, 2003). *POL* encodes a nuclear-localised protein phosphatase 2C (PP2C) that has a broad domain of expression and may function in other pathways (Yu *et al.*, 2003).

of *CLV1* is also dependent on WUS activity and at least one other factor (see below; Gallois *et al.*, 2002).

The extensive molecular and genetic characterisation of the CLV–WUS pathway has thus provided insight into the complex signalling involved in maintaining the integrity of the meristem (see Fig. 6.2B). The pathway is essentially a negative feedback loop in which the *WUS*-expressing organising centre generates a non-cell autonomous signal that, in conjunction with other factors (see below and Table 6.2), promotes stem cell identity and *CLV3* expression in the overlying layers of the central zone. The secreted CLV3 ligand either diffuses or is actively transported through the apoplast of the meristem where it activates the putative CLV1/2 receptor complex in underlying cells. Activation of the CLV pathway ultimately leads to transcriptional repression of *WUS*, which in turns limits the size of the organising centre and the strength of the stem-cell-promoting signal.

This model does not explain why CLV3 fails to activate CLV signalling in the organising centre, a region expressing *CLV1*. As ectopic expression of *CLV3* in this region causes *WUS* repression (Brand *et al.*, 2000), there must be some mechanism that prevents it from occurring normally. A recent study has shown that the putative

CLV1/2 receptor complex is not only activated by CLV3 but also sequesters it in the region above the organising centre, thus preventing CLV signalling in underlying cells (Lenhard & Laux, 2003). This work also shows that fluorescently labelled CLV3 protein accumulates in regions of the meristem that do not express *CLV3*. The non-cell autonomous accumulation of CLV3 apparently serves two purposes: it prevents lateral expansion of the organising centre and promotes the transition of central zone cells into the peripheral zone where they eventually differentiate (Lenhard & Laux, 2003).

Despite the progress in understanding how stem cells are maintained in the meristem, fundamental questions remain. For instance, how does *WUS* signal stem cell identity in overlying cell layers? How many and what function do other RLKs play in meristem homeostatis? In addition, what is the role of the recently identified *CLV3*-like (*CLE*; Cock & McCormick, 2001) and *WUS*-like (*WOX*; Haecker et al., 2004) genes in meristem function? To date, only one *CLE* gene has been examined in detail (*CLE40*; Hobe et al., 2003) and this showed that targeted expression of *CLE40* in the shoot apex activates the CLV pathway. However, as *CLE40* expression is broader than *CLV3* expression, and does not accumulate to high levels in the shoot apex, it is likely that *CLE40* and *CLV3* have different functions. Interestingly, ectopic expression of *CLE40* and *CLV3* results in a loss of the root apical meristem, suggesting that CLV signalling may occur in the root (see also Chapter 7, this volume). Of the 14 *WOX* genes that have recently been reported in the *Arabidopsis* genome, a subset is expressed in developing embryos and may therefore be involved in early patterning events (Haecker et al., 2004). This idea is supported by the finding that *wox2* mutants have aberrant divisions in the apical region of the embryo. Mutations affecting other *WOX* genes do not affect embryo development, perhaps suggesting that this family is highly redundant.

6.5 Maintaining indeterminate cells in the meristem requires homeobox genes

Superimposed on the CLV signalling pathway are other pathways that keep meristem cells in an indeterminate state until they reach the peripheral zone, where they are recruited into organ formation. A highly conserved family of homeodomain transcription factors encoded by the *KNOX* (*KNOTTED1*-like homeobox) genes perform this function in the meristems of diverse species (see Reiser et al., 2000). The founding member of this family is the maize *KNOTTED1* gene, which is expressed throughout the meristem but excluded from founder cells and organ primordia (Vollbrecht et al., 1991; Jackson et al., 1994). The close correlation between undifferentiated cells of the meristem and *KN1* expression suggests that *KN1* may function by repressing organ identity in the meristem and/or keeping meristem cells in an undifferentiated state. Consistent with this is the finding that ectopic expression of *KN1* or other members of the *KNOX* family in leaves results in a less determinate pattern of leaf development and may, if transcript levels are sufficiently high, induce the formation of ectopic shoots (reviewed in Reiser et al., 2000).

Loss-of-function mutations in *KN1* and its functional orthologue in *Arabidopsis*, *SHOOT MERISTEMLESS* (*STM*), are characterised by a failure to develop and/or maintain a shoot apical meristem (Barton & Poethig, 1993; Clark *et al.*, 1996; Endrizzi *et al.*, 1996; Long *et al.*, 1996; Kerstetter *et al.*, 1997; Vollbrecht *et al.*, 2000). As both *stm* and *wus* mutants impart similar mutant phenotypes, their functional relationship has been the focus of several studies. Genetic analysis places *WUS* downstream of *STM* (Endrizzi *et al.*, 1996), which is consistent with an inability to maintain *WUS* expression in developing *stm* mutant embryos (Mayer *et al.*, 1998). However, this hierarchy of gene activity is at odds with the observation that *WUS* is expressed before *STM* in embryogenesis (Long & Barton, 1998; Mayer *et al.*, 1998). The precise regulatory relationship between *STM* and *WUS* has therefore been difficult to determine. Using slightly different approaches, several recent studies have shown that *STM* and *WUS* regulate distinct genetic pathways (Brand *et al.*, 2002; Gallois *et al.*, 2002; Lenhard *et al.*, 2002). For instance ectopic expression of *STM* in organs promotes the expression of several other *KNOX* genes but not *CLV3*, whereas *CLV3* is detected in organs expressing *WUS* (Lenhard *et al.*, 2002). Ectopic expression of both *WUS* and *STM* in organs has a synergistic effect, resulting in elevated levels of *CLV3* expression (Brand *et al.*, 2002; Gallois *et al.*, 2002; Lenhard *et al.*, 2002). Based on these observations it has been proposed that one function of *STM* is to prevent cell differentiation within the meristem, rather than directly promoting stem cell identity (Lenhard *et al.*, 2002). *WUS* on the other hand promotes stem cell identity, but in the absence of *STM*, these cells eventually differentiate. Thus the combined activity of *STM* and *WUS* is required to maintain a self-perpetuating population of stem cells. However, although meristem-like structures arise from organs expressing both *STM* and *WUS*, they do not develop into fully formed shoots, suggesting that other factors are required for meristem maintenance (see Tables 6.1 and 6.2).

STM maintains stem cell fate in part by restricting the expression of the MYB-domain transcription factor *ASYMMETRIC LEAVES1* (*AS1*) and the leucine zipper domain transcription factor *AS2* to organ founder cells and primordia in the periphery of the meristem. Presence of both *AS1* and *AS2* transcript throughout the apex of *stm* mutant embryos suggests that the failure to develop a meristem may be a direct consequence of *AS1* and *AS2* misexpression (Byrne *et al.*, 2000, 2002; Iwakawa *et al.*, 2002). This was elegantly demonstrated by showing that *as1 stm* and *as2 stm* double mutants form functional meristems (Byrne *et al.*, 2000, 2002). One of the functions of *AS1* and *AS2* is to prevent *KNOX* gene expression in incipient organ primordia. Both *as1* and *as2* mutants have lobed leaves that are similar in appearance to the leaves of plants constitutively expressing the *KNOX* gene, *BREVIPEDICELLUS* (Lincoln *et al.*, 1994; Byrne *et al.*, 2000; Ori *et al.*, 2000; Semiarti *et al.*, 2001). Transcripts of several *KNOX* genes were subsequently found to accumulate in *as1* and *as2* leaves, suggesting the leaf phenotype is caused by ectopic *KNOX* expression (Ori *et al.* 2000; Semiarti *et al.*, 2001). The similarity between *as1* and *as2* mutant phenotypes implies that they may function in a common pathway to regulate *KNOX* expression. Consistent with this model is the recent finding that AS1 and AS2 proteins physically interact *in vitro*, suggesting they may form a complex *in planta*

(Xu et al., 2003). Similar interactions between *KNOX* and *AS1*-like genes occur in maize and *Antirrhinum*, showing the importance of this pathway in defining cell identities within the shoot apex (Timmermans et al., 1999; Tsiantis et al., 1999).

In addition to *STM*, there are three other class I *KNOX* genes in *Arabidopsis*. These genes – *BREVIPEDICELLUS* [*BP*, formerly *KNOTTED-like from ARABIDOPSIS THALIANA1* (*KNAT1*)], *KNAT2* and *KNAT6* – are expressed in the shoot apical meristem; however unlike *STM*, their expression is restricted to specific domains (for example see Lincoln et al., 1994). *BP/KNAT1* is expressed in the basal regions of the meristem around the sites of incipient organ formation, where one of its functions is to promote internode growth (Douglas et al., 2002; Venglat et al., 2002). The function of *KNAT2* has yet to be determined, as *knat2* mutants lack a discernable phenotype, possibly because of redundancy with the closely related *KNAT6* gene (Byrne et al., 2002).

The surprising finding that *as1 stm* plants have relatively normal vegetative meristem raised the possibility that in the absence of *STM*, other *KNOX* genes might acquire *STM*-like functions (Byrne et al., 2000). Consistent with this hypothesis is the finding that *BP/KNAT1* promotes meristem formation in *as1 stm* mutant plants (Byrne et al., 2002). Based on these observations, a model depicting the likely interactions between *KNOX* genes and *AS1/AS2* has been proposed (see Fig. 6.3). While the downregulation of *KNOX* genes is closely associated with organ formation and

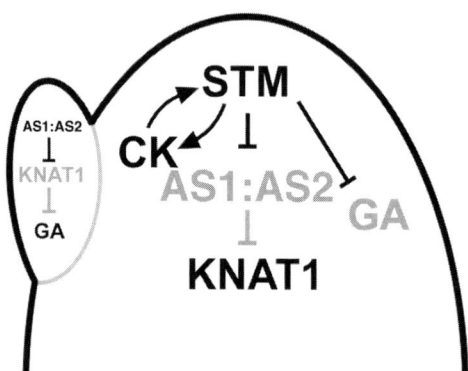

Figure 6.3 Factors regulating meristem maintenance and the patterning of lateral organs. This model shows the likely relationship between *KNOX* genes (*STM*, *KNAT1*), *AS1*, *AS2* and growth regulators such as gibberellin (GA) and cytokinin (CK) in the meristems and organs of *Arabidopsis*. *STM* is expressed throughout the meristem but excluded from organ founder cells and initiating primordia. The function of *STM* is to keep meristems cell in an undifferentiated state, which is achieved in two ways. *STM* restricts *AS1* and *AS2* expression to organ initials and in doing so allows *KNAT1* expression in the meristem. Both *KNOX* genes prevent an accumulation of GA in the meristem by directly repressing the expression of GA biosynthetic genes. In contrast, *KNOX* genes promote the accumulation of CK, which in turn promotes *KNOX* gene expression and meristem activity. *AS1* and *AS2* accumulate in organ primordia, where they likely form a complex that represses *KNOX* gene expression and allows, either directly or indirectly, accumulation of GA.

expression of *AS1*, the signals responsible for the patterned expression of these genes have yet to be identified. Recent work has suggested that auxin might be involved in this process (see Section 6.7).

An intriguing feature of the KN1 protein is its ability to move between cells of the shoot apex (Jackson *et al.*, 1994; Kim *et al.*, 2002). Similarly, STM and BP/KNAT1 proteins also move between cell layers of the *Arabidopsis* apex, suggesting that this is a general property of KNOX proteins and is important for their function (Kim *et al.*, 2003). Further analysis of KNOX movement showed that these proteins form steep concentration gradients over several cell layers (Kim *et al.*, 2003). These transcription factors might therefore be involved in short-range signalling in specific regions of the meristem. The presence of developmentally regulated symplastic networks within the meristem lends weight to this hypothesis (Rinne & van der Schoot, 1998; Gisel *et al.*, 1999). Interestingly, movement is not limited to KNOX proteins as RNA of the tomato *KN1* orthologue, *LeT6*, was recently shown to move over much larger distances through the phloem (Kim *et al.*, 2001). However, the relevance of this movement for KNOX function has yet to be resolved.

6.6 Interactions between *KNOX* genes and hormones regulate meristem activity

Classic studies have shown that phytohormones play an important role in the meristem. For instance, a high cytokinin-to-auxin ratio is required to generate shoot meristems from callus. When levels of endogenous cytokinin are reduced, plants display a number of defects associated with an aberrant meristem, including a reduction in meristem size (Werner *et al.*, 2001). The formation of ectopic shoots on leaves with elevated cytokinins levels is consistent with this hormone-promoting meristem activity (Estruch *et al.*, 1991). Interestingly, ectopic *KNOX* gene expression was also detected in leaves with elevated cytokinins levels, suggesting that one function of cytokinins is to promote *KNOX* gene expression (Rupp *et al.*, 1999). However, other work points to a different relationship, with ectopic *KNOX* gene expression causing an accumulation of cytokinins (Tamaoki *et al.*, 1997; Ori *et al.*, 1999; Frugis *et al.*, 2001). Taken together, these results show that cytokinins and *KNOX* genes promote each other's accumulation within the meristem (see Fig. 6.3).

KNOX genes also function to repress biosynthesis of gibberellin (GA), a hormone associated with cell differentiation. Misexpression of the tobacco *KNOX* gene *NTH15* in leaves results in direct repression *Ntc12*, a key GA biosynthetic gene (Sakamoto *et al.*, 2001). Thus, *KNOX* genes may prevent GA accumulation in the meristem by repressing genes involved in GA biosynthesis. Several lines of evidence support this model. Firstly, expression of *Ntc12* and *NTH15* is mutually exclusive, with *NTC12* transcript accumulating in organs and *NTH15* in the meristem (Tamaoki *et al.*, 1997; Tanaka-Ueguchi *et al.*, 1998; Sakamoto *et al.*, 2001). And secondly, exogenous application of GA or increasing the level of GA signalling is sufficient to overcome the affects of *KNOX* gene misexpression in leaves

(Tanaka-Ueguchi *et al.*, 1998; Hay *et al.*, 2002). Conversely, increased GA signalling causes a weak *stm* mutant to phenocopy a strong *stm* mutant (Hay *et al.*, 2002). Thus the antagonistic interaction between GA and *KNOX* genes mediates the balance between indeterminate (meristem) and determinate (organ) cell fates in the shoot apex (see Fig. 6.3).

6.7 Signals involved in organ formation

Lateral organs arise at regular positions in the periphery of the meristem, a pattern termed *phyllotaxis* (from the Greek word for 'leaf arrangement'). The position of an organ on the stem is often referred to as the node and the space between adjacent nodes – the internode. Nodes may either contain a single organ, a pair of organs or a group of organs. In some cases, each adjacent organ arises directly (180°) opposite to the earlier formed organ, called *distichous phyllotaxis*. An example of this pattern is the arrangement of leaves along the maize stem. More common is *spiral phyllotaxis*, where adjacent nodes are frequently offset by $\sim$137°. For example *Arabidopsis* leaves and flowers are arranged spirally around the stem in either a left- or right-handed pattern. When two organs arise directly opposite each other at a single node, the arrangement is called *opposite phyllotaxis*. In the majority of cases nodes are spirally arranged, although occasionally they may be separated by 90°, termed *decussate* (for example see Fig. 6.1A). The term *whorled* is often used to describe more than two organs arising at a node, as seen in the flowers of many species.

6.7.1 Models of phyllotaxis

How phyllotaxis is established and maintained has puzzled generations of biologists and mathematicians (Steeves & Sussex, 1989). Early observations of phyllotaxis recognised that organs tend to initiate in the greatest space available and at a point farthest from established primordia and the central zone of the meristem – termed Hofmeister rule (Snow & Snow, 1962). Indeed, a spontaneous or induced change in the position of a single organ is sufficient to change the position of all subsequent organs and may even cause a shift in the handedness of the spiral. This shows that existing primordia somehow influence the position at which lateral organs will subsequently form. Classic surgical and pharmacological experiments suggest that organs produce some kind of chemical or physical signal that determines the position of subsequent organs (extensively reviewed in Steeves & Sussex, 1989; Lyndon, 1990). Based on these observations an inhibitory field theory was proposed, in which existing primordia and the central zone of the meristem produce a diffusible signal that inhibits the formation of new organs. It follows that new primordia will arise only at a point farthest from the source of this signal in a region of least inhibition (Schoute, 1913; Wardlaw, 1949). An alternative model proposes that the signal promotes organ formation and that it is depleted from tissue surrounding developing primordia. Accordingly, organ initiation will occur only when the levels of the activator have reached a sufficient level, at a point farthest from the existing organs (reviewed in Lyndon, 1998).

Changing the size of the meristem without altering the strength of the signal is expected to alter phyllotaxis, as the region experiencing least inhibition or greatest activation will be greatly expanded. Several *Arabidopsis* mutants display such alterations including the *clv* and *fasciata* mutants (Leyser & Furner, 1992; Clark *et al.*, 1993) as well as the maize *abphyll* (*abph1*) mutant (Greyson *et al.*, 1978; Jackson & Hake, 1999). Similarly, reducing the size of organ primordia allows more organs to form from a meristem, as seen in the whorled arrangement of the smaller needle-like organs of the *Antirrhinum phantastica* (*phan*) mutants (Waites & Hudson, 1995).

An alternative, but not necessarily exclusive, mechanism for phyllotaxis is based on physical forces (Green, 1996). According to this model, tensile and compressive forces within the meristem determine the site of organ formation. Computer modelling has shown that such a process is capable of accounting for many aspects of phyllotaxis, including the regular pattern of organ formation and changes in phyllotaxis that arise following disruptions in organ formation (Green, 1992, 1996). As this mechanism is unlikely to require the action of specific genes, it is hard to test genetically. However, experiments that alter the physical forces acting within the meristem have provided some supporting evidence for this model. The sunflower inflorescence (capitulum) is a large flat meristem that initiates primordia in multiple spirals. Each primordium develops into a flower and subtending bract. Hernandez and Green (1993) showed that lateral compression of a developing capitulum led to the formation of ridges that run parallel to the applied pressure. These ridges fail to separate into primordia, developing instead as extended (united) bracts that lack flowers. Thus, applied physical forces can change phyllotaxis in predictable ways as well as cause changes to primordia development.

6.7.2 *The role of auxin in phyllotaxis*

Much attention has recently been focused on auxin as a key regulator of phyllotaxis. It is synthesised in the developing tissue of the shoot, particularly young leaves (Davies, 1995), and actively moved throughout the plant by a polar transport system. When polar auxin transport is inhibited chemically, phyllotaxis is significantly altered and in many cases organs fail to form, leaving a naked pin-like stem (Okada *et al.*, 1991; Reinhardt *et al.*, 2000). A number of *Arabidopsis* mutants mimic these effects, including *pin-formed1* (*pin1*; Okada *et al.*, 1999), *pinoid* (*pid*; Bennett *et al.*, 1995) and *monopterous* (*mp*; Berleth & Jürgens, 1993; Przemeck *et al.*, 1996), which are all implicated in either auxin transport or signalling.

PIN1 encodes an auxin efflux carrier that is expressed in the meristem, developing primordia and vascular tissue (Gälweiler *et al.*, 1998; Vernoux *et al.*, 2000; see also Chapter 2). The lack of organs on *pin1* stems may reflect the need either to remove auxin from lateral organ initials or to accumulate auxin, depending on whether auxin inhibits or promotes primordium formation. Several lines of evidence point to auxin as a lateral organ promoter. The *pin1* mutant phenotype can be rescued by exogenous application of auxin (Reinhardt *et al.*, 2000). Similarly, local application of auxin onto wild-type tomato and *Arabidopsis* inflorescences can induce organ outgrowth, with the size of the primordium being proportional to the amount of applied auxin (Reinhardt *et al.*, 2000, 2003b). However, application of auxin to the central zone does not lead to organ outgrowth, showing that auxin is only active in the peripheral zone of the apex. These observations suggest that a patterned distribution of auxin might largely account for phyllotaxis.

Although the loss of polar auxin transport and auxin signalling in the *pin1* and *pid* mutants does not affect the expression of genes involved in meristem maintenance, genes regulating organ identity are no longer expressed in a phyllotactic pattern (Christensen *et al.*, 2000; Vernoux *et al.*, 2000; Reinhardt *et al.*, 2003b). These genes are instead expressed in concentric rings around the naked stem, suggesting that one function of auxin is to limit their expression to sites of organ formation. Based on these and other observations, a model of phyllotaxis involving two signalling processes has been proposed (Reinhardt *et al.*, 2000; Vernoux *et al.*, 2000; Reinhardt & Kuhlemeier, 2002). One, that does not involve auxin, promotes the formation of evenly spaced rings of cells that are competent to assume lateral organ identity in the periphery of the meristem. And a second auxin-dependent mechanism that partitions each ring into lateral organ and non-lateral organ initials. According to this model, partitioning is achieved by the initiating organ primordia generating a localised zone of auxin depletion which prevents neighbouring cells from forming organs, as originally proposed by Sachs (1991). A higher auxin concentration at more distant positions leads to primordium formation (and perhaps promotes increased auxin accumulation).

This model assumes that the distribution of auxin is patterned within the meristem, and that this pattern determines future sites of organ initiation. Two elegant studies have recently used the distributions of PIN1 protein to infer the likely movement and distribution of auxin within the meristem (Benková *et al.*, 2003; Reinhardt *et al.* 2003b). PIN1 protein is present in the plant membranes and accumulates in the side of the cell that is actively involved in auxin efflux (Gälweiler *et al.*, 1998). In the meristem, PIN1 accumulates in the apical side of L1 and L2 cells, suggesting that auxin moves towards the summit of the apex through the epidermal and subepidermal tissue. However, around the sites of future organ formation, PIN1 accumulation is redirected to membranes facing the incipient primordium, showing that a change in auxin movement is one of the earliest markers of organ formation. Following primordium outgrowth, PIN1 distribution indicates that auxin flows from the surrounding tissue into the centre of the primordium; then later as the organ enlarges, auxin moves through the epidermis towards the apex of the organ before being redirected inwards towards the centre. Auxin movement in maturing organs as inferred by PIN1 localisation is consistent with the process of canalisation, the selective channelling of auxin that leads to the formation of the vasculature (Sachs, 1991).

The use of PIN1 as a marker for auxin movement clearly shows that auxin is patterned within the meristem, and that auxin accumulation is closely associated with organ formation (see Fig. 6.4). However if auxin is the signal underlying phyllotaxis,

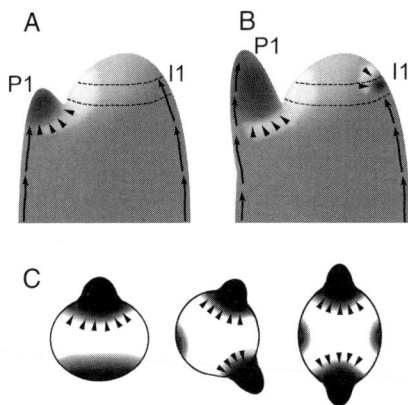

Figure 6.4 Model for the role of polar auxin transport in phyllotaxis. The shoot apical meristem with an expanding primordium (P1) is shown in (A) and (B). The region between the dashed lines depicts the peripheral organogenic zone of the meristem, and I1 marks the site of organ initiation. Arrows indicate polar auxin transport through the epidermis and sub-epidermis, and arrowheads indicate the flow of auxin from surrounding tissue into the developing organ. Shading represents the likely distribution of auxin (darker – more auxin, lighter – less auxin). (A) Following organ initiation, the primordium (P1) becomes an auxin sink, diverting both the flow of auxin away from the apex and draining auxin from the surrounding meristem tissue. As a consequence, a zone of auxin depletion forms around the expanding organ, with the lowest concentration of auxin being in region above the primordium. Auxin accumulates in the region of the meristem farthest from P1, which in this case is 180° from P1 (distichous). (B) Once the concentration of auxin passes a threshold within organogenic region, organ formation is initiated. As the organ forms it becomes a new sink for auxin, resulting in the local depletion of auxin from surrounding tissue. This pattern of auxin depletion and accumulation is self-perpetuating and can largely account for the maintenance of phyllotaxis once it has been established. Redrawn from Reinhardt *et al.* (2003b). (C) The type of phyllotaxis displayed by a plant is largely determined by factors that influence the distribution of auxin, such as relative sink strengths, and size and growth rates of the central and peripheral zones. This panel, which illustrates how different types of phyllotaxis arise, shows transverse sections through the meristems of plants displaying distichous (left), spiral (middle) and decussate (right) phyllotaxis. In distichous phyllotaxis, the youngest primordium (P1) absorbs auxin whereas the older primordium (P2, not shown) has either stopped absorbing auxin or is too far from the meristem to affect subsequent organ initiation. As a result the newly initiating organ (I1) forms opposite P1. In spiral phyllotaxis, P1 and P2 both compete for auxin, with P1 being a large sink than P2. As a consequence I1 forms between P1 and P2, but closer to P2. In decussate phyllotaxis the two opposing primordia are weaker sinks for auxin than in the previous examples, which allows organ initiation to occur at two opposing positions equidistant from the P1 organs. Redrawn from Reinhardt and Kuhlemeier (2002).

it must somehow generate a self-perpetuating pattern. The above studies suggest how this might be achieved. Initiating organs generate a zone of auxin depletion by both deflecting the flow of auxin away from the apex and draining auxin from surrounding tissue (Fig. 6.4A). As a consequence, auxin accumulates in regions farthest from developing organs (Fig. 6.4B). Organ formation is initiated only when the concentration of auxin has passed a certain threshold in the organogenic zone. This in turn is likely to promote the expression of *PIN1*, further enhancing the flow of auxin to sites of organ formation (Reinhardt *et al.*, 2003b). According to this model, different auxin distributions within the apex largely determine the plant's phyllotaxis (Fig. 6.4C). The distribution of auxin may be influenced by a number of factors including the relative auxin sink strengths within the apex, as well as the size and growth rates of the central and peripheral zones of the shoot apical meristem.

6.7.3 Organ outgrowth involves physical forces

Given the likely role of auxin in phyllotaxis, it seems unlikely that physical forces have a direct role in determining the position of organ formation. However, there is evidence that physical forces are involved in the process of organ outgrowth. Work in which the flexibility of meristem cell walls was altered using exogenously applied expansins, proteins that weaken the bonds between cell wall polysaccharides (Cosgrove, 2000), showed that an outgrowth with leaf-like qualities could be induced to form on tomato apicies (Fleming *et al.*, 1997). Similar studies in tobacco, where expansin activity was induced in all three layers of the meristem, rather than just the epidermis, resulted in an outgrowth that formed a near perfect leaf (Pien *et al.*, 2001). In both studies, formation of an organ at the site of expansin application affected phyllotaxis, in many cases reversing the spiral arrangement of organs. These studies imply that changes to cell wall flexibility, and by inference the physical forces acting on these cells, are sufficient to induce the formation of organs. However, expansin application generated outgrowths only at sites that would normally form organs during the course of development. This suggests that cell wall loosening resulted in the early outgrowth of an existing initial rather than the *de novo* formation of an organ. Interestingly, localised changes to the pattern of cell division and proliferation within the tobacco apex did not lead to organ formation (Wyrzykowska *et al.*, 2002; Wyrzykowska & Fleming, 2003), suggesting that cell wall loosening is the first step in organ outgrowth.

The process of organ formation and outgrowth may therefore involve a combination of signalling and physical forces. According to this model, accumulation of auxin determines the site of organ formation, triggering the localised expression of expansin and cell wall loosening. It should also be noted that auxin itself can induce cell expansion. The pressure from underlying layers subsequently triggers cell division and organ outgrowth. Consistent with this model is the observation that certain isoforms of expansin are expressed specifically at sites of organ formation prior to any sign of organ development (Fleming *et al.*, 1997; Reinhardt *et al.*, 1998). Localised expansin expression also occurs at sites where auxin is applied exogenously to the meristem (Reinhardt & Kuhlemeier, 2002).

6.8 Signalling between organ primordia and the meristem

As organs emerge from the shoot apical meristem, they typically flatten laterally before developing unique cell types along the adaxial (facing the meristem) – abaxial (away from the meristem) axis. A connection between the ad-abaxial patterning of a lateral organ and the shoot apical meristem was first established by surgical experimentation (Sussex, 1955; Snow & Snow, 1959; Hanawa, 1961). When an incision or impermeable barrier was placed between a young organ primordium and the meristem, formation of adaxial cells was impaired and the resulting organ had a radial symmetry (i.e. needle-like). These findings gave the first tantalising evidence that a meristem-derived signal promotes adaxial cell identity in developing organs. In addition, it showed that abaxial cells form in the absence of adaxial cells and that the presence of adaxial cell types was required for lateral growth, as in their absence organs failed to form a lamina. Similar surgical treatment to slightly older primordia had no affect on adaxial cell identity and leaf morphology, suggesting that there is only a transitory requirement for the meristem signal during the early stages of organ development.

Supporting evidence for a meristem-derived adaxialising morphogen comes from the characterisation of gain-of-function mutations in two closely related *Arabidopsis* genes, *PHABULOSA* (*PHB*) and *PHAVOLUTA* (*PHV*). These mutations cause a dose-dependent adaxialisation of organs, with severely affected organs having a needle-like morphology when abaxial cell types fail to form (McConnell & Barton, 1998). The loss of lateral growth when either adaxial or abaxial cells are missing from a developing organ implies that signalling between these cell types may be required to initiate lateral growth at the point where they meet, as first proposed by Waites and Hudon (1995). *PHB* and *PHV* belong to a family of likely transcription factors with homeodomain, leucine zipper and START domain motifs (*HD-ZIP*). The gain-of-function mutations affecting *PHB* and *PHV* arise from nucleotide changes in a small region of the START domain, a motif that has been implicated in the binding of sterol lipid ligands in a number of diverse proteins (McConnell *et al.*, 2001). This raises the possibility that PHB and PHV proteins are activated by a meristem-derived sterol, which is presumably more abundant in the adaxial domain (side closest to the meristem) than the abaxial domain of the incipient primordium. It has been proposed that the proteins encoded by the dominant *phb* and *phv* mutant alleles are constitutively active and are no longer responsive to the meristem-derived signal. The accumulation of *phb* and *phv* transcripts in the abaxial domain of mutant organs is consistent with activated PHB and PHV proteins promoting their own RNA as well as adaxial cell identity (McConnell *et al.*, 2001).

Normally *PHB* and *PHV* expression is restricted to the adaxial domain, although within this domain there is a gradient of expression, with the strongest expression being closest to the meristem. The gradient of *PHB* and *PHV* expression, and by inference activation, may therefore reflect the gradient of activating signal from the meristem. Similarly, accumulation of *PHB* and *PHV* transcripts in narrow strips that extend from the initiating organ into the centre of the meristem might also reflect the likely path of the signal. Based on these observations it is tempting to speculate that the signal promoting PHB and PHV activation is the same as that identified in the surgical experiments. It is produced in the centre of the meristem and flows towards the periphery forming a concentration gradient, with adaxial fate arising from a high concentration of the signal, and therefore closest to the meristem.

An alternative model for ad-abaxial patterning of organs has recently been proposed following the discovery of two short (~22 bp) microRNA (miRNA165, 166) complementary to RNA from the *PHB* and *PHV* loci and a third member of this gene family, *REVOLUTA* (*REV*; Reinhart et al., 2002; Rhoades et al., 2002). This model proposes that the miRNAs accumulate in the abaxial domain of the incipient primordium where they promote the degradation of *HD-ZIP* RNA. Absence of the miRNA from the adaxial domain allows PHB and PHV to promote adaxial cell identity and REV to promote vascular development. This model does not exclude the possibility that HD-ZIPs are activated by a meristem-derived adaxialising signal. Experimental support for this model comes from the observation that *PHV* RNA is targeted for miRNA-mediated degradation, and that RNA produced by the *PHV* gain-of-function mutant allele is resistant to this degradation because of an imperfect match with the miRNA (Tang et al., 2003). The lack of degradation may explain why *PHV*, and by inference *PHB*, persists in the abaxial domain of mutant organs, where it could promote ectopic adaxial fate. Recent work has also shown that *REV* is the likely target of miRNA degradation (Emery et al., 2003).

Also consistent with this model is the recent finding that the *Arabidopsis* and maize miRNA166 accumulates in the abaxial domain of developing organ primordia (Juarez et al., 2004; Kidner & Martienssen, 2004). This implies that the organ is already patterned or that the miRNA itself acts as the polarising signal. There is some support for the latter scenario, as the maize miRNA166 is first expressed below the initiating organ primordium and subsequently accumulates, possibly by movement through the phloem, in the abaxial domain of the growing organ (Juarez et al., 2004).

Two gene families promote abaxial cell identities in developing organs, *YABBY*s and *KANADI*s (reviewed by Bowman et al., 2002). The *KANADI* family were first identified as mutants that enhance the polarity defects of the *crabs claw* (*CRC*, the founding member of the *YABBY* family) mutant and were later shown to encode likely transcription factors with a GARP domain (Eshed et al., 1999, 2001; Kerstetter et al., 2001). Mutant and ectopic expression studies suggest that *KAN*s and *HD-ZIP*s function as mutual repressors in developing organs. For instance, *kan1 kan2* mutants have adaxialised organs with a broader domain of *PHB* expression, whereas ubiquitous *KAN* expression in developing seedlings leads to abaxialisation of cotyledons and a loss of the shoot apical meristem, phenotypes similar to plants triply mutant for *phb phv rev* loss-of-function mutation (Eshed et al., 2001; Kerstetter et al., 2001; Emery et al., 2003). It is not clear whether *KAN*s repress HD-ZIPs directly, or indirectly by promoting miRNA expression in the abaxial domain (Fig. 6.5).

As lateral growth of organs requires a juxtaposition of adaxial and abaxial cell types, it is likely that the *HD-ZIP*s and *KAN*s regulate a signalling pathway that directly controls cell proliferation at the junction between the different cell types (see Fig. 6.5). Signalling from the adaxial domain of a developing organ is also required to promote meristem formation and maintenance. The *Antirrhinum phantastica* mutant initially has leaves with both adaxial and abaxial cell types, but as successive leaves arise from the meristem there is an increasing loss of

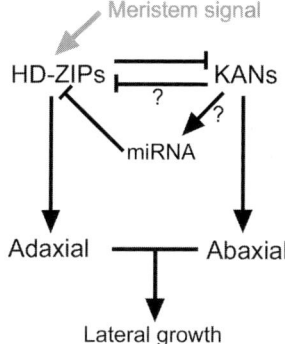

Figure 6.5 Factors regulating the patterning of lateral organs. This figure shows genetic interactions that are likely to pattern the emerging organ primordium. The Class III *HD-ZIP* genes – *PHABULOSA*, *PHAVOLUTA* and *REVOLUTA* – are expressed in the adaxial domain of a developing organ and in the meristem. The adaxial expression pattern in organs is established in part by the miRNA-mediated degradation of *HD-ZIP* RNA in the abaxial domain. HD-ZIPs, which also have a sterol-binding domain (START domain), promote adaxial cell identity perhaps following activation by a meristem-derived sterol (indicated with an arrow). Loss of *HD-ZIP* activity results in an abaxialisation of organs, a phenotype similar to that seen when the abaxially expressed *KANADI* (*KAN*) genes are expressed throughout a developing organ. Thus one function of *HD-ZIPs* may be to limit *KAN* expression to the abaxial domain. *KAN*s promote abaxial identity as in their absence organs become adaxialised. As *HD-ZIP* expression is detected in the abaxial domain of *kan* mutant organs, it is likely that *KAN*s restrict *HD-ZIPs* to the adaxial domain. They may do this directly or perhaps by promoting miRNA expression. The juxtaposition of cells with adaxial and abaxial identity triggers lateral growth, a process that presumably requires signalling between cells. Redrawn from Emery *et al.* (2003).

adaxial cell types, ultimately leading to the formation of completely abaxialised needle-like organs (Waites & Hudson, 1995). Sensitivity of the *phan* mutant phenotype towards cold suggested that *PHAN*, which is the orthologue of *AS1*, promotes adaxial cell identity redundantly with cold-sensitive factors (Waites & Hudson, 1995; Waites *et al.*, 1998). At the most severe, cold-grown *phan* mutants fail to produce any organs and the meristem becomes inactive. A link between adaxial cell identity and meristem activity is also apparent in *Arabidopsis*, as lines ectopically expressing *KAN*s in the adaxial domain of the cotyledons fail to develop a shoot apical meristem (Eshed *et al.*, 2001).

The characterisation of mutants with defects in adaxial–abaxial patterning has revealed extensive signalling between the developing organ and meristem. The nature of these signals is presently not well understood, but clearly the ability of the meristem to impart adaxial identity on organs argues for the existence of a morphogen.

6.9 Conclusion

This chapter has described how diverse signalling pathways, which operate over distances of several cells to many hundreds of cells, are integrated into a functioning shoot apical meristem. Analysis of the *Arabidopsis* genome has shown that there are a staggering number of RLKs with no known function, suggesting that receptor–ligand interactions are likely to be the most common form of signalling in plants. It would therefore not be surprising to find many more CLV-like pathways operating in the meristem. However as can be seen from this and other chapters, not all signalling in plants involves the perception of ligands. A number of transcriptions factors, such as the *KNOX* genes, move through the meristem and may form concentration gradients that activate different sets of target genes. This type of signalling may rely on the developmental regulation of symplastic fields within the meristem, an area that is still poorly understood. The formation of organs also requires the movement and perception of a signalling molecule, which in this case is auxin. How cells perceive and respond to increased levels of auxin remains to be determined. Similarly, understanding how auxin-regulated pathways interact with other signalling pathways in the meristem will be particularly interesting. The addition of microRNAs to the list of potential signalling molecules is intriguing, but not necessarily unexpected. RNA movement in plants is well known and has been extensively characterised. However it remains to be seen whether localised movement of RNA or microRNAs within the meristem, or between the meristem and emerging organs, has a role in regulating meristem functions. Thus studies of the shoot apical meristem have provided unique insights into various types of signalling systems that operate within plants, and are likely to do so well into the future.

Acknowledgements

The author thanks Gwyneth Ingram, David Smyth and members of the Smyth laboratory for thoughtful discussions and comments on this manuscript.

References

Aida, M., Ishida, T., Fukaki, H., Fujisawa, H. & Tasaka, M. (1997) Genes involved in organ separation in *Arabidopsis*: an analysis of the *cup-shaped cotyledon* mutant. *Plant Cell*, **9**, 841–857.

Aida, M., Ishida, T. & Tasaka, M. (1999) Shoot apical meristem and cotyledon formation during *Arabidopsis* embryogenesis: interaction among the *CUP-SHAPED COTYLEDON* and *SHOOT MERISTEMLESS* genes. *Development*, **119**, 823–831.

Asai, T., Tena, G., Plotnikova, J., Willmann, M.R., Chiu, W.-L., Gomez-Gomez, L., Boller, T., Ausubel, F.M. & Sheen, J. (2002) MAP kinase signalling cascade in *Arabidopsis* innate immunity. *Nature*, **415**, 977–983.

Autran, D., Jonak, C., Belcram, K., Beemster, G.T.S., Kronenberger, J., Grandjean, O., Inze, D. & Traas, J. (2002) Cell numbers and leaf development in *Arabidopsis*: a functional analysis of the *STRUWWELPETER* gene. *EMBO J.* **21**, 6036–6049.

Banno, H., Ikeda, Y., Niu, Q.W. & Chua, N.H. (2001) Overexpression of *Arabidopsis* ESR1 induces initiation of shoot regeneration. *Plant Cell*, **13**, 2609–2618.

Barton, M.K. & Poethig, R.S. (1993) Formation of the shoot apical meristem in *Arabidopsis thaliana*: an analysis of development in the wildtype and in the *shoot meristemless* mutant. *Development*, **119**, 823–831.

Benková, E., Michniewicz, M., Sauer, M., Teichmann, T., Seifertová, D., Jürgens, G. & Friml, J. (2003) Local, efflux-dependent auxin gradients as a common module for plant organ formation. *Cell*, **115**, 591–602.

Bennett, S.R.M., Alvarez, J., Bossinger, G. & Smyth, D.R. (1995) Morphogenesis in *pinoid* mutants of *Arabidopsis thaliana*. *Plant J.*, **8**, 505–520.

Berleth, T. & Jürgens, G. (1993) The role of the *monopterous* gene in organising the basal body region of *Arabidopsis* embryo. *Development*, **118**, 575–587.

Bowman, J.L., Eshed, Y. & Baum, S.F. (2002) Establishment of polarity in angiosperm lateral organs. *Trend Genet.*, **18**, 134–141.

Brand, U., Fletcher, J.C., Hobe, M., Meyerowitz, E.M. & Simon, R. (2000) Dependence of stem cell fate in *Arabidopsis* on a feedback loop regulated by *CLV3* activity. *Science*, **289**, 617–619.

Brand, U., Grunewald, M., Hobe, M. & Simon, R. (2002) Regulation of *CLV3* by two homeobox genes in *Arabidopsis*. *Plant Physiol.*, **129**, 565–575.

Byrne, M.E., Barley, R., Curtis, M., Arroyo, J.M., Dunham, M., Hudson, A. & Martienssen, R.A. (2000) *Asymmetric leaves1* mediates leaf patterning and stem cell function in *Arabidopsis*. *Nature*, **408**, 967–971.

Byrne, M.E., Simorowski, J. & Martienssen, R.A. (2002) *ASYMMETRIC LEAVES1* reveals *knox* gene redundancy in *Arabidopsis*. *Development*, **129**, 1957–1965.

Byrne, M.E., Groover, A.T., Fontana, J.R. & Martienssen, R.A. (2003) Phyllotatctic pattern and stem cell fate determined by the *Arabidopsis* homeobox gene *BELLRINGER*. *Development*, **130**, 3941–3950.

Callos, J.D., DiRado, M., Xu, B., Behringer, F.J., Link, B.M. & Medford, J.I. (1994) The forever young gene encodes an oxidoreductase required for proper development of the *Arabidopsis* vegetative shoot apex. *Plant J.*, **6**, 835–847.

Carlos, C.C. & Fletcher, J.C. (2003) Shoot apical meristem maintenance: the art of a dynamic balance. *Trends Plant Sci.*, **8**, 394–401.

Clark, S.E., Jacobsen, S.E., Levin, J.Z. & Meyerowitz, E.M. (1996) The *CLAVATA* and *SHOOT MERISTEMLESS* loci competitively regulate meristem activity in *Arabidopsis*. *Development*, **122**, 1567–1575.

Clark, S.E., Running, M.P. & Meyerowitz, E.M. (1993) *CLAVATA1*, a regulator of meristem and flower development in *Arabidopsis*. *Development*, **119**, 397–418.

Clark, S.E., Running, M.P. & Meyerowitz, E.M. (1995) *CLAVATA3* is a specific regulator of shoot and floral meristem development affecting the same processes as CLAVATA1. *Development*, **121**, 2057–2067.

Clark, S.E., Williams, R.W. & Meyerowitz, E.M. (1997) The *CLAVATA1* gene encodes a putative receptor kinase that controls shoot and floral meristem size in *Arabidopsis*. *Cell*, **89**, 575–585.

Cock, J.M. & McCormick, S. (2001) A large family of genes that share homology with *CLAVATA3*. *Plant Physiol.*, **126**, 939–942.

Cosgrove, D. (2000) Loosening of plant cell walls by expansins. *Nature*, **407**, 321–326.

Christensen, S.K., Dagenais, N., Chory, J. & Weigel, D. (2000) Regulation of auxin response by the protein kinase PINOID. *Cell*, **100**, 469–478.

Cutler, S., Ghassemian, M., Bonetta, D., Cooney, S. & McCourt, P. (1996) A protein farnesyl transferase involved in abscisic acid signal transduction in *Arabidopsis*. *Science*, **273**, 1239–1241.

Davies, P.J. (1995) *Plant Hormones – Physiology, Biochemistry and Molecular Biology*. Kluwer Academic Publishers, Dordrecht, The Netherlands.

Dermen, H. (1953) Periclinal cytochimaeras and the origin of tissues in stem and leaf of peach. *Am. J. Bot.*, **40**, 154–168.

Dievart, A., Dala, M., Tax, F.E., Lacey, A.D., Huttly, A., Li, J. & Clark, S.E. (2003) CLAVATA1 dominant-negative alleles reveal functional overlap between multiple receptor kinases that regulate meristem and organ development. *Plant Cell*, **15**, 1198–1211.

Douglas, S.J., Chuck, G., Dengler, R.E., Pelecanda, L. & Riggs, C.D. (2002) *KNAT1* and *ERECTA* regulate inflorescence architecture in *Arabidopsis*. *Plant Cell*, **14**, 547–558.

Emery, J.F., Floyd, S.K., Alvarez, J., Eshed, Y., Hawker, N.P., Izhaki, A., Baum, S.F. & Bowman, J.L. (2003) Radial patterning of *Arabidopsis* shoots by class III *HD-ZIP* and *KANADI* genes. *Curr. Biol.*, **13**, 1768–1774.

Endrizzi, K., Moussian, B., Haecker, A., Levin, J.Z. & Laux, T. (1996) The *SHOOT MERISTEMLESS* gene is required for maintenance of undifferentiated cells in *Arabidopsis* shoot and floral meristems and acts at a different regulatory level than the meristem genes *WUSCHEL* and *ZWILLE*. *Plant J.*, **10**, 967–979.

Eshed, Y., Baum, S.F. & Bowman, J.L. (1999) Distinct mechanisms promote polarity establishment in carpels of *Arabidopsis*. *Cell*, **99**, 199–209.

Eshed, Y., Baum, S.F., Perea, J.V. & Bowman, J.L. (2001) Establishment of polarity in lateral organs of plants. *Curr. Biol.*, **11**, 1251–1260.

Estruch, J.J., Prinsen, E., Van Onckelen, H., Schell, J. & Spena, A. (1991) Viviparous leaves produced by somatic activation of an inactive cytokinin-synthesizing gene. *Science*, **254**, 1364–1367.

Fleming, A.J., McQueen-Mason, S., Madel, T. & Kuhlemeier, C. (1997) Induction of leaf primordia by the cell wall protein expansin. *Science*, **276**, 1415–1418.

Fletcher, J.C. (2001) The *ULTRAPETALA* gene controls shoot and floral meristem size in *Arabidopsis*. *Development*, **128**, 1323–1333.

Fletcher, J.C., Brand, U., Running, M.P., Simon, R. & Meyerowitz, E.M. (1999) Signalling of cell fate decisions by *CLAVATA3* in *Arabidopsis* shoot meristems. *Science*, **283**, 1911–1914.

Frugis, G., Giannino, D., Mele, G., Nicolodi, C., Chiappetta, A., Bitonti, M.B., Innocenti, A.M., Dewitte, W., Van Onckelen, H. & Mariotti, D. (2001) Overexpression of *KNAT1* in lettuce shifts leaf determinate growth to a shoot-like indeterminate growth associated with accumulation of isopentenyl-type cytokinins. *Plant Physiol.*, **126**, 1370–1380.

Furner, I.J. & Pumfrey, J.E. (1992) Cell fate in the shoot apical meristem of *Arabidopsis thaliana*. *Development*, **115**, 755–764.

Gallois, J-L., Woodward, C., Reddy, V. & Sablowski, R. (2002) Combined SHOOT MERISTEMLESS and WUSCHEL trigger ectopic organogenesis in *Arabidopsis*. *Development*, **129**, 3207–3217.

Gälweiler, L., Guan, C., Müller, A., Wismen, E., Mendgen, K., Yephremov, A. & Palme, K. (1998) Regulation of polar auxin transport by AtPIN1 in *Arabidopsis* vascular tissue. *Science*, **282**, 2226–2230.

Gisel, A., Barella, S., Hempel, F.D. & Zambryski, P.C. (1999) Temporal and spatial regulation of symplastic trafficking during development in *Arabidopsis thaliana* apices. *Development*, **126**, 1879–1889.

Green, P.B. (1992) Pattern formation in shoots: a likely role for minimal energy configurations of the tunica. *Int. J. Plant Sci.*, **153** (Suppl.), 59–75.

Green, P.B. (1996) Expression of form and pattern in plants – a role for biophysical fields. *Semin. Cell Dev. Biol.*, **7**, 903–911.

Greyson, R.I., Walden, D.B., Humes, J.A. & Erickson, R.O. (1978) The ABPHYL syndrome in *Zea mays*, II: Patterns of leaf initiation and the shape of the shoot meristem. *Can. J. Bot.*, **56**, 1545–1550.

Haecker, A., Grob-Hardt, R., Geiges, B., Sarkar, A., Breuninger, H., Herrmann, M. & Laux, T. (2004) Expression dynamics of *WOX* genes mark cell fate decisions during early embryonic patterning in *Arabidopsis thaliana*. *Development*, **131**, 657–668.

Hanawa, J. (1961) Experimental studies of leaf dorsiventrality in *Sesamum indicum* L. *Bot. Mag. Tokyo*, **74**, 303–309.

Hay, A., Kaur, H., Phillips, A., Hedden, P., Hake, S. & Tsiantis, M. (2002) The gibberellin pathway mediates KNOTTED1-type homeobox function in plants with different body plans. *Curr. Biol.*, **12**, 1557–1565.

Haywood, V., Kragler, F. & Lucas, W. (2002) Plasmodesmata: pathways for protein and riboprotein signaling. *Plant Cell*, **14** (Suppl.), 303–325.

Hernandez, L.F. & Green, P.B. (1993) Tranductions for the expression of structural pattern: analysis in sunflower. *Plant Cell*, **5**, 1725–1738.

Hobe, M., Muller, R., Grunewald, M., Brand, U. & Simon, R. (2003) Loss of CLE40, a protein functionally equivalent to the stem cell restricting signal CLV3, enhances root waving in *Arabidopsis*. *Dev. Genes Evol.*, **213**, 371–381.

Irish, V.F. & Sussex, I.M. (1992) A fate map of the *Arabidopsis* embryonic shoot apical meristem. *Development*, **115**, 745–753.

Ishiguro, S., Watanabe, Y., Ito, N., Nonaka, H., Takeda, N., Sakai, T., Kanaya, H. & Okada, K. (2002) SHEPHERD is the *Arabidopsis* GRP94 responsible for the formation of functional CLAVATA proteins. *EMBO J.*, **21**, 898–908.

Iwakawa, H., Ueno, Y., Semiarti, E., Onouchi, H., Kojima, S., Tsukaya, H., Hasebe, M., Soma, T., Ibezaki, M., Machida, C. & Machida, Y. (2002) The *ASYMMETRIC LEAVES2* gene of *Arabidopsis thaliana*, required for formation of a symmetric flat leaf lamina, encodes a member of a novel family of proteins characterized by cysteine repeats and a leucine zipper. *Plant Cell Physiol.*, **43**, 467–478.

Jackson, D. & Hake, S. (1999) Control of phyllotaxy in maize by the *abphyl1* gene. *Development*, **126**, 315–323.

Jackson, D., Veit, B. & Hake, S (1994) Expression of maize *KNOTTED1* related homeobox genes in the shoot apical meristem predicts patterns of mophogenesis in the vegetative shoot. *Development*, **120**, 405–413.

Jeong, S., Trotochaud, A.E. & Clark, S.E. (1999) The *Arabidopsis CLAVATA2* gene encodes a receptor-like protein required for the stability of the CLAVATA1 receptor-like kinase. *Plant Cell*, **11**, 1925–1933.

Jorgenson, C.A. & Crane, M.B. (1927) Formation and morphology of *Solanum* chimaeras. *J. Genet.*, **18**, 247–273.

Juarez, M.T., Kui, J.S., Thomas, J., Heller, B.A. & Timmermans, M.C.P. (2004) microRNA-mediated repression of rolled leaf1 specifies maize leaf polarity. *Nature*, **428**, 84–88.

Kaya, H., Shibahara, K.-I., Taoka, K.-I., Iwabuchi, M., Stillman, B. & Araki, T. (2001) *FASCIATA* gene for chromatin assembly factor-1 in *Arabidopsis* maintain the cellular organisation of apical meristems. *Cell*, **104**, 131–142.

Kayes, J.M. & Clark, S.E. (1998) *CLAVATA2*, a regulator of meristem and organ development in *Arabidopsis*. *Development*, **125**, 3843–3851.

Kerstetter, R.A., Bollman, K., Taylor, R.A., Bomblies, K. & Poethig, R.S. (2001) *KANADI* regulates organ polarity in *Arabidopsis*. *Nature*, **411**, 706–709.

Kerstetter, R.A., Laudencia-Chingcuanco, D., Smith, L.G. & Hake, S. (1997) Loss-of-function mutations in the maize homeobox gene, *knotted1*, are defective in shoot meristem maintenance. *Development*, **124**, 3054–3054.

Kidner, C.A. & Martienssen, R.A. (2004) Spatially restricted microRNA directs leaf polarity through ARGONAUTE1. *Nature*, **428**, 81–84.

Kirch, T., Simon, R., Grunewald, M. & Werr, W. (2003) The *DORNROSCHEN/ENHANCER OF SHOOT REGENERATION1* gene of *Arabidopsis* acts in the control of meristem cell fate and lateral organ development. *Plant Cell*, **15**, 694–705.

Kim, J.Y., Yuan, Z., Cilia, M., Khalfan-Jagani, Z. & Jackson, D. (2002) Intercellular trafficking of a KNOTTED1 green fluorescent protein fusion in the leaf and shoot meristem of *Arabidopsis*. *Proc. Natl. Acad. Sci. U.S.A.*, **99**, 4103–4108.

Kim, J.Y., Yuan, Z. & Jackson, D. (2003) Developmental regulation and significance of KNOX protein trafficking in *Arabidopsis*. *Development*, **130**, 4351–4362.

Kim, M., Canio, W., Kessler, S. & Sinha, N. (2001) Developmental changes due to long-distance movement of a homeobox fusion transcript in tomato. *Science*, **293**, 287–289.

Laufs, P., Dockx, J., Kronenberger, J. & Traas, J. (1998a) *MGOUN1* and *MGOUN2*: two genes required for primordium initiation at the shoot apical and floral meristem in *Arabidopsis thaliana*. *Development*, **125**, 1253–1260.

Laufs, P., Grandjean, O., Jonak, C., Kieu, K. & Traas, J. (1998b) Cellular parameters of the shoot apical meristem in *Arabidopsis*. *Plant Cell*, **10**, 1375–1389.

Laux, T., Mayer, K.F.X., Berger, J. & Jürgens, G. (1996) The *WUSCHEL* gene is required for shoot and floral meristem integrity in *Arabidopsis*. *Development*, **122**, 87–96.

Lenhard, M. & Laux, T. (2003) Stem cell homeostasis in the *Arabidopsis* shoot meristem is regulated by intercellular movement of CLAVATA3 and its sequestration by CLAVATA1. *Development*, **130**, 3163–3173.

Lenhard, M., Jurgen, G. & Laux, T. (2002) The *WUSCHEL* and *SHOOTMERISTEMLESS* genes fulfil complementary roles in *Arabidopsis* shoot meristem regulation. *Development*, **129**, 3195–3206.

Leyser, H.M.O. & Furner, I.J. (1992) Characterisation of three shoot apical meristem mutants of *Arabidopsis thaliana*. *Development*, **116**, 397–403.

Lincoln, C., Long, J., Yamaguchi, J., Serikawa, K. & Hake, S. (1994) A *knotted1*-like homeobox gene in *Arabidopsis* is expressed in the vegetative meristem and dramatically alters leaf morphology when overexpressed in transgenic plants. *Plant Cell*, **6**, 1859–1876.

Long, J.A. & Barton, M.K. (1998) The development of apical embryonic pattern in *Arabidopsis*. *Development*, **126**, 3027–3035.

Long, J.A., Moan, E.I., Medford, J.I. & Barton, M.K. (1996) A member of the KNOTTED class of homeodomain proteins encoded by the *SHOOT MERISTEMLESS* gene of *Arabidopsis*. *Nature*, **379**, 66–69.

Lyndon, R.F. (1990) *Plant Development – The Cellular Basis*. Unwin Hyman, London.

Lyndon, R.F. (1998) *The Shoot Apical Meristem: Its Growth and Development*. Cambridge University Press, Cambridge.

Lynn, K., Fernandez, A., Aida, M., Sedbrook, J., Tasaka, M., Masson, P. & Barton, M.K. (1999) The *PINHEAD/ZWILLE* gene acts pleiotropically in *Arabidopsis* development and has overlapping functions with the *ARGONAUTE1* gene. *Development*, **126**, 469–481.

Mayer, K.F.X, Schoof, H., Haecker, A., Lenhard, M., Jurgens, G. & Laux, T. (1998) Role of *WUSCHEL* in regulating stem cell fate in the *Arabidopsis* shoot apical meristem. *Cell*, **95**, 805–815.

McConnell, J.R. & Barton, M.K. (1995) Effect of mutations in the *PINHEAD* gene of *Arabidopsis* on the formation of shoot apical meristems. *Dev. Genet.*, **16**, 358–366.

McConnell, J.R. & Barton, M.K. (1998) Leaf polarity and meristem formation in *Arabidopsis*. *Development*, **125**, 2935–2942.

McConnell, J.R., Emery, J., Eshed, Y., Bao, N., Bowman, J. & Barton, M.K. (2001) Role of *PHABULOSA* and *PHAVOLUTA* in determining radial patterning in shoots. *Nature*, **411**, 709–713.

Moussian, B., Schoof, H., Haecker, A., Jurgens, G. & Laux, T. (1998) Role of the *ZWILLE* gene in the regulation of central shoot meristem cell fate during *Arabidopsis* embryogenesis. *EMBO J.*, **17**, 1799–1809.

Okada, K., Ueda, J., Komaki, M.K., Bell, C.J. & Shimura, Y. (1991) Requirement of the auxin polar transport system in early stages of *Arabidopsis* floral bud formation. *Plant Cell*, **3**, 677–684.

Ori, N., Eshed, Y., Chuck, G., Bowman, J.L. & Hake, S. (2000) Mechanisms that control *knox* gene expression in the *Arabidopsis* shoot. *Development*, **127**, 5523–5532.

Ori, N., Juarez, M.T., Jackson, D., Yamaguchi, J., Banowetz, G.M. & Hake, S. (1999) Leaf senescence is delayed in tobacco plants expressing the maize homeobox gene *knotted1* under the control of a senescence-activated promoter. *Plant Cell*, **11**, 1073–1080.

Pien, S., Wyrzykowska, J., McQueen-Mason, S., Smart, C. & Fleming, A. (2001) Local expression of expansin induces the entire process of leaf development and modifies leaf shape. *Proc. Natl. Acad. Sci. U.S.A.*, **98**, 11812–11817.

Przemeck, G.K.H., Mattsson, J., Hardtke, C.S., Sung, Z.R. & Berleth, T. (1996) Studies on the role of the *Arabidopsis* gene *MONOPTEROS* in vascular development and plant cell axialization. *Planta*, **200**, 229–237.

Reinhardt, D. & Kuhlemeier, C. (2002) Phyllotaxis in higher plants. In: *Meristematic Tissue in Plant Growth and Development*, 1st edn (eds M.T. McManus & B.E. Veit), pp. 172–212. Sheffield Academic Press, Sheffield.

Reinhardt, D., Frenz, M., Mandel, T. & Kuhlemeier, C. (2003a) Microsurgical and laser ablation analysis of interactions between the zones and layers of the tomato shoot apical. *Development*, **130**, 4073–4083.

Reinhardt, D., Mandel, T. & Kuhlemeier, C. (2000) Auxin regulates the initiation and radial position of plant lateral organs. *Plant Cell*, **12**, 507–518.

Reinhardt, D., Pesce, E.-R., Stieger, P., Mandal, T., Baltensperger, K., Bennett, M., Traas, J., Friml, J. & Kuhlemeier, C. (2003b) Regulation of phyllotaxis by polar auxin transport. *Nature*, **426**, 255–260.

Reinhardt, D., Wittwer, F., Mandel, T. & Kuhlemeier, C. (1998) Localized upregulation of a new expansin gene predicts the site of leaf formation in the tomato meristem. *Plant Cell*, **10**, 1427–1437.

Reinhart, B.J., Weinstein, E.G., Rhoades, M.W., Bartel, B. & Bartel, D.P. (2002) MicroRNAs in plants. *Genes Dev.*, **16**, 1616–1626.

Reiser, L., Sánchez-Baracaldo, P. & Hake, S. (2000) Knots in the family tree: evolutionary relationships and functions of *knox* homeobox genes. *Plant Mol. Biol.*, **42**, 151–166.

Rhoades, M.W., Reinhart, B.J., Lim, L.P., Burge, C.B., Bartel, B. & Bartel, D.P. (2002) Prediction of plant microRNA targets. *Cell*, **110**, 513–520.

Rinne, P.L. & van der Schoot, C. (1998) Symplasmic fields in the tunica of the shoot apical meristem coordinate morphogenetic events. *Development*, **125**, 1477–1485.

Rojo, E., Sharma, V.K., Kovaleva, V., Raikhel, N.V. & Fletcher, J.C. (2002) CLV3 is localized to the extracellular space, where it activates the *Arabidopsis* CLAVATA stem cell signalling pathway. *Plant Cell*, **14**, 969–977.

Running, M.P., Fletcher, J.C. & Meyerowitz, E.M. (1998) The *WIGGUM* gene is required for proper regulation of floral size in *Arabidopsis*. *Development*, **125**, 2545–2553.

Rupp, H.M., Frank, M., Werner, T., Strnad, M. & Schmulling, T. (1999) Increased steady state mRNA levels of the STM and KNAT1 homeobox genes in cytokinins overproducing *Arabidopsis thaliana* indicate a role for cytokinins in the shoot apical meristem. *Plant J.*, **18**, 557–563.

Sachs, T. (1991) *Pattern Formation in Plant Tissues*. Cambridge University Press, Cambridge.

Sakamoto, T., Kamiya, N., Ueguchi-Tanaka, M., Iwahori, S. & Matsuoka, M. (2001) KNOX homeodomain protein directly suppresses the expression of a gibberellin biosynthetic gene in the tobacco shoot apical meristem. *Genes Dev.*, **15**, 581–590.

Satina, S., Blakeslee, A.F. & Avery, A.G. (1940) Demonstration of the three germ layers in the shoot apex of *Datura* by means of induced polyploidy in periclinal chimaeras. *Am. J. Bot.*, **27**, 895–905.

Schoof, H., Lenhard, M., Haecker, A., Mayer, K.F.X., Jurgen, G. & Laux, T. (2000) The stem cell population of *Arabidopsis* shoot meristems is maintained by a regulatory loop between *CLAVATA* and *WUSCHEL* genes. *Cell*, **100**, 635–644.

Schoute J.C. (1913) Beiträge zur Blattstellungslehre. *Ré Trav. Bot. Néerl.*, **10**, 153–235.

Semiarti, E., Ueno, Y., Tsukaya, H., Iwakawa, H., Machida, C. & Machida, Y. (2001) The *ASYMMETRIC LEAVES2* gene of *Arabidopsis thaliana* regulates formation of a symmetric lamina, establishment of venation and repression of meristem-related homeobox genes in leaves. *Development*, **128**, 1771–1783.

Shiu, S.H. & Bleecker, A.B. (2003) Expansion of the receptor-like kinase/Pelle gene family and receptor-like family in *Arabidopsis*. *Plant Physiol.*, **132**, 530–543.

Smith, H.M. & Hake, S. (2003) The interaction of two homeobox genes, *BREVIPEDICELLUS* and *PENNYWISE*, regulates internode patterning in the *Arabidopsis* inflorescence. *Plant Cell*, **15**, 1717–1727.

Snow, M. & Snow, R. (1959) The dorsiventrality of leaf primordia. *New Phytol.*, **58**, 188–207.

Snow, M. & Snow, R. (1962) A theory of the regulation of phyllotaxis based on *Lupinus albus*. *Philos. Trans. R. Soc. Lond. B Biol. Sci.*, **244**, 483–513.

Souer, E., van Houwelingen, A., Kloos, D., Mol, J. & Koes, R. (1996) The *No Apical Meristem* gene of Petunia is required for pattern formation in embryos and flowers and is expressed at meristem and primordia boundaries. *Cell*, **85**, 159–170.

Steeves, T.A. & Sussex, I.M. (1989) *Patterns in Plant Development*. Cambridge University Press, Cambridge.

Stewart, R.N. & Burk, L.G. (1970) Independence of tissues derived from apical layers in ontogeny of the tobacco leaf and ovary. *Am. J. Bot.*, **57**, 1010–1016.

Stewart, R.N. & Dermen, H. (1970) Determination of number and mitotic activity of shoot apical initial cells by analysis of mericlinal chimaeras. *Am. J. Bot.*, **57**, 816–826.

Stewart, R.N. & Dermen, H. (1975) Flexibility in ontogeny as shown by the contribution of the shoot apical layers to leaves of periclinal chimaeras. *Am. J. Bot.*, **62**, 935–947.

Stewart, R.N. & Dermen, H. (1979) Ontogeny in monocotyledons as revealed by studies of the developmental anatomy of periclinal chloroplast chimaeras. *Am. J. Bot.*, **66**, 47–58.

Stewart, R.N., Meyer, F.G. & Dermen, H. (1972) Camellia + 'Daisy Eagleson', a graft chimera of *Camellia sasanqua* and *C. japonica*. *Am. J. Bot.*, **59**, 515–524.

Stone, J.M., Trotochaud, A.E., Walker, J.C. & Clark, S.E. (1998) Control of meristem development by CLAVATA1 receptor kinase and kinase-associated protein phosphatase interactions. *Plant Physiol.*, **117**, 1217–1235.

Stuurman, J., Jaggi, F. & Kuhlemeier, C. (2002) Shoot meristem maintenance is controlled by a *GRAS*-gene mediated signal from differentiating cells. *Genes Dev.*, **16**, 2213–2218.

Sussex, I.M. (1955) Experiments on the cause of dorsiventrality in leaves. *Nature*, **174**, 351–352.

Sussex, I.M. (1989) Developmental programming of the shoot meristem. *Cell*, **56**, 225–229.

Szymkowiak, E.J. & Sussex, I.M. (1996) What chimaeras can tell us about plant development. *Annu. Rev. Plant Physiol. Plant Mol. Biol.*, **47**, 351–376.

Takada, S., Hibara, K-I., Ishida, T. & Tasaka, M. (2001) The *CUP-SHAPED COTYLEDON1* gene of *Arabidopsis* regulates shoot apical meristem formation. *Development*, **128**, 1127–1135.

Tamaoki, M., Kusaba, S., Kano-Murakami, Y. & Matsuoka, M. (1997) Ectopic expression of a tobacco homeobox gene, *NTH15*, dramatically alters leaf morphology and hormones levels in transgenic tobacco. *Plant Cell Physiol.*, **38**, 917–927.

Tanaka-Ueguchi, M., Itoh, H., Oyama, N., Koshioka, M. & Matsuoka, M. (1998) Over-expression of a tobacco homeobox gene, *NTH15*, decreases the expression of a giberellin biosynthetic gene encoding GA 20-oxidase. *Plant J.*, **15**, 391–400.

Tang, G., Reinhart, B.J., Bartel, D.P. & Zamore, P.D. (2003) A biochemical framework for RNA silencing in plants. *Genes Dev.*, **17**, 49–63.

Timmermans, M.C.P., Hudson, A., Becraft, P.W. & Nelson, T. (1999) ROUGH SHEATH2: a Myb protein that represses *knox* homeobox genes in maize lateral organ primordia. *Science*, **284**, 151–153.

Trotochaud, A.E., Hao, T., Wu, G., Yang, Z. & Clark, S.E. (1999) The CLAVATA1 receptor-like kinase requires CLAVATA3 for its assembly into a signaling complex that includes KAPP and a Rho-related protein. *Plant Cell*, **11**, 393–406.

Tsiantis, M., Schneeberger, R., Golz, J.F., Freeling, M. & Langdale, J.A. (1999) The maize *rough sheath2* gene and leaf development programs in monocots and dicot plants. *Science*, **284**, 154–156.

Venglat, S.P., Dumonceaux, T., Rozwadowski, K., Parnell, L., Babic, V., Keller, W., Martienssen, R., Selvaraj, G. & Datla, R. (2002) The homeobox gene *BREVIPEDICELLUS* is a key regulator of inflorescence architecture in *Arabidopsis*. *Proc. Natl. Acad. Sci. U.S.A.*, **99**, 4730–4735.

Vernoux, T., Kronenberger, J., Grandjean, O., Laufs, P. & Traas, J. (2000) PIN-FORMED 1 regulates cell fate at the periphery of the shoot apical meristem. *Development*, **127**, 5157–5165.

Vollbrecht, E., Reiser, L. & Hake, S. (2000) Shoot meristem size is dependent on inbred background and presence of the maize homeobox gene, *knotted1*. *Development*, **127**, 3161–3172.

Vollbrecht, E., Veit, B., Sinha, N. & Hake, S. (1991) The developmental gene *Knotted-1* is a member of a maize homeobox gene family. *Nature*, **350**, 241–243.

Vroemen, C.W., Mordhorst, A.P., Albrecht, C., Kwaaitaal, M.A.C.J. & de Vries, S.C. (2003) The *CUP-SHAPED COTYLEDON3* gene is required for boundary and shoot meristem formation in *Arabidopsis*. *Plant Cell*, **15**, 1563–1577.

Waites, R. & Hudson, A. (1995) *Phantastica*: a gene required for dorsoventrality in leaves of *Antirrhinum majus*. *Development*, **121**, 2143–2154.

Waites, R., Selvadurai, H.R., Oliver, I.R. & Hudson, A. (1998) The *PHANTASTICA* gene encodes a MYB transcription factor involved in growth and dorsoventrality of lateral organs in *Antirrhinum*. *Cell*, **93**, 779–789.

Wardlaw, C.W. (1949) Further experimental observations on the shoot apex of *Dryopteris aristata* Druce. *Philos. Trans. R. Soc. Lond. B Biol. Sci.*, **233**, 415–451.

Weir, I., Lu, J., Cook, H., Causier, B., Schwarz-Sommer, Z. & Davis, B. (2004) *CUPULIFORMIS* establishes lateral organ boundaries in *Antirrhinum*. *Development*, **131**, 915–922.

Werner, T., Motyka, V., Strnad, M. & Schmülling, T. (2001) Regulation of plant growth by cytokinins. *Proc. Natl. Acad. Sci. U.S.A.*, **98**, 10487–10492.

Williams, R.W., Wilson, J.M. & Meyerowitz, E.M. (1997) A possible role for kinase-associated protein phosphatase in the *Arabidopsis* CLAVATA1 signalling pathway. *Proc. Natl. Acad. Sci. U.S.A.*, **94**, 10467–10472.

Wyrzykowska, J. & Fleming, A. (2003) Cell division pattern influences gene expression in the shoot apical meristem. *Proc. Natl. Acad. Sci. U.S.A.*, **100**, 5561–5566.

Wyrzykowska, J., Pien, S. Shen, W.H. & Fleming, A. (2002) Manipulation of leaf shape by modulating of cell division. *Development*, **129**, 957–964.

Xu, L., Xu, Y., Dong, A., Sun, Y., Pi, L., Xu, Y. & Huang, H. (2003) Novel *as1* and *as2* defects in leaf adaxial-abaxial polarity reveal the requirement for *ASYMMETRIC LEAVES1* and 2 and *ERECTA* functions in specifying leaf adaxial identity. *Development*, **130**, 4097–4107.

Yalovsky, S., Kulukian, A., Rodriguez-Concepcion, M., Young, C.A. & Gruissem, W. (2000) Functional requirement of plant farnesyltransferase during development in *Arabidopsis*. *Plant Cell*, **12**, 1267–1278.

Yu, L.P., Miller, A.K. & Clark, S.E. (2003) *POLTERGEIST* encodes a protein phosphatase 2C that regulates CLAVATA pathways controlling stem cell identity in *Arabidopsis* shoot and flower meristems. *Curr. Biol.*, **13**, 179–188.

Yu, L.P., Simon, E.J., Trotochaud, A.E. & Clark, S.E. (2000) *POLTERGEIST* functions to regulate meristem development downstream of the *CLAVATA* loci. *Development*, **127**, 1661–1670.

Ziegelhoffer, E.C., Medrano, L.J. & Meyerowitz, E.M. (2000) Cloning of the *Arabidopsis WIGGUM* gene identifies a role for farnesylation in meristem development. *Proc. Natl. Acad. Sci. U.S.A.*, **97**, 7633–7638.

7 Intercellular communication during floral initiation and development

George Coupland

7.1 Introduction

The adult organs of a plant are derived from groups of stem cells, or meristems, that are formed during embryo development. These organs develop from primordia that are formed repetitively on the flanks of the shoot apical meristem (SAM). The identity of the organs formed from these primordia can be influenced by environmental conditions, illustrating the extreme flexibility of plant development. Furthermore, analysis of genetic mosaics, in which the development of a tissue consisting of two phenotypically distinct genotypes is followed (Huala & Sussex, 1993), or the effect of ablation of single cells within the developing root (Vandenberg et al., 1995) indicated that positional information produced by signaling from neighboring cells can determine the fate of plant cells late in the development of an organ. Similarly, the identity of the organ formed from a primordium or the behavior of axillary meristems present at the junction between the shoot and leaf can be influenced by long-distance signals formed in other tissues and transported to their site of action (Bernier et al., 1993; Booker et al., 2003).

Short- and long-distance signaling appear to have important roles in the transition from vegetative development to flowering. This transition requires a change in the behavior of meristems (Huala & Sussex, 1993). For example, in *Arabidopsis* the morphology of the SAM is altered irreversibly by increased rates of cell division generating a larger, taller meristem called the inflorescence meristem (Vaughn, 1955), and inflorescences form from axillary meristems present in the axils of the last few leaves to be formed during vegetative growth. New primordia formed on the flanks of the inflorescence meristem give rise to flowers. In many species, including *Arabidopsis*, the timing of the transition from vegetative to reproductive development is determined by environmental conditions such as temperature, day length and light quality. The environmental conditions that trigger flowering can be perceived in different organs, so that changes in day length are detected in the leaves (Zeevaart, 1976), while vernalization (extended exposures to low temperatures that mimic winter conditions) is detected in the meristem (Michaels & Amasino, 2000). Since flower development occurs at the meristem, then at least in the case of the day-length response, long-distance signaling must be required to induce flower development. More recently, analysis of developing flowers that are mosaics of mutant and wild-type tissues demonstrated that the effect of mutations in regulatory genes can be corrected by short-distance signaling from wild-type cells (Wu et al., 2002).

In the following sections, I describe how molecular genetic and cell biology based approaches are being used to elucidate the molecular mechanisms underlying these signaling processes.

7.2 Long-distance signaling during the induction to flowering

7.2.1 Discovery of a role for long-distance signaling in the induction of flowering

The involvement of long-distance signaling in initiating the transition to flowering was originally proposed based on the demonstration that differences in day length are perceived in the leaves. The first experiments that demonstrated this were based on exposure of different parts of the plant to distinct day lengths. James Knott showed using spinach plants that exposing the whole plant or only the foliage to days containing 15 h light would induce rapid flowering, while exposing the apex of the shoot to such long day lengths while the foliage is exposed to shorter day lengths would not induce flowering (Knott, 1934). He concluded that the role of the leaves in the induction of flowering in response to day length is 'in the production of some substance, or stimulus, that is transported to the growing point'.

The conclusion that leaves are the source of a floral stimulus was strengthened by generating grafts between plants exposed to different day lengths (Zeevaart, 1976). In particular, grafting of leaves of the short-day plant *Perilla* to shoots that had been exposed only to long days was sufficient to induce flowering (King & Zeevaart, 1973). These experiments confirmed that leaves are the source of the floral stimulus, and that a single leaf produces sufficient stimulus to induce floral development at the apex of the plant. Indeed in *Perilla*, successively grafting the same leaves to seven different shoots over a period of 97 days demonstrated that these leaves were stably induced to produce floral stimulus and produced enough to trigger flowering of all seven recipient shoots (Zeevaart, 1985).

Physiological experiments indirectly indicated that the stimulus is transported from the leaf to the apex through the phloem. This was done in *Perilla* by comparing the effect of the donor leaf on flowering with the movement of radioactively labeled assimilates from the donor leaf through the phloem and into the axillary shoot (King & Zeevaart, 1973). Assimilates and the floral stimulus are both transported over long distances in the shoot, and there is a close correlation between the translocation of radioactively labeled assimilates into the axillary shoot of *Perilla* and the movement of the floral stimulus into the shoot, as detected by the induction of floral development.

Although transport of the floral stimulus across graft junctions could be followed indirectly by its effect on flowering, its identity has been difficult to establish. In the mustard *Sinapis alba*, flowering can be induced by exposure to a single long day (Bernier *et al.*, 1993). Sucrose was the earliest detected signal transported from the leaves in response to long-day treatment, and was transported both up to the apex and down to the roots (Havelange *et al.*, 2000). Transport of sucrose to the root was correlated with increased transport of cytokinin from the root up the shoot to the apex. Similarly, nitrogenous compounds, such as the amino acids glutamine and asparagine, were transported from the leaf up the shoot to the apex (Corbesier *et al.*, 2001). When movement through the phloem from the shoot to the root was prevented by removing living tissues, including phloem, from the surface of the plant (girdling), or when movement from the root to the shoot through the xylem was prevented by increasing relative humidity, the flowering response of the plant to a single long day was reduced (Perilleux & Bernier, 2002). However, application of sucrose and cytokinins to vegetative plants was not sufficient to induce flowering, and no genetic evidence clearly indicates a signaling role for these substances in the transition to flowering. This led to the suggestion that the floral stimulus may be a complex mixture of compounds (Bernier *et al.*, 1993), and to the perception that genetic-based approaches are required to supplement physiology to define the floral stimulus (Colasanti & Sundaresan, 2000; Perilleux & Bernier, 2002).

7.2.2 *Mutations that impair long-distance signaling in pea and maize*

Mutations or natural-genetic variations that alter flowering-time have been described in many species. Studies in pea plants were enhanced by the availability of extensive genetic stocks, and the ability to readily graft different genotypes so that the effect of this variation on long-distance signaling could be assessed (Weller *et al.*, 1997b). For example, *gigas* (*gi*) mutants flower later than wild-type plants, but their flowering is accelerated by grafting a *gi* shoot onto a wild-type stock (Beveridge & Murfet, 1996). This suggests that in wild-type plants the *GI* gene may be involved in the synthesis or transport of the floral stimulus. Flowering of wild-type pea plants is accelerated in response to long days and is delayed by exposure to short days. The *gi* mutant flowers later under both conditions, and often never flowers under short days. This suggests that the floral stimulus controlled by *GI* is not part of the response to day length, but is expressed under all environmental conditions tested. The *LATE FLOWERING* (*LF*) gene is proposed to act at the apex and encode a target of the floral stimulus; dominant alleles at this gene delay flowering and the effect is not influenced by grafting of an *LF* shoot onto a wild-type stock (Murfet, 1971, 1985). In addition to the floral stimulus controlled by *GI*, there is evidence for a long-distance inhibitory signal regulating flowering-time of pea plants. Mutations in the *STERILE NODES* (*SNE*), *DIE NEUTRALIS* (*DNE*) or *PHOTOPERIOD* (*PPD*) genes cause early flowering, and flowering of the shoots of these plants can be delayed by grafting onto a root stock of a wild-type plant (King & Murfet, 1985; Weller *et al.*, 1997a,b). Plants in which these genes are mutated are almost day-length insensitive, flowering at the same time under both long and short days, indicating that the photoperiod response is largely caused by production of an inhibitor under short days. These experiments suggested a model in which the timing of the transition to flowering at the apex of pea plants is determined by a balance between long-distance promotive and inhibitory signals, so that when the ratio of stimulus to inhibitor exceeds a certain level flowering occurs (Weller *et al.*, 1997b). Further progress in understanding the identity of these signals has been delayed by a lack of molecular information, since none of the genes associated with production of the floral stimulus or the inhibitor of flowering have been cloned.

Analysis of the *INDETERMINATE* (*ID*) gene of maize provided the first molecular information on a gene that regulates the floral stimulus. Mutations in the *ID* gene dramatically delay the transition to flowering, so that many more leaves are formed than in wild-type plants (Colasanti *et al.*, 1998). Eventually *id* mutants do flower, but the reproductive structures develop abnormally and show vegetative characteristics. The male tassel formed at the apex of a wild-type plant forms vegetative plantlets on its flanks, while the female inflorescence, which is generated from axillary meristems in the wild-type plant, either does not form or is converted into vegetative branches. The *ID* gene was cloned and shown to encode a protein containing predicted zinc fingers that may act as a transcriptional regulator (Colasanti *et al.*, 1998). Analysis of the expression of *ID* detected the mRNA in young, immature leaves, but not in the SAM or in mature leaves. The expression of *ID* in the leaves, but not the SAM, indicated that it acts to regulate long-distance signals that influence the transition to flowering of the meristem. The expression of *ID* appears to occur in sink leaves, which receive nutrients from photosynthetically active source tissues, and not to be expressed in source leaves (Colasanti & Sundaresan, 2000). This observation led to the suggestion that ID may not promote the production of the floral stimulus, but rather act in the developing leaves to regulate its flow. However, the mechanism by which ID regulates flowering requires further knowledge of the identity and function of the genes whose expression it regulates.

7.2.3 *Molecular genetic analysis of flowering-time control in Arabidopsis places the long-distance signal within a regulatory hierarchy*

7.2.3.1 *A network of pathways controls flowering of Arabidopsis*
The genetic control of flowering has been most extensively studied in *Arabidopsis*. The behavior of mutants exhibiting a severe delay in flowering was first described in detail by Redei (Redei, 1962), and this analysis was later broadened and extended by Koornneef (Koornneef *et al.*, 1991, 1998). More recently, a large number of mutants showing either later or earlier flowering have been described (Mouradov *et al.*, 2002). Furthermore, study of natural genetic variation between *Arabidopsis* accessions identified loci that were not detected by extensive mutagenesis of standard laboratory accessions (Alonso-Blanco & Koornneef, 2000).

Environmental conditions influence flowering-time of *Arabidopsis*. Flowering is promoted by exposure to long days and delayed under short days, whereas vernalization treatments promote flowering (Martinez-Zapater *et al.*, 1994). In addition to these seasonal cues, less dramatic changes in ambient conditions also strongly influence flowering-time. Exposure to lower temperatures (16°C) delays flowering compared to the effect of growing plants at typical growth temperatures of 20–24°C, and exposure to the high ratios of far-red to red light associated with shading conditions accelerates flowering (Blazquez et al., 2003; Cerdan & Chory, 2003). The genes identified by mutagenesis and by allelic variation between accessions were placed in pathways based on genetic criteria and their effect on the response of flowering-time to different environmental cues (Koornneef et al., 1998). The major features of this model were later confirmed by the cloning of the genes and analysis of their expression patterns in wild-type and mutant plants (Mouradov et al., 2002; Simpson & Dean, 2002).

Within this model, four major pathways control flowering-time and converge to regulate the expression of genes that integrate the information received from the different pathways (Fig. 7.1). One pathway controls the response to day-length, and specifically promotes flowering in response to long days (Hayama & Coupland, 2003; Yanovsky & Kay, 2003; Searle & Coupland, 2004). Mutations in this pathway can either delay flowering under long days or accelerate flowering under short days. The last gene that is specifically involved in this pathway is *CONSTANS* (*CO*), which encodes a zinc finger protein that promotes transcription of downstream

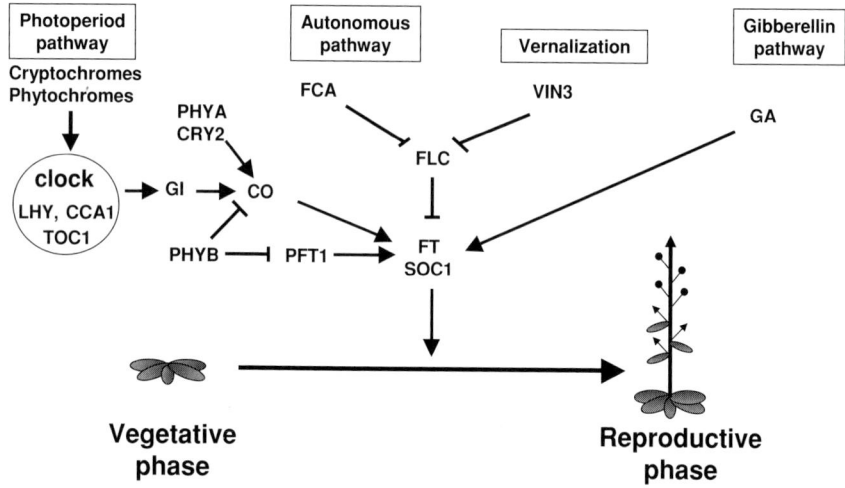

Figure 7.1 A network of four major pathways controls flowering-time in *Arabidopsis*. The photoperiod pathway promotes flowering specifically under long days. The transcription of the *GI* and *CO* genes is regulated by the circadian clock, whereas the photoreceptors PHYA, CRY2 and PHYB regulate CO protein abundance. Flowering is also influenced by light quality; in particular, low ratios of red to far-red light resembling the shaded conditions formed by vegetation promote flowering. This response probably acts partly by stabilization of CO protein and partly independently of CO through PFT1. The autonomous pathway negatively regulates the abundance of the mRNA of the floral repressor FLC, and FCA is included as a representative of this pathway. Vernalization also promotes flowering by repressing *FLC* mRNA levels and acts independently of the autonomous pathway. VERNALIZATION INSENSITIVE 3 (VIN3) is shown as a representative of this pathway (Sung & Amasino, 2004). Finally, gibberellin promotes flowering of *Arabidopsis*, particularly under short days. All four pathways appear to converge on the transcriptional regulation of the *FT* and *SOC1* genes, which are often referred to as floral integrators and promote flowering. These pathways are described in more detail in the text and in recent reviews (Michaels & Amasino, 2000; Mouradov et al., 2002; Simpson & Dean, 2002; Searle & Coupland, 2004).

flowering-time genes (Putterill *et al.*, 1995; Robson *et al.*, 2001), and is discussed in more detail in Section 7.2.3.2. This pathway probably also plays a role in the effect of light quality on flowering, because high ratios of far-red to red light promote flowering and stabilize the CO protein (Valverde *et al.*, 2004), although the flowering response to light quality also involves a CO-independent pathway (Cerdan & Chory, 2003).

A second genetic pathway controls the response to vernalization. In response to extended exposures to low temperature, this pathway reduces the abundance of the mRNA encoding the MADS box transcription factor FLC, which is a potent repressor of flowering (Michaels & Amasino, 1999; Sheldon *et al.*, 1999). Therefore, vernalization accelerates flowering by reducing *FLC* expression. The autonomous pathway also regulates *FLC* expression (Sheldon *et al.*, 2000; Michaels & Amasino, 2001). Mutations in a third pathway, the autonomous pathway, delay flowering under both long and short days, and cause an increase in *FLC* mRNA levels. This pathway regulates *FLC* expression independently of vernalization, so that the late flowering caused by mutations in this pathway and the high *FLC* mRNA levels observed in these mutants can be corrected by vernalization. Mutants affected in this pathway also show an altered flowering-time in response to ambient temperatures (Blazquez *et al.*, 2003). The autonomous pathway appears to represent protein complexes involved in histone modification and RNA processing (Simpson *et al.*, 2003; Ausin *et al.*, 2004), and probably also has a more general role than the regulation of *FLC* expression. Finally, the growth regulator gibberellic acid (GA) promotes flowering of *Arabidopsis*, and mutations that affect genes required for GA biosynthesis delay flowering, particularly under short days (Wilson *et al.*, 1992). This general framework of three interacting pathways that promote flowering of *Arabidopsis* in the absence of vernalization is supported by the observation that a triple mutant carrying the *co* mutation (impairs the photoperiod pathway), the *ga1* mutation (reduces GA synthesis) and the *fca* mutation (impairs the autonomous pathway) never flowered under long days (Reeves & Coupland, 2001).

These distinct genetic pathways converge to regulate the expression of a small group of downstream genes, sometimes described as floral integrators (Mouradov *et al.*, 2002; Simpson & Dean, 2002). This group contains three genes, two of which promote flowering, *FLOWERING LOCUS T* (*FT*) and *SUPPRESSOR OF OVEREXPRESSION OF CO 1* (*SOC1*), and a third, *LEAFY*, that encodes a transcription factor required to confer floral identity on developing floral primordia. *FT* encodes a protein with similarity to RAF kinase inhibitors of animals (Kardailsky *et al.*, 1999; Kobayashi *et al.*, 1999), and is discussed in more detail in the following section, whereas *SOC1* encodes a MADS box transcription factor (Borner *et al.*, 2000; Lee *et al.*, 2000; Samach *et al.*, 2000). Mutations in each of these genes delay flowering, whereas their overexpression from the viral CaMV 35S promoter causes extreme early flowering. The expression of *SOC1* and *FT* is increased by CO and reduced by FLC, indicating that they are downstream of the point of convergence of the vernalization and photoperiod pathways (Samach *et al.*, 2000; Hepworth *et al.*, 2002). Furthermore, the expression of *SOC1* is increased by treating plants with GA, suggesting that it acts downstream of all three pathways (Moon *et al.*, 2003).

7.2.3.2 Spatial regulation of flowering-time control

The question of in which tissues the separate flowering-time pathways or their individual components act to regulate flowering was not addressed until recently. Therefore in most cases, it remains unclear whether these pathways regulate the function of long-distance signals expressed in the leaves or respond to these signals in the meristem. Simply analyzing the spatial pattern of expression of these genes did not help to address this problem since many flowering-time genes are expressed broadly.

Classical physiological experiments suggested that vernalization acts in the meristem to promote flowering (Michaels & Amasino, 2000). Initial observations were based on exposing only the leaves or only the apices of celery plants to vernalization treatments, and demonstrating that vernalization of the meristem was sufficient to induce flowering. The vernalization pathway is therefore likely to act in the meristem to reduce *FLC* expression and thereby induce flowering, and this may also be true for the autonomous pathway. Consistent with vernalization acting in the meristem, *FLC* is expressed specifically in the shoot and root meristems in young seedlings, although in older plants it is also expressed in expanded leaves (Sheldon *et al.*, 2002; Noh & Amasino, 2003; Bastow *et al.*, 2004). FLC protein binds to the promoter of the *SOC1* gene *in vitro* and the binding sites are required for repression of *SOC1* expression by FLC (Hepworth *et al.*, 2002). The expression pattern of *FLC* and its role in repressing *SOC1* expression suggests that it may act directly in the meristem to repress transcription of target genes. Such a role would suggest that FLC does not repress flowering by altering the synthesis or transport of the floral stimulus, but the response of the meristem to the stimulus. Nevertheless, analysis of somatic sectors predicted to express FLC suggested that its repressive effect on flowering can be overcome by non-cell autonomous signaling (Furner *et al.*, 1996). Somatic sectors homozygous for the *fca* mutation were created in an otherwise wild-type plant, and since this mutation impairs the autonomous pathway these sectors would be predicted to cause localized expression of *FLC* at high levels. Mutant *fca* sectors within the L2 and L3 layers of the meristem did not affect the morphology of the meristem or of the mutant cells, suggesting that the effect of *FLC* expression in those cells is overcome by non-cell autonomous signaling from wild-type cells (Furner *et al.*, 1996). This signal may originate in cells present in the L1, or in L2 or L3 cells neighboring the mutant sector or even from other organs such as the leaves.

Existence of the floral stimulus was originally demonstrated by inducing flowering with appropriate day lengths, as described in Section 7.2.1. Therefore the photoperiod pathway of *Arabidopsis* might be expected to include a long-distance signaling component analogous to the floral stimulus. A molecular hierarchy within the photoperiod pathway has been defined. Two flowering-time genes specific to this pathway are *GIGANTEA* (*GI*) and *CO*. The *GI* gene encodes a large protein of 1180 amino acids that is present in the nucleus and is highly conserved among the angiosperms but has no animal homologues (Fowler *et al.*, 1999; Park *et al.*, 1999). The biochemical function of GI is unknown, but *gi* mutations cause severe late flowering (Redei, 1962), while overexpression of *GI* causes early flowering (L. Wright and G. Coupland, unpublished results, 2004). GI regulates flowering-time at least in part by the regulation of *CO* mRNA abundance; *gi* mutants contain less *GI* mRNA (Suarez-Lopez *et al.*, 2001) while *GI* overexpressors show higher *CO* mRNA abundance. The abundance of *GI* and *CO* mRNAs is circadian clock regulated. Under days containing 16 h light, in which these genes promote early flowering, *GI* mRNA abundance peaks around 10–12 h after dawn, whereas *CO* mRNA abundance rises around 12 h after dawn and stays high throughout the night until the following dawn (Fowler *et al.*, 1999; Park *et al.*, 1999; Suarez-Lopez *et al.*, 2001). *CO* mRNA abundance is therefore high when plants are exposed to light at the end of a long day. *CO* expression is also regulated at the post-transcriptional level, so that the cryptochrome and phytochrome A photoreceptors act at the end of the day to stabilize the CO protein (Valverde *et al.*, 2004), whereas in darkness the protein is rapidly degraded, probably as a consequence of being ubiquitinated. Under short days, the *CO* mRNA is expressed only in the dark, and so the protein would be predicted never to accumulate. In agreement with these data, in wild-type plants *FT* is activated by CO under long days, but not under short days (Suarez-Lopez *et al.*, 2001; Yanovsky & Kay, 2002). Therefore, the combination of circadian clock mediated regulation of *CO* mRNA abundance and stabilization of CO protein by exposure to light can explain why CO promotes *FT* expression and thus flowering only under long days.

The grafting experiments performed in *Perilla* and many other species indicated that day length is perceived in the leaves. The observation that CO is a major part of the molecular mechanism by which *Arabidopsis* discriminates between long and short days suggests that CO may act in the leaf to regulate the transition to flowering at the apex. The *CO* mRNA is present at very low abundance, but is expressed widely. *In situ* hybridizations and RT-PCR detected the *CO* mRNA in the meristem, young leaf primordia and whole seedling RNA. A more refined expression pattern was identified using fusions of the *CO* promoter to the GUS marker gene (Takada & Goto, 2003; An *et al.*, 2004). In CO:GUS plants, GUS expression was most strongly detected in the phloem of cotyledons, leaves and stems, but also in the protoxylem, young leaves and meristem (see Plate 7.1, following page 146). Several recent observations suggest that CO acts in the vascular tissue and not the meristem to promote flowering. The pattern of expression of the CO target gene, *FT*, has not been described in wild-type plants, because of its low level of expression. However, *FT* expression is increased in the early flowering *terminal flower 2 (tfl2)* mutant, and is present in the vascular tissue, suggesting that CO may activate its target gene in these tissues (Takada & Goto, 2003). Consistent with this conclusion, *FT* expression was reduced in *tfl2 co-2* plants compared to *tfl2* mutants. In an independent approach, expression of *CO* from heterologous promoters specific to the phloem, such as that of the *SUCROSE TRANSPORTER 2 (SUC2)* gene, complemented the *co-2* mutation, but expression of CO from meristem-specific promoters had no effect on flowering (An *et al.*, 2004). Therefore, *CO* appears to act in the vascular tissue to regulate the synthesis or transport of a long-distance signal that initiates floral development at the apex.

Proteins have been shown to move through the phloem from source to sink tissues, but CO protein itself is unlikely to be the long-distance signal. The *SUC2* promoter is specific to the companion cells of the phloem in source leaves, and expression of GUS enzyme from this promoter produced staining specifically in the vascular tissue of these leaves (Truernit *et al.*, 1996). In contrast, expression of GFP from the same promoter produced fluorescence both in source and sink leaves, indicating that GFP can be downloaded from the companion cells into the phloem sieve elements and transported to sink leaves. However, CO protein is approximately 20 kDa larger than GFP and expression of CO:GFP from the *SUC2* promoter complemented the *co* mutation, but GFP fluorescence was detected only in the vascular tissue and not in the meristem or leaf epidermal cells (An *et al.*, 2004). This, together with the observation that *CO* does not promote flowering when expressed in the meristem, suggests that CO protein acts in the phloem to promote flowering and is not transported to other cells.

The mechanism by which CO acts to promote flowering in the phloem involves the *FT* gene. FT mRNA abundance was increased in the phloem of *SUC2::CO* plants, and the *ft* mutation strongly suppressed the early flowering of *SUC2::CO* (An *et al.*, 2004). Furthermore, expression of *FT* in the phloem from the *SUC2* promoter complemented the *co* mutation. However, in contrast to CO, FT promoted flowering when expressed in the meristem and the epidermal layer, as well as the phloem (An *et al.*, 2004). No data are available on the movement of the FT protein between cells. However, FT is a small protein of 23 kD (Kardailsky *et al.*, 1999; Kobayashi *et al.*, 1999) and is, therefore, smaller than GFP, suggesting that it may be able to move freely between cells. Also, the observation that *FT* can promote flowering when expressed in the meristem is consistent with the idea that the protein could move from the phloem to the meristem where it acts to promote flower development. However, these data could also be explained if FT can act in almost any cell type to trigger the synthesis of a small molecule that induces flowering and is able to move freely between cells.

The biochemical function of FT is unknown. It is a member of a small protein family in *Arabidopsis* and shares homology with characterized proteins in other species. These proteins are referred to as CETS, after CENTRORADIALIS (CEN) of *Antirrhinum*, TERMINAL FLOWER 1 (TFL1) of *Arabidopsis* and SELF PRUNING (SP) of tomato (Bradley *et al.*, 1997; Pnueli *et al.*, 1998; Kardailsky *et al.*, 1999; Kobayashi *et al.*, 1999). CETS proteins share homology to RAF kinase inhibitor proteins of mammals (Kardailsky *et al.*, 1999; Pnueli *et al.*, 2001), and the crystal structure of CEN is similar to that of RAF kinase inhibitors (Banfield & Brady, 2000). In the yeast two-hybrid system, SP interacted with a NIMA-like kinase, bZIP transcription factors and a 14-3-3 protein, which led to the suggestion that CETS proteins act as adapters in a variety of signaling pathways (Pnueli *et al.*, 2001). How these interactions relate to the role of FT in promoting flowering is unknown. Combining *ft* mutations with mutations affecting flower development identified a strong genetic interaction between *ft* and mutations in the floral organ identity gene *LEAFY* (*LFY*) (Ruiz-Garcia *et al.*, 1997). The *ft lfy* double mutant failed to produce any mature floral organs, and resembled a *lfy apetala1* (*ap1*) double mutant. This suggested that the role of FT in promoting flowering may involve the activation of *AP1*, which is expressed exclusively at the meristem in floral primordia. This suggests that in *SUC2::FT* plants, FT might activate a long-distance signal in the phloem that leads to *AP1* activation in the meristem or that FT protein might move to the meristem where it activates *AP1*. An alternative possibility is that in wild-type plants *FT* mRNA is also expressed in the meristem, although so far this has not been detected.

7.2.3.3 Identifying the floral stimulus: a perspective from Arabidopsis molecular genetics

Analysis of the spatial regulation of components of the photoperiod pathway of *Arabidopsis* showed that a long-distance, graft-transmissible signal similar to the floral stimulus acts within this pathway (Takada & Goto, 2003; An *et al.*, 2004). CO acts cell autonomously in the phloem to activate *FT* expression, and therefore the signal does not seem to comprise the CO protein itself, but rather is regulated by CO. In turn, *FT* is still activated by CO, and promotes flowering when expressed in the phloem, but also does so when expressed in other tissues, including the meristem. The small size of the FT protein suggests that it may move from the phloem to the meristem and directly trigger changes in gene expression. Symplastic downloading of proteins from the sieve elements into the sink tissues of the apex through plasmodesmata has been proposed (Ruiz-Medrano *et al.*, 2001; see also Chapter 5), suggesting that FT may move directly by this mechanism into apical cells and induce flowering. Furthermore, the size exclusion limit or selectivity of plasmodesmata that allow downloading from the sieve elements into the meristem have been shown to change around the time of flowering (Gisel *et al.*, 2002), suggesting that this might be an important regulatory step. If FT does move from the companion cells of the leaf to the meristem, nothing is known of the mechanisms underlying its export from the leaf and import to the crucial cells of the meristem.

An alternative possibility is that FT acts in the phloem to induce the expression of enzymes that generate small molecules that are transported to the apex, or of small RNAs that induce flowering. Such small molecules might be metabolites or growth regulators, such as sucrose or cytokinin, as suggested from physiological studies. Alternatively, RNA molecules can be transmitted through the phloem (see Chapter 3). Strikingly, small RNAs that induce gene silencing have been demonstrated to cross graft junctions (Palauqui *et al.*, 1997). Furthermore, endogenous microRNAs that decrease the expression of transcription factors related to APETALA2 were recently shown to promote flowering of *Arabidopsis* (Aukerman & Sakai, 2003). Although these particular microRNAs appear to be expressed at the shoot apex, other microRNAs expressed in the leaf might be transported to the apex where they influence flowering-time. In support of the biological significance of long-distance transport of RNA, the mRNA of the *mouse ears* mutant gene was recently shown to cross graft junctions in tomato plants (Kim *et al.*, 2001). The *mouse ears* mutation is dominant and caused by a complex fusion of a gene encoding an enzyme in the glycolytic pathway and *LeT6*, which encodes a homeobox transcription factor. The mutant shows more compound leaves than in wild-type plants. Grafting of wild-type shoots onto stocks of the dominant *mouse ears* mutant caused the morphology of leaves at the apex of the plant to appear similar to those of *mouse ears* mutants, and this correlated with the presence of increased abundance of *mouse ears* mRNA in the wild-type shoot (Kim *et al.*, 2001).

Positioning the floral stimulus within a regulatory hierarchy will allow it to be approached using the molecular genetic tools available in *Arabidopsis*. Screening for mutations that suppress the early-flowering phenotype caused by expressing *CO* or *FT* in the phloem, or screening randomly for further proteins and RNAs that influence flowering-time when expressed in the phloem, is likely to identify further components of the system. Similarly, use of full genome microarrays will enable identification of all of the genes whose expression is induced by overexpression of *CO* in the phloem, and therefore provide leads as to the identity of mobile components.

7.3 Intercellular communication during floral development

Dramatic changes in gene expression occur at the apex of the plant on the transition to flowering. Recently, these changes were analyzed using microarrays containing oligonucleotides derived from almost all *Arabidopsis* genes, and within 7 days of a shift from short to long day lengths, the expression of 332 genes was induced or repressed at least twofold in both the Landsberg *erecta* and Columbia accessions (Schmid *et al.*, 2003). Perhaps surprisingly, of these 332 genes, the majority (231) showed reduced expression during floral induction, whereas the remainder were increased. At least one of the repressed genes, encoding an AP2-like transcription factor, was shown to actively repress flowering when overexpressed, suggesting that its downregulation during flowering may play an active role in floral induction (Schmid *et al.*, 2003). Similarly, some of the earliest genes that show increased expression during floral induction appear to promote flowering. For example, the *SOC1* gene of *Arabidopsis* is induced at the meristem within a few hours of shifting plants from short to long days (Borner *et al.*, 2000; Samach *et al.*, 2000), and the orthologous gene in *Sinapis alba* behaves similarly (Bonhomme *et al.*, 2000). Mutations in *SOC1* delay flowering, but do not have an effect on floral development (Borner *et al.*, 2000; Lee *et al.*, 2000; Samach *et al.*, 2000). In contrast, *AP1* and *LFY*, two genes that confer floral meristem identity on the developing floral primoridum, are among the first genes with established roles in floral development whose transcripts increase in abundance at the meristem (Mandel *et al.*, 1992; Weigel *et al.*, 1992). Mutations in these genes cause shoots or flowers with shoot characters to develop from primordia that would normally form flowers, and the proteins encoded by these genes are involved in activation of downstream transcription factors that confer organ identity on floral organs (Weigel & Meyerowitz, 1993). Recent work indicates that *LFY* and orthologous genes in other species, as well as some of the transcription factors involved in the development of floral organs, can act non-cell autonomously to influence floral organ development (Wu et al., 2003). In this section, I review these data and describe the mechanisms thought to underlie the non-cell autonomy.

7.3.1 Some of the transcription factors that control floral meristem or organ identity act non-cell autonomously in the developing flower

The first molecular evidence for non-cell autonomy of genes involved in floral development came from studying *floricaula* (*flo*) mutants in *Antirrhinum majus*. The *FLO* and *LFY* genes are orthologues and have similar functions in *Antirrhinum* and *Arabidopsis*, respectively (Coen et al., 1990; Weigel et al., 1992). *FLO* was originally cloned based on the insertion of a transposon so that excision of the transposon restores *FLO* activity (Coen et al., 1990). Excision of the transposon can generate somatic sectors so that a single plant contains both mutant and wild-type cells. The shoot meristem and flowers that form on the flanks of the meristem consist of three cell layers, the L1, L2 and L3, which represent separate clones formed during embryogenesis (Huala & Sussex, 1993; Wu et al., 2003). *FLO* mRNA is expressed in all three layers, but a transposon excision in the *flo* mutant can generate a wild-type clone of cells that restores gene function in a single layer. Restoration of *FLO* gene function in the L1 allowed the formation of phenotypically wild-type flowers, whereas FLO function in the L2 or L3 caused the development of flowers that still showed some aspects of the mutant phenotype (Carpenter & Coen, 1995; Hantke et al., 1995). In *flo* mutants, the expression of the MADS box transcription factors PLE, which is involved in stamen and carpel development, and DEF, which is involved in petal stamen and development, is reduced. Analysis of expression of these genes in flowers in which *FLO* mRNA is expressed in only one layer showed that *FLO* can induce signaling events that affect other layers. For example, in plants expressing *FLO* mRNA only in the L1, *DEF* mRNA expression occurred in all layers in a similar pattern to wild-type plants, although its expression occurred later and at lower levels (Carpenter & Coen, 1995; Hantke et al., 1995).

In *Arabidopsis*, *LFY* acts non-cell autonomously in a similar way to *FLO* in *Antirrhinum*, but in contrast another floral meristem identity gene, *AP1*, acts almost exclusively cell autonomously. Expression of *LFY* in the L1 layer using the *ML1* epidermal-specific promoter was sufficient to fully complement the *lfy* mutant phenotype (Sessions et al., 2000), and some plants showed a phenotype similar to that of plants overexpressing *LFY* from the CaMV 35S promoter (Weigel & Nilsson, 1995). Also, as shown in *Antirrhinum* for *FLO*, expression of *LFY* in the L1 layer was sufficient to activate expression of the gene encoding the MADS box transcription factors APETALA3 (AP3), required for petal and stamen development, and AGAMOUS (AG), required for stamen and carpel development, in all three layers. However, a quite different effect was observed when *AP1* was expressed in the L1 layer in *ap1* mutants (Sessions et al., 2000). In these plants rescue of sepal and petal identity occurred in the first and second whorls, but only in the L1 layer. The L2 and L3 layers of the sepals of these plants were more typical of bracts. Similarly, the activation of *AP3* and *AG* in *ap1 ML1::AP1* plants was restricted to the L1 layer, indicating that *AP1* acts cell autonomously. These observations suggest that LFY, but not AP1, triggers signaling between cell layers in the developing flower.

Non-cell autonomy in floral development is not restricted to floral meristem identity genes that act early in floral development, such as *AP1* and *LFY*, but is also shown by genes that specify the identity of individual organs. An example of this comes from studies with the *DEF* and *GLOBOSA* (*GLO*) genes of *Antirrhinum*, which are required for normal petal and stamen development (Sommer *et al.*, 1990; Troebner *et al.*, 1992). Transposon-induced reversion of *def* mutations generated somatic sectors in which activity of the gene occurred only in the L1 (Perbal *et al.*, 1996). This was sufficient to cause expansion of second whorl organs, similar to the expansion shown by petal lobes, and accumulation of pigment in L1 cells similar to the epidermis of petals. However, L2 and L3 cells accumulated chlorophyll and did not show petal identity. Expression of *DEF* mRNA in the L1 is therefore not sufficient to confer full petal identity on inner layers. These observations were confirmed using transgenic *Antirrhinum* plants expressing *DEF* only in the L1 from the *ANTIRRHINUM FIDDLEHEAD* (*AFI*) promoter (Efremova *et al.*, 2001). In contrast, sectors in which *DEF* mRNA was expressed in the L2 and L3 were sufficient to confer petal identity on the L1 layer, but not to promote petal expansion. This observation indicates that *DEF* can act non-cell autonomously in the inner layers to confer petal identity on the epidermis. Further examples of the analysis of cell autonomy in floral organ identity gene function are those performed in *Arabidopsis* on *AG*, which is required for stamen and carpel expression, and on the *PISTILLATA* (*PI*) and *APETALA 3* (*AP3*) genes that are required for petal and stamen identity (Bouhidel & Irish, 1996; Sieburth *et al.*, 1998; Jenik & Irish, 2001).

7.3.2 *Movement of transcription factors between cells defines one mechanism for short-distance signaling in the developing flower*

Signaling between animal cells typically involves activation of a receptor located in the membrane of one cell by a ligand formed in a nearby cell (Pires-daSilva & Sommer, 2003). Until recently short-distance signaling between plant cells during flower development was assumed to be exclusively based on similar processes. However, analysis of KNOTTED (KN), a homeobox transcription factor of maize, demonstrated that transcription factors can move between plant cells (Jackson *et al.*, 1994) and that this may provide an alternative to membrane-bound receptor-based systems in plant cells. The original observation was that although *KN* mRNA is detected only in the L2 and L3 layers of maize SAM, the protein is found in the L1 (Jackson *et al.*, 1994). The authors proposed that although KN is a nuclear transcription factor, it may nevertheless move between plant cells. This was later confirmed using fluorescently labeled forms of the KN protein (Lucas *et al.*, 1995; Kim *et al.*, 2002, 2003).

The mechanism by which KN moves between plant cells has not been established, but is likely to involve movement through plasmodesmata. These are channels that connect neighboring cells, and consist of an outer membrane derived from plasma membrane and a segment of the endoplasmic reticulum that is continuous between the two cells (Lucas et al., 2001; see Chapter 5). Plasmodesmata are either made at the cell plate during cytokinesis (primary plasmodesmata) or are formed between existing cells (secondary plasmodesmata) (Lucas et al., 2001; Wu et al., 2002). The types of molecule that move through plasmodesmata appear to be tightly regulated by the developmental stage of the plant and the environmental signals to which it is exposed (Ding et al., 1992; Oparka et al., 1999; Kim et al., 2003). Also, the size exclusion limit of plasmodesmata, which defines the size of the molecules that can move through particular plasmodesmata, appears to change depending on the cell type in which they are present and the stage in development at which they are studied. Evidence that KN moves through plasmodesmata is indirect and is based on the observation that expression of the KN protein increases the size exclusion limit of plasmodesmata, allowing the *KN* mRNA to move to neighboring cells (Lucas et al., 1995). Also, RNAs and proteins encoded by other genes have been shown to move between plant cells through plasmodesmata (Lucas et al., 2001; Wu et al., 2002; see also Chapter 5).

DEF was the first protein involved in flower development that was shown to move between cells (Perbal et al., 1996). In chimeric *Antirrhinum* plants that expressed the *DEF* gene only in the L2 and L3 layers of petals, the mRNA was detected specifically in these layers, whilst the DEF protein was present in all three layers. This observation demonstrated that the DEF protein moves from inner layers into the L1 and probably explains the wild-type L1 phenotype shown by plants expressing DEF mRNA only in the L2 and L3. In contrast, in sectored plants expressing the *DEF* mRNA only in the L1, DEF protein was detected in the L1, but not the L2 or L3 layers (Perbal et al., 1996). This indicated that the DEF protein could not move through plasmodesmata from the L1 to underlying layers, consistent with the cell autonomy of the complementation of the *def* mutant phenotype. Taken together, these data indicate that trafficking of the DEF MADS box transcription factor occurs between layers but that this movement is polar; it occurs from the L2 or L3 layers to the L1, but not from the L1 to inner layers. Movement of DEF from the L2 to the L1 is likely to occur through secondary plasmodesmata, since primary plasmodesmata, which are formed during cell division, cannot be formed between cells that are not clonally related. However, DEF is also unlikely to be trafficked through primary plasmodesmata, since small, cell autonomous revertant sectors were previously described on *Antirrhinum def* mutant petals (Carpenter & Coen, 1990).

Trafficking of the LFY transcription factor was tested in transgenic plants in which expression of *LFY* mRNA from the epidermal *ML1* promoter complemented the *lfy* mutation (Sessions et al., 2000). Although in these plants *LFY* mRNA was detected only in the L1 layer, the protein was detected in all layers. LFY protein is therefore able to traffick from the L1 to inner layers. Movement of this protein was tested more extensively using fusions with GFP (Wu et al., 2003). Expression of several LFY-GFP fusion proteins from the *ML1* promoter complemented the *lfy* mutation, as shown for the wild-type LFY protein. GFP fluorescence was detected strongly in the L1 but also up to four cell layers deeper into the meristem. A gradient of fluorescence was detected from the L1 into the deeper layers. Previously, proteins were proposed to move through plasmodesmata either using a specific mechanism that utilizes signals within the protein or through nontargeted movement that involves only diffusion (Crawford & Zambryski, 1999). The gradient of fluorescence observed with LFY-GFP fusion proteins suggests that the movement may be nontargeted because proteins that move through a targeted mechanism, such as viral movement proteins, are able to move further and do not show a gradient in abundance in deeper layers. LFY-GFP also did not appear to move laterally within a layer, but moved effectively between layers, suggesting that movement might be restricted to secondary plasmodesmata and not occur between primary plasmodesmata.

During flower development, not all transcription factors move between cells. Expression of *AP1* mRNA from the *ML1* promoter did not rescue the *ap1* mutant phenotype in underlying layers, and AP1:GFP fusion proteins remained in the L1 cells when expressed from the *ML1* promoter (Wu et al., 2003). If LFY protein moves passively between cells through secondary plasmodesmata, then why does AP1, which is a smaller protein than LFY, not move between cells? Several possibilities have been suggested (Wu et al., 2003). Efficient nuclear localization of transcription factors may reduce the possibilities for movement, and there is evidence that AP1:GFP is more efficiently targeted to the nucleus than LFY:GFP fusions. Alternatively, the formation of higher order complexes, similar to those formed by AP1 and other MADS box transcription factors during floral development, might effectively increase the size of the protein and thereby restrict its movement by diffusion.

Short-range signaling between plant cells during floral development may not only involve trafficking of organ and meristem identity proteins. Expression of DEF in the L1 of *Antirrhinum* partially restores the petal identity of underlying cells, but the protein itself apparently does not move to the L2 (Efremova et al., 2001). This suggests that DEF activates signaling processes in the L1 that influence the identity of the subepidermal cells, but the nature of these signals is not known. These signals may be even more effective at promoting petal and stamen identity in *Arabidopsis* since epidermal expression of DEF in an *ap3* mutant almost completely rescued the effect of the *ap3* mutation on petal and stamen identity, but also in this species the AP3 protein was not able to move between cell layers.

How important is transcription factor movement between cells in the development of a wild-type flower? During root development, the *SHORTROOT* (*SHR*) mRNA, which encodes a transcription factor of the GRAS family, is expressed only in the stele, but is required for the development of the adjacent endodermis (Helariutta et al., 2000; see also Chapter 8). Analysis of the SHR protein showed that it is present both in the stele and endodermis, indicating that the protein must move from the stele to the endodermis, where it is required for the development of this cell layer and for the activation of the *SCARECROW* gene (Nakajima et al., 2001).

In contrast, *LFY* mRNA is expressed in all layers of the flower primordium and within their domains of expression the floral organ identity genes are also expressed in all layers, therefore there is no obvious requirement for protein movement. One possibility is that the induction of transcription of floral meristem identity and floral organ identity genes does not occur reliably within their domain of expression and that short-distance transcription factor movement ensures that gene activity occurs reliably in every cell.

7.4 Perspectives

Application of molecular genetic approaches in *Arabidopsis* is likely to lead to the identification of the long-distance signal involved in leaf-apex signaling during floral induction, whereas the existence of short-distance signaling between cell layers in the floral meristem has been established and shown to involve movement of particular transcription factors. However, there are also processes of short-distance signaling between cell layers of the flower that have not yet been described and appear not to involve movement of the primary floral organ identity transcription factors (Efremova *et al.*, 2001). Furthermore, the mechanisms by which these signals move between cells are unknown. Plasmodesmata appear to play important roles both in the movement of transcription factors in the floral primordium and in the downloading of the long-distance floral induction signal from the phloem companion cells into the sieve elements. Explaining how selectivity and size exclusion limits of plasmodesmata are determined, and identification of the structural components of plasmodesmata, will be essential in understanding the mechanisms of signaling between cells during flowering.

In addition, there are likely to be relationships between the long-distance signals that induce flowering and later processes during floral development that have not been described. Normally, the long-distance induction of flowering from the leaf is assumed to be involved in the initial induction of flowering and to set in train a series of events that results in floral development. However, there is evidence in some species and particular environmental conditions that the long-distance signal has a continued role in the later events of floral development. This is most evident in the process of floral reversion. For example, if *Arabidopsis* plants are grown under short days, exposed to several long days and then returned to short days, they are induced to flower by the long days but subsequently revert to vegetative growth (Laibach, 1951; Martinez-Zapater *et al.*, 1994). This phenomenon is more pronounced in *Impatiens*. Floral reversion in *Impatiens* can cause floral organs within a developing flower to revert to vegetative structures, suggesting that the floral inductive signal is required to maintain expression of floral organ identity genes (Pouteau *et al.*, 1997). The two processes described here of long-distance signaling to induce flowering and signaling within the developing flower may therefore be interrelated at levels which are not yet described, and of which we have no mechanistic information.

References

Alonso-Blanco, C. & Koornneef, M. (2000) Naturally occurring variation in *Arabidopsis*: an underexploited resource for plant genetics. *Trends Plant Sci.*, **5**, 22–29.

An, H., Roussot, C., Suarez-Lopez, P., Corbesier, L., Vincent, C., Pineiro, M., Hepworth, S., Mouradov, A., Justin, S., Turnbull, C.G.N. & Coupland, G. (2004) CONSTANS acts in the phloem to regulate a systemic signal that induces photoperiodic flowering of *Arabidopsis*. *Development*, **131**, 3615–3626.

Aukerman, M.J. & Sakai, H. (2003) Regulation of flowering time and floral organ identity by a microRNA and its APETALA2-like target genes. *Plant Cell*, **15**, 2730–2741.

Ausin, I., Alonso-Blanco, C., Jarillo, J. A., Ruiz-Garcia, L. & Martinez-Zapater, J.M. (2004) Regulation of flowering time by FVE, a retinoblastoma-associated protein. *Nat. Genet.*, **36**, 162–166.

Banfield, M.J. & Brady, R.L. (2000) The structure of *Antirrhinum* centroradialis protein (CEN) suggests a role as a kinase regulator. *J. Mol. Biol.*, **297**, 1159–1170.

Bastow, R., Mylne, J.S., Lister, C., Lippman, Z., Martienssen, R.A. & Dean, C. (2004) Vernalization requires epigenetic silencing of FLC by histone methylation. *Nature*, **427**, 164–167.

Bernier, G., Havelange, A., Houssa, C., Petitjean, A. & Lejeune, P. (1993) Physiological signals that induce flowering. *Plant Cell*, **5**, 1147–1155.

Beveridge, C.A. & Murfet, I.C. (1996) The gigas mutant in pea is deficient in the floral stimulus. *Physiol. Plant.*, **96**, 637–645.

Blazquez, M.A., Ahn, J.H. & Weigel, D. (2003) A thermosensory pathway controlling flowering time in *Arabidopsis thaliana*. *Nature Genet.*, 33, 168–171.

Bonhomme, F., Kurz, B., Melzer, S., Bernier, G. & Jacqmard, A. (2000) Cytokinin and gibberellin activate SaMADS A, a gene apparently involved in regulation of the floral transition in *Sinapis alba*. *Plant J.*, **24**, 103–111.

Booker, J., Chatfield, S. & Leyser, O. (2003) Auxin acts in xylem-associated or medullary cells to mediate apical dominance. *Plant Cell*, **15**, 495–507.

Borner, R., Kampmann, G., Chandler, J., Gleissner, R., Wisman, E., Apel, K. & Melzer, S. (2000) A MADS domain gene involved in the transition to flowering in *Arabidopsis*. *Plant J.*, **24**, 591–599.

Bouhidel, K. & Irish, V.F. (1996) Cellular interactions mediated by the homeotic PISTILLATA gene determine cell fate in the *Arabidopsis* flower. *Dev. Biol.*, **174**, 22–31.

Bradley, D., Ratcliffe, O., Vincent, C., Carpenter, R. & Coen, E. (1997) Inflorescence commitment and architecture in *Arabidopsis*. *Science*, **275**, 80–83.

Carpenter, R. & Coen, E. (1990) Floral homeotic mutations produced by transposon-mutagenesis in *Antirrhinum majus*. *Genes Dev.*, **4**, 1483–1493.

Carpenter, R. & Coen, E.S. (1995) Transposon induced chimaeras show that floricaula, a meristem identity gene, acts non-autonomously between cell-layers. *Development*, **121**, 19–26.

Cerdan, P.D. & Chory, J. (2003) Regulation of flowering time by light quality. *Nature*, **423**, 881–885.

Coen, E.S., Romero, J.M., Doyle, S., Elliott, R., Murphy, G. & Carpenter, R. (1990) Floricaula – a homeotic gene required for flower development in *Antirrhinum majus*. *Cell*, **63**, 1311–1322.

Colasanti, J. & Sundaresan, V. (2000) 'Florigen' enters the molecular age: long-distance signals that cause plants to flower. *Trends Biochem. Sci.*, **25**, 236–240.

Colasanti, J., Yuan, Z. & Sundaresan, V. (1998) The indeterminate gene encodes a zinc finger protein and regulates a leaf-generated signal required for the transition to flowering in maize. *Cell*, **93**, 593–603.

Corbesier, L., Havelange, A., Lejeune, P., Bernier, G. & Perilleux, C. (2001) N content of phloem and xylem exudates during the transition to flowering in *Sinapis alba* and *Arabidopsis thaliana*. *Plant Cell Environ.*, **24**, 367–375.

Crawford, K.M. & Zambryski, P.C. (1999) Plasmodesmata signaling: many roles, sophisticated statutes. *Curr. Opin. Plant Biol.*, **2**, 382–387.

Ding, B., Haudenshield, J.S., Hull, R.J., Wolf, S., Beachy, R.N. & Lucas, W.J. (1992) Secondary plasmodesmata are specific sites of localization of the tobacco mosaic-virus movement protein in transgenic tobacco plants. *Plant Cell*, **4**, 915–928.

Efremova, N., Perbal, M.-C., Yephremov, A., Hofmann, W.A., Saedler, H. & Schwarz-Sommer, Z. (2001) Epidermal control of floral organ identity by class B homeotic genes in *Antirrhinum* and *Arabidopsis*. *Development*, **128**, 2661–2671.

Fowler, S., Lee, K., Onouchi, H., Samach, A., Richardson, K., Coupland, G. & Putterill, J. (1999) *GIGANTEA*: a circadian clock-controlled gene that regulates photoperiodic flowering in *Arabidopsis* and encodes a protein with several possible membrane-spanning domains. *EMBO J.*, **18**, 4679–4688.

Furner, I.J., Ainscough, J.F.X., Pumfrey, J.A. & Petty, L.M. (1996) Clonal analysis of the late flowering fca mutant of *Arabidopsis thaliana*: cell fate and cell autonomy. *Development*, **122**, 1041–1050.

Gisel, A., Hempel, F.D., Barella, S. & Zambryski, P. (2002) Leaf-to-shoot apex movement of symplastic tracer is restricted coincident with flowering in *Arabidopsis*. *Proc. Natl. Acad. Sci. U.S.A.*, **99**, 1713–1717.

Hantke, S.S., Carpenter, R. & Coen, E.S. (1995) Expression of floricaula in single cell layers of periclinal chimaeras activates downstream homeotic genes in all layers of floral meristems. *Development*, **121**, 27–35.

Havelange, A., Lejeune, P. & Bernier, G. (2000) Sucrose/cytokinin interaction in *Sinapis alba* at floral induction: a shoot-to-root-to-shoot physiological loop. *Physiol. Plant.*, **109**, 343–350.

Hayama, R. & Coupland, G. (2003) Shedding light on the circadian clock and the photoperiodic control of flowering. *Curr. Opin. Plant Biol.*, **6**, 13–19.

Helariutta, Y., Fukaki, H., Wysocka-Diller, J., Nakajima, K., Jung, J., Sena, G., Hauser, M.T. & Benfey, P.N. (2000) The SHORT-ROOT gene controls radial patterning of the *Arabidopsis* root through radial signaling. *Cell*, **101**, 555–567.

Hepworth, S.R., Valverde, F., Ravenscroft, D., Mouradov, A. & Coupland, G. (2002) Antagonistic regulation of flowering-time gene SOC1 by CONSTANS and FLC via separate promoter motifs. *EMBO J.*, **21**, 4327–4337.

Huala, E. & Sussex, I.M. (1993) Determination and cell-interactions in reproductive meristems. *Plant Cell*, **5**, 1157–1165.

Jackson, D., Veit, B. & Hake, S. (1994) Expression of maize KNOTTED1 related homeobox genes in the shoot apical meristem predicts patterns of morphogenesis in the vegetative shoot. *Development*, **120**, 405–413.

Jenik, P.D. & Irish, V.F. (2001) The *Arabidopsis* floral homeotic gene APETALA3 differentially regulates intercellular signaling required for petal and stamen development. *Development*, **128**, 13–23.

Kardailsky, I., Shukla, V.K., Ahn, J.H., Dagenais, N., Christensen, S.K., Nguyen, J.T., Chory, J., Harrison, M.J. & Weigel, D. (1999) Activation tagging of the floral inducer *FT*. *Science*, **286**, 1962–1965.

Kim, J.Y., Yuan, Z., Cilia, M., Khalfan-Jagani, Z. & Jackson, D. (2002) Intercellular trafficking of a KNOTTED1 green fluorescent protein fusion in the leaf and shoot meristem of *Arabidopsis*. *Proc. Natl. Acad. Sci. U.S.A.*, **99**, 4103–4108.

Kim, J.Y., Yuan, Z. & Jackson, D. (2003) Developmental regulation and significance of KNOX protein trafficking in *Arabidopsis*. *Development*, **130**, 4351–4362.

Kim, M., Canio, W., Kessler, S. & Sinha, N. (2001) Developmental changes due to long-distance movement of a homeobox fusion transcript in tomato. *Science*, **293**, 287–289.

King, R.W. & Zeevaart, J.A.D. (1973) Floral stimulus movement in *Perilla* and flower inhibition caused by noninduced leaves. *Plant Physiol.*, **51**, 727–738.

King, W.M. & Murfet, I.C. (1985) Flowering in *Pisum*: a sixth locus, *DNE*. *Ann. Bot.*, **56**, 835–846.

Knott, J.E. (1934) Effect of a localized photoperiod on spinach. *Proc. Soc. Horticult. Sci.*, **31**, 152–154.

Kobayashi, Y., Kaya, H., Goto, K., Iwabuchi, M. & Araki, T. (1999) A pair of related genes with antagonistic roles in mediating flowering signals. *Science*, **286**, 1960–1962.

Koornneef, M., Alonso-Blanco, C., Vries, H.B.-D., Hanhart, C.J. & Peeters, A.J.M. (1998) Genetic interactions among late-flowering mutants of *Arabidopsis*. *Genetics*, **148**, 885–892.

Koornneef, M., Hanhart, C.J. & Van Der Veen, J.H. (1991) A genetic and physiological analysis of late flowering mutants in *Arabidopsis thaliana*. *Mol. Gen. Genet.*, **229**, 57–66.

Laibach, F. (1951) Uber sommer und winterannuelle Rasse von *Arabidopsis thaliana* (L.) Heynh. Ein Beitrag zur Atiologie der Blutenbildung. *Beitr. Biol. Pflanz.*, **28**, 173–210.

Lee, H., Suh, S.-S., Park, E., Cho, E., Ahn, J.H., Kim, S.-G., Lee, J.S., Kwon, Y.M. & Lee, I. (2000) The AGAMOUS-LIKE 20 MADS domain protein integrates floral inductive pathways in *Arabidopsis*. *Genes Dev.*, **14**, 2366–2376.

Lucas, W.J., Bouche-Pillon, S., Jackson, D.P., Nguyen, L., Baker, L., Ding, B. & Hake, S. (1995) Selective trafficking of KNOTTED1 homeodomain protein and its mRNA through plasmodesmata. *Science*, **270**, 1980–1983.

Lucas, W.J., Yoo, B.C. & Kragler, F. (2001) RNA as a long-distance information macromolecule in plants. *Nat. Rev. Mol. Cell Biol.*, **2**, 849–857.

Mandel, M.A., Gustafson-Brown, C., Savidge, B. & Yanofsky, M.F. (1992) Molecular characterization of the *Arabidopsis* floral homeotic gene APETALA1. *Nature*, **360**, 273–277.

Martinez-Zapater, J.M., Coupland, G., Dean, C. & Koornneef, M. (1994) The transition to flowering in *Arabidopsis*. In: *Arabidopsis, Cold Spring Harbor Monograph Series* (eds E.M. Meyerowitz & C.R. Somerville), pp. 403–433. Cold Spring Harbor Laboratory Press, Plainview, NY.

Michaels, S.D. & Amasino, R.M. (1999) FLOWERING LOCUS C encodes a novel MADS domain protein that acts as a repressor of flowering. *Plant Cell*, **11**, 949–956.

Michaels, S.D. & Amasino, R.M. (2000) Memories of winter: vernalization and the competence to flower. *Plant Cell Environ.*, **23**, 1145–1153.

Michaels, S.D. & Amasino, R.M. (2001) Loss of FLOWERING LOCUS C activity eliminates the late-flowering phenotype of FRIGIDA and autonomous pathway mutations but not responsiveness to vernalization. *Plant Cell*, **13**, 935–941.

Moon, J., Suh, S.S., Lee, H., Choi, K.R., Hong, C.B., Paek, N.C., Kim, S.G. & Lee, I. (2003) The SOC1 MADS-box gene integrates vernalization and gibberellin signals for flowering in *Arabidopsis*. *Plant J.*, **35**, 613–623.

Mouradov, A., Cremer, F. & Coupland, G. (2002) Control of flowering time: interacting pathways as a basis for diversity. *Plant Cell*, **14** (Suppl.), S111–S130.

Murfet, I.C. (1971) Flowering in *Pisum*: reciprocal grafts between known genotypes. *Aust. J. Biol. Sci.*, **24**, 1089–1101.

Murfet, I.C. (1985) *Pisum sativum*. In: *Handbook of Flowering*, Vol. 4 (ed. A.H. Halevy), pp. 97–126. CRC Press, Boca Raton, FL.

Nakajima, K., Sena, G., Nawy, T. & Benfey, P.N. (2001) Intercellular movement of the putative transcription factor SHR in root patterning. *Nature*, **413**, 307–311.

Noh, Y.S. & Amasino, R.M. (2003) PIE1, an ISWI family gene, is required for FLC activation and floral repression in *Arabidopsis*. *Plant Cell*, **15**, 1671–1682.

Oparka, K.J., Roberts, A.G., Boevink, P., Santa Cruz, S., Roberts, L., Pradel, K.S., Imlau, A., Kotlizky, G., Sauer, N. & Epel, B. (1999) Simple, but not branched, plasmodesmata allow the nonspecific trafficking of proteins in developing tobacco leaves. *Cell*, **97**, 743–754.

Palauqui, J.C., Elmayan, T., Pollien, J.M. & Vaucheret, H. (1997) Systemic acquired silencing: transgene-specific post-transcriptional silencing is transmitted by grafting from silenced stocks to non-silenced scions. *EMBO J.*, **16**, 4738–4745.

Park, D.H., Somers, D.E., Kim, Y.S., Choy, Y.H., Lim, H.K., Soh, M.S., Kim, H.J., Kay, S.A. & Nam, H.G. (1999) Control of circadian rhythms and photoperiodic flowering by the *Arabidopsis GIGANTEA* gene. *Science*, **285**, 1579–1582.

Perbal, M.C., Haughn, G., Saedler, H. & SchwarzSommer, Z. (1996) Non-cell-autonomous function of the *Antirrhinum* floral homeotic proteins DEFICIENS and GLOBOSA is exerted by their polar cell-to-cell trafficking. *Development*, **122**, 3433–3441.

Perilleux, C. & Bernier, G. (2002) The control of flowering: do genetical and physiological approaches converge? *Annu. Plant Rev.*, **6**, 1–32.

Pires-daSilva, A. & Sommer, R.J. (2003) The evolution of signalling pathways in animal development. *Nat. Rev. Genet.*, **4**, 39–49.

Pnueli, L., Carmel-Goren, L., Hareven, D., Gutfinger, T., Alvarez, J., Ganal, M., Zamir, D. & Lifschitz, E. (1998) The *SELF-PRUNING* gene of tomato regulates vegetative to reproductive switching of sympodial meristems and is the ortholog of *CEN* and *TFL1*. *Development*, **125**, 1979–1989.

Pnueli, L., Gutfinger, T., Hareven, D., Ben-Naim, O., Ron, N., Adir, N. & Lifschitz, E. (2001) Tomato SP-interacting proteins define a conserved signaling system that regulates shoot architecture and flowering. *Plant Cell*, **13**, 2687–2702.

Pouteau, S., Nicholls, D., Tooke, F., Coen, E. & Battey, N. (1997) The induction and maintenance of flowering in Impatiens. *Development*, **124**, 3343–3351.

Putterill, J., Robson, F., Lee, K., Simon, R. & Coupland, G. (1995) The *CONSTANS* gene of *Arabidopsis* promotes flowering and encodes a protein showing similarities to zinc finger transcription factors. *Cell*, **80**, 847–857.

Redei, G.P. (1962) Supervital mutants of *Arabidopsis*. *Genetics*, **47**, 443–460.

Reeves, P.H. & Coupland, G. (2001) Analysis of flowering time control in *Arabidopsis* by comparison of double and triple mutants. *Plant Physiol.*, **126**, 1085–1091.

Robson, F., Costa, M.M.R., Hepworth, S., Vizir, I., Pineiro, M., Reeves, P.H., Putterill, J. & Coupland, G. (2001) Functional importance of conserved domains in the flowering-time gene CONSTANS demonstrated by analysis of mutant alleles and transgenic plants. *Plant J.*, **28**, 619–631.

Ruiz-Garcia, L., Madueno, F., Wilkinson, M., Haughn, G., Salinas, J. & Martinez-Zapater, J.M. (1997) Different roles of flowering-time genes in the activation of floral initiation genes in *Arabidopsis*. *Plant Cell*, **9**, 1921–1934.

Ruiz-Medrano, R., Xoconostle-Cazares, B. & Lucas, W.J. (2001) The phloem as a conduit for inter-organ communication. *Curr. Opin. Plant Biol.*, **4**, 202–209.

Samach, A., Onouchi, H., Gold, S.E., Ditta, G.S., Schwarz-Sommer, Z., Yanofsky, M.F. & Coupland, G. (2000) Distinct roles of *CONSTANS* target genes in reproductive development of *Arabidopsis*. *Science*, **288**, 1613–1616.

Schmid, M., Uhlenhaut, N.H., Godard, F., Demar, M., Bressan, R., Weigel, D. & Lohmann, J.U. (2003) Dissection of floral induction pathways using global expression analysis. *Development*, **130**, 6001–6012.

Searle, I. & Coupland, G. (2004) Induction of flowering by seasonal changes in photoperiod. *EMBO J.*, **23**, 1217–1222.

Sessions, A., Yanofsky, M.F. & Weigel, D. (2000) Cell-cell signaling and movement by the floral transcription factors *LEAFY* and *APETALA1*. *Science*, **289**, 779–781.

Sheldon, C.C., Burn, J.E., Perez, P.P., Metzger, J., Edwards, J.A., Peacock, W.J. & Dennis, E.S. (1999) The FLF MADS box gene: a repressor of flowering in *Arabidopsis* regulated by vernalization and methylation. *Plant Cell*, **11**, 445–458.

Sheldon, C.C., Conn, A.B., Dennis, E.S. & Peacock, W.J. (2002) Different regulatory regions are required for the vernalization-induced repression of FLOWERING LOCUS C and for the epigenetic maintenance of repression. *Plant Cell*, **14**, 2527–2537.

Sheldon, C.C., Rouse, D.T., Finnegan, E.J., Peacock, W.J. & Dennis, E.S. (2000) The molecular basis of vernalization: the central role of FLOWERING LOCUS C (FLC). *Proc. Natl. Acad. Sci. U.S.A.*, **97**, 3753–3758.

Sieburth, L.E., Drews, G.N. & Meyerowitz, E.M. (1998) Non-autonomy of *AGAMOUS* function in flower development: use of a Cre/loxP method for mosaic analysis in *Arabidopsis*. *Development*, **125**, 4303–4312.

Simpson, G.G. & Dean, C. (2002) *Arabidopsis*, the Rosetta stone of flowering time? *Science*, **296**, 285–289.

Simpson, G.G., Dijkwel, P.P., Quesada, V., Henderson, I. & Dean, C. (2003) FY is an RNA 3′ end-processing factor that interacts with FCA to control the *Arabidopsis* floral transition. *Cell*, **113**, 777–787.

Sommer, H., Beltran, J.P., Huijser, P., Pape, H., Lonnig, W.E., Saedler, H. & Schwarzsommer, Z. (1990) Deficiens, a homeotic gene involved in the control of flower morphogenesis in *Antirrhinum majus* – the protein shows homology to transcription factors. *EMBO J.*, **9**, 605–613.

Suarez-Lopez, P., Wheatley, K., Robson, F., Onouchi, H., Valverde, F. & Coupland, G. (2001) *CONSTANS* mediates between the circadian clock and the control of flowering in *Arabidopsis*. *Nature*, **410**, 1116–1120.

Sung, S.B. & Amasino, R.M. (2004) Vernalization in *Arabidopsis thaliana* is mediated by the PHD finger protein VIN3. *Nature*, **427**, 159–164.

Takada, S. & Goto, K. (2003) TERMINAL FLOWER2, an *Arabidopsis* homolog of HETEROCHROMATIN PROTEIN1, counteracts the activation of FLOWERING LOCUS T by CONSTANS in the vascular tissues of leaves to regulate flowering time. *Plant Cell*, **15**, 2856–2865.

Troebner, W., Ramirez, L., Motte, P., Hue, I., Huijser, P., Loennig, W.E., Saedler, H., Sommer, H. & Schwarz-Sommer, Z. (1992) Globosa – a homeotic gene which interacts with deficiens in the control of *Antirrhinum* floral organogenesis. *EMBO J.*, **11**, 4693–4704.

Truernit, E., Schmid, J., Epple, P., Illig, J. & Sauer, N. (1996) The sink-specific and stress-regulated *Arabidopsis STP4* gene: enhanced expression of a gene encoding a monosaccharide transporter by wounding, elicitors, and pathogen challenge. *Plant Cell*, **8**, 2169–2182.

Valverde, F., Mouradov, A., Soppe, W., Ravenscroft, D., Samach, A. & Coupland, G. (2004) Photoreceptor regulation of CONSTANS protein and the mechanism of photoperiodic flowering. *Science*, **303**, 1003–1006.

Vandenberg, C., Willemsen, V., Hage, W., Weisbeek, P. & Scheres, B. (1995) Cell fate in the *Arabidopsis* root-meristem determined by directional signaling. *Nature*, **378**, 62–65.

Vaughn, J.E. (1955) The morphology and growth of the vegetative and reproductive apices of *Arabidopsis thaliana* (L.) Heynh., *Capsella bursapastoris* (L.) Medic. & *Anagallis arvensis*. *J. Linn. Soc. Bot.*, **55**, 279–301.

Weigel, D. & Meyerowitz, E.M. (1993) Activation of floral homeotic genes in *Arabidopsis*. *Science*, **261**, 1723–1726.

Weigel, D. & Nilsson, O. (1995) A developmental switch sufficient for flower initiation in diverse plants. *Nature*, **377**, 495–500.

Weigel, D., Alvarez, J., Smyth, D.R., Yanofsky, M.F. & Meyerowitz, E.M. (1992) Leafy controls floral meristem identity in *Arabidopsis*. *Cell*, **69**, 843–859.

Weller, J.L., Murfet, I.C. & Reid, J.B. (1997a) Pea mutants with reduced sensitivity to far-red light define an important role for phytochrome A in day-length detection. *Plant Physiol.*, **114**, 1225–1236.

Weller, J.L., Reid, J.B., Taylor, S.A. & Murfet, I.C. (1997b) The genetic control of flowering in pea. *Trends Plant Sci.*, **2**, 412–418.

Wilson, R.N., Heckman, J.W. & Somerville, C.R. (1992) Gibberellin is required for flowering in *Arabidopsis thaliana* under short days. *Plant Physiol.*, **100**, 403–408.

Wu, X., Dinneny, J.R., Crawford, K.M., Rhee, Y., Citovsky, V., Zambryski, P.C. & Weigel, D. (2003) Modes of intercellular transcription factor movement in the *Arabidopsis* apex. *Development*, **130**, 3735–3745.

Wu, X.L., Weigel, D. & Wigge, P.A. (2002) Signaling in plants by intercellular RNA and protein movement. *Genes Dev.*, **16**, 151–158.

Yanovsky, M.J. & Kay, S.A. (2002) Molecular basis of seasonal time measurement in *Arabidopsis*. *Nature*, **419**, 308–312.

Yanovsky, M.J. & Kay, S.A. (2003) Living by the calendar: how plants know when to flower. *Nat. Rev. Mol. Cell Biol.*, **4**, 265–275.

Zeevaart, J.A.D. (1976) Physiology of flower formation. *Annu. Rev. Plant Physiol.*, **27**, 321–348.

Zeevaart, J.A.D. (1985) Perilla. In: *CRC Handbook of Flowering*, Vol. 5 (ed. A.H. Halevy), pp. 239–252. CRC Press, Boca Raton, FL.

8 Lessons from the root apex

Martin Bonke, Sari Tähtiharju and Ykä Helariutta

8.1 Introduction

The *Arabidopsis* root with its simple structure serves as an excellent model for understanding plant organogenesis. In this chapter, we shall review the recent progress in understanding intercellular communication underlying pattern formation, cell division and cell differentiation during root development.

8.2 Organization of the root

8.2.1 Anatomy of the root meristem and procambium in the apex of a growing root

The principal tissues of *Arabidopsis* primary root are arranged simply, and in concentric layers (see Plate 8.1, following page 146). In the mature region of the root, single layers of epidermis (surrounded by the lateral root cap at the most distal region of the root tip), cortex, endodermis and pericycle surround a small number of vascular cells (Dolan *et al.*, 1993; Scheres *et al.*, 1994). All these layers of the *Arabidopsis* root arise from files of cells that originate from a small number of initial cells at the tip of the root, defined as a meristem. Internal to and contacting all the initials are a small number of central cells that are mitotically inactive and are known as the quiescent centre (QC) (Clowes, 1956; Dolan *et al.*, 1993). On the proximal (shoot) side of the QC, a tier of cells comprises the pericycle initials and the vascular tissue initials, the latter dividing to form the youngest xylem and phloem cells. At the radial flanks are the initials that give rise to the endodermal and cortical layers. At the distal (root apex) side of the QC are the initials for both the columella and lateral root cap as well as the epidermis, with the lateral root cap and epidermis appearing to originate from the same initials (Dolan *et al.*, 1993). All initial cells together are defined as the promeristem, the minimal group of cells that is capable of making all tissues by ordered divisions. With each division the initials add one cell to the plant body, while the initial cell retains its position within the meristem. This division of initials can be either solely anticlinal (resulting in a single file of cells) or first periclinal then anticlinal (resulting in two or more cell layers). The columella initials divide only anticlinally and their progeny undergo rapid cell expansion and then differentiate. The other three types of initials undergo both periclinal and anticlinal divisions, resulting in cell lineages that acquire different identities (Dolan *et al.*, 1993). In contrast to other cell lineages, vascular cell lineages appear to arise slightly proximally to the promeristem. In serial sections of the primary root meristem, Mähönen et al. (2000) and Baum et al. (2002) observed that xylem cell lineages form an axis composed of four to five cell files very close to the underlying QC. Two domains of initials that give rise to the phloem cell lineages and to the undifferentiated procambial cell lineages through asymmetric cell divisions flank this axis. The number and exact pattern of these procambial divisions show some variability between individual seedlings, which is in contrast to the invariant pattern of cell lineages in the endodermis and outer layers (Scheres et al., 1994).

Upon germination, cells of the meristem begin to divide and the root expands axially as a result of cell expansion. As the root continues to grow, the number of cells in the meristem increases and the rate of cell production increases. Cells are organized along the longitudinal axis in the root in distinct developmental zones (Dolan et al., 1993). Cells are produced at the basal region of the meristem, the meristematic/division zone, which is overlaid by the root cap. Proximal to this zone is the elongation zone. The next zone is the differentiation zone, in which elongated cells from different tissues mature into fully differentiated cells. Because root growth is indeterminate, these processes are continual, resulting in all developmental stages being present at all times. The radial symmetry of the root combined with a lack of cell movement means that clonally related cells are frequently found in cell files.

8.2.2 Cellular organization of the root is established during embryonic development

The basic layout of the mature plant, including root, is established during embryogenesis. In *Arabidopsis*, the zygote elongates approximately threefold and subsequently divides asymmetrically, resulting in a large basal and a smaller apical cell (Mansfield & Briarty, 1991). The octant stage (Fig. 8.1A) is reached after three rounds of cell division of the apical cell. At this stage the embryo constitutes of the upper and lower tiers, which will produce the apical and basal areas of the later stages of embryo and seedling respectively. Meanwhile, the basal cell divides only horizontally and forms the suspensor structure; only the top cell (hypophysis) of the suspensor will be part of the later seedling as the QC and the columella root cap (Scheres et al., 1994).

Subsequent tangential cell divisions of the octant result in the formation of the epidermal (protoderm) layer: dermatogen stage (Fig. 8.1B). Succeeding cell divisions lead to the globular stage (Figs. 8.2C and 8.2D), where the inner cell divisions are oriented in the apical–basal plane, and axis formation can be seen. After the next set of divisions, clear polarity of the embryo is established when in the lower tiers the procambium is formed. In the upper tiers, the cell divisions start the formation of two cotyledon primordia, which leads to the triangular shape embryo (Fig. 8.1E). Cell number is now increased to more than a hundred and in the lower tiers the ground tissue, epidermis and vascular initials are visible. The hypophysis also starts to divide at this stage.

At the heart stage (Fig. 8.1F), most tissue types are present, cotyledon primordia start to grow and the shoot apical meristem is initiated. Also the QC and columella

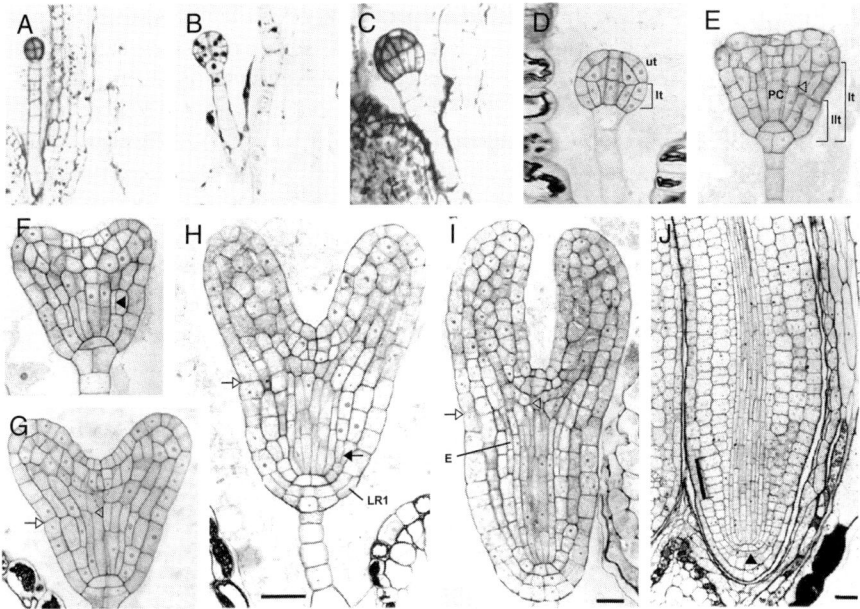

Figure 8.1 Stages of *Arabidopsis* embryogenesis: (A) octant stage; four of eight cells; (B) dermatogen stage, the inner cells are separated from the epidermis by a set of tangential cell divisions; (C) early globular stage, the embryonic axis is visible for the first time; (D) globular stage; (E) triangular stage; (F) early heart stage; (G) mid heart stage; (H) late heart stage; (I) mid torpedo stage; (J) mature embryo stage; ut, upper tier; lt, lower tier; llt, lower region of lt; PC, procambium; LR1, 1st lateral root cap layer; E, endodermis; Closed arrows and arrowheads: cell walls marking divisions. Adapted from Scheres *et al.* (1994) and from Berleth and Chatfield (2002).

root cap initials are formed. During the next subsequent heart, torpedo and bent cotyledon (Figs. 8.2F and 8.2H) stages, development of the embryo is concluded and all tissue types are set in place, leading to a final structure that has similar structural complexity as that of a seedling. The actual differentiation of most tissues is delayed until after germination.

During embryogenesis the different tissues form a radially symmetrical pattern around the apical–basal axis. These layers are established in stages. In *Arabidopsis*, they start forming just after the octant stage when the epidermal cell layer is created through periclinal cell divisions. During the succeeding stages of embryogenesis the other tissues are formed from the inner cells. These cells undergo periclinal cell divisions, which increase cell number and establish the ground tissue layer and provascular bundle. At the transition of globular to triangular stage, the provascular cells divide once more periclinally to separate the pericycle layer from rest of the vascular bundle. The ground tissue initial cells will also undergo an asymmetric periclinal cell division to form the endodermis (inside) and cortex (outside) layers. During the last stage of embryogenesis, cell number in the vascular bundle increases and the phloem–xylem pattern is completed.

As the cellular organization of the meristem in an embryonic root is very similar to that in the growing root, an important question was to determine if the cellular organization of an embryo serves as the functional basis for that of the growing root. To answer this question the fate of the descendents of *Arabidopsis* embryo cells was determined using the clonal analysis technique. This technique depends on the possibility to use a heritable genetic marker that can be traced in the descendents of that cell. Since all descendents will possess this marker, they are considered clones. Once the embryo has matured and germinated it can thus be determined how much the original cell has contributed to the mature seedling. Factors such as the size and distribution of the clonal area can be used to determine developmental parameters such as cell division rates, founder cell number, etc. For shoot tissues several markers are available, mostly involving intense staining with anthocyanins or pigment loss due to chlorophyll deficiency; however, they do not function in root tissues. Scheres *et al.* (1994) solved this problem by transforming a disrupted β-glucuronidase (GUS) gene under the 35S promoter into the *Arabidopsis* genome. The GUS gene was disrupted with a maize *Ac* transposable element. Once this excised an intact GUS, coding sequence was restored and blue staining could be monitored in those cells where excision occurred. Using this technique Scheres *et al.* (1994) analysed large blue-stained clones that would demarcate the early embryonic divisions. The larger the sector, the earlier the excision event must have occurred. Since most of the sector boundaries either coincided or were closely associated with embryonic cell divisions, they were able to establish a 'fate map', which predicted the probable fate of embryonic cells (Fig. 8.2).

The embryonic boundary between root and hypocotyl was determined and this separation was mapped to divisions at the early heart stage. However, since some of

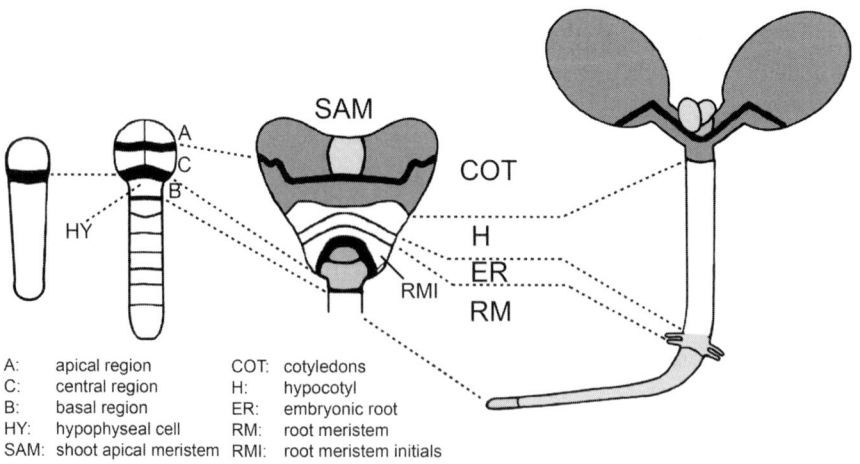

Figure 8.2 Embryonic origin of the *Arabidopsis* root. From left to right: first zygotic division; octant stage embryo; heart stage embryo; seedling. Adapted from Scheres *et al.* (2002).

these sectors sometimes crossed over between root and hypocotyl, it was concluded that individual cells can change hypocotyl/root fate depending on yet unidentified positional cues until the end of embryonic development. It was also determined that the promeristem (the QC with the surrounding initials) is established during the heart stage and maintained thereafter. Unlike the root–hypocotyl sectors, the lineage of the hypophysis showed no variation and its descendants always give rise to the QC and columella initials. The cells in the small part of the embryonic root that is not a part of the meristem will give rise to the 'collet', the junction between root and hypocotyl with typical epidermal morphology and early emerging root hairs. Clonal and anatomical analyses also show that the most basal epidermal cell will become the initial of the lateral root cap. This takes place at the heart stage of embryogenesis.

8.2.3 Development of secondary roots

In most plant species, secondary (such as lateral) roots are initiated from the pericycle cells that are associated with protoxylem. These particular pericycle cells are distinguishable from the other pericycle cells by the increased number of cell divisions in the axial plane, causing them to be shorter. When a secondary root is initiated, the pericycle cells undergo a set of periclinal cell divisions, increasing cell layer number to between 2 and 4. These cells then increase in size, causing the bud to emerge from the root. The radial patterning of endodermis, cortex and epidermis of the prospective lateral root is then already in place (Malamy & Benfey, 1997). After the lateral root primordium emergences, cell division rate increases and a new meristem is formed. Laskowski *et al.* (1995) showed by excision experiments that a functional meristem has developed at the 3–5 cell layer stage. Auxin treatments indicated that the lateral root formation is a two-step process where first a set of cells is initiated to divide, followed by activation of those cells to form an organized meristem. Dubrovsky *et al.* (2000) proposed that by maintaining cell proliferation the root is prepared for environmental changes and can activate the formation of secondary roots once the need for them is high. As the anatomy of a secondary root is very similar to that of the primary root, the emerging picture is that secondary root development recapitulates genetic processes controlling development of a primary root.

8.3 Cell fate studies of the growing root

As the root meristem gives rise to clonal cell files, it was originally proposed that clonal relationships are important for patterning of the root tissues. This view was already challenged by the observation of the mitotically inactive QC in the focal point of the clonal cell files by thymidine radiolabelling studies (Clowes, 1956). Experimental studies on young *Arabidopsis* seedlings during 1990s resolved the question of the clonal relationships, initial cells and the role of QC (van den Berg *et al.*, 1995, 1997; Berger *et al.*, 1998; Kidner *et al.*, 2000).

In the classical study by van den Berg *et al.* (1995), each of the four cells of the QC was laser ablated. This resulted in expansion and division of the cells in the proximal vascular bundle to replace the dead cells. Furthermore, the study of cell-specific markers showed that these invading cells expressed the root cap marker instead of vascular marker in their new position (van den Berg *et al.*, 1995). This was the first demonstration that cell fates in the root are not permanent but can be changed based on their position in relation to other cells.

Although this study revealed the important principle of position-dependent cell fates, the role of QC itself remained open. That was revealed by a subsequent study in which only one of the QC cells was ablated at a time (van den Berg *et al.*, 1997). Ablation of one QC cell resulted in cessation of cell division of the columella initials, which were in direct contact with the ablated cell but not in those that were contacting the remaining intact QC cells. This demonstrated that contact with QC cells keeps cells in an initial-specific, less differentiated state. Thus, central cells control cell differentiation possibly in a cell-contact-dependent manner. Furthermore, examination of the fates of QC cells and initials by inducing clones in young meristems using a heat-inducible transposon-based marker indicated that the central cells (QC) divide at a low frequency and that their daughter cells end up in various cell files in a relatively random fashion (Kidner *et al.*, 2000).

Positional information also determines cell fate in the radial plane. Pericycle cells were able to replace ablated cortical initials, indicating that pericycle cells switch fate when moved over radial clonal boundaries. Cortical cells also respond to changes in position (van den Berg *et al.*, 1995). When epidermal initials were ablated, a neighbouring cortical cell occupied its position and took up its fate (van den Berg *et al.*, 1995, Berger *et al.*, 1998). This indicated that cell fate in the epidermis is determined by position relative to the underlying cell layer, the cortex (Berger *et al.*, 1998). Laser ablation studies indicated that all cell types can adopt position-dependent fates (van den Berg *et al.*, 1995). This was also shown by clonal analysis studies (Berger *et al.*, 1998, Kidner *et al.*, 2000) in which a rare class of sectors was observed that spanned the tissue radially, indicating that also in a growing root sometimes cells that have a clonal origin end up in adjacent radial cell files.

In order to understand the direction of information flow underlying the radial patterning in the root apex, van den Berg *et al.* (1995) uncoupled the initial cells from the older tissue. Isolation of ground tissue initial cells from more mature ground tissue cells by ablation of their apically located daughters interferes with the change in orientation of the cell division plane required to form new endodermal and cortical cells. Thus, positional signals can be perpetuated from more mature to initial cells to guide the pattern of meristem cell differentiation (van den Berg *et al.*, 1995). On the other hand, since the QC delays differentiation or promotes stem cell fate, these observations may also reflect an accumulation of stem cell promoting factors. Laser ablation and genetic data have further shown that it is a balance between short-range signals inhibiting differentiation and signals that reinforce cell fate decisions which control pattern formation in the root meristem (van den Berg *et al.*, 1997). Therefore, root development involves cell-to-cell signalling.

Taken together, in each of these sets of experiments, cells had effectively undergone a post-embryonic change in their position and in response exhibited a change in their developmental fate (van den Berg et al., 1995, 1997; Berger et al., 1998; Kidner et al., 2000). This suggests that positional information is provided both embryonically and post-embryonically to ensure appropriate cell specification in the *Arabidopsis* root (Berger et al., 1998; Kidner et al., 2000). Below we will review the molecular nature of such intercellular signalling.

8.4 Molecular genetics of root development

8.4.1 Distal patterning

Classical experiments already implicated auxin as a growth substance promoting development of the whole root system. More recently, Sabatini *et al.* (1999) showed that a promoter consisting of 7 tandem repeats of an auxin response element fused to the GUS gene led to a maximum of GUS activity at the meristematic region where auxin also accumulates, that is in the columella initials under the QC (Fig. 8.3). They also showed that treatment with the chemical auxin transport inhibitor naphtylphtalamic acid resulted in patterning defects, while introduction of a new auxin maximum elsewhere in the root resulted in the formation of a new meristem. These results strongly suggested that auxin is an important determinant of root meristem formation.

In mature plants, auxin is mainly produced in the shoot tissues and is most likely transported to the site of action in two different manners: polar and non-polar transport. Polar auxin transport is an active process involving specific efflux carriers. Long-distance transport through the phloem, however, is thought to be non-polar. This assumption is based on three arguments. Polar auxin transport is known to have a velocity of approximately 10 mm/h, which would make it too slow to be an efficient signalling method in large plant species. Also high concentrations of free auxin have been measured in phloem exudates (Baker, 2000). Finally, *aux1* mutants are both unable to load auxin into the phloem in the shoot and unload auxin from the phloem at the root. This means they cannot transport auxin between shoot and root (Swarup et al., 2001; Marchant et al., 2002), indicating that AUX1, a putative auxin permease, acts at both ends of the non-polar auxin transport route.

In embryos on the other hand, polar auxin transport seems to be the main mechanism of achieving an auxin gradient. Throughout the formation of the embryo, auxin seems to have an important role in defining the pattern formation. Already during the earliest stages of embryogenesis, auxin is actively transported into the embryo by the PIN7 efflux carrier (Friml et al., 2003). PIN7 is localized at the apical membrane of the basal cell just after the division of the zygote. Since in *pin7* mutants the apical cell fate is disrupted, it can be concluded that the actively maintained auxin accumulation in the apical cell leads directly or indirectly to correct specification of this apical cell (see also Chapter 1 for a more detailed exploration of the role of auxin in signalling).

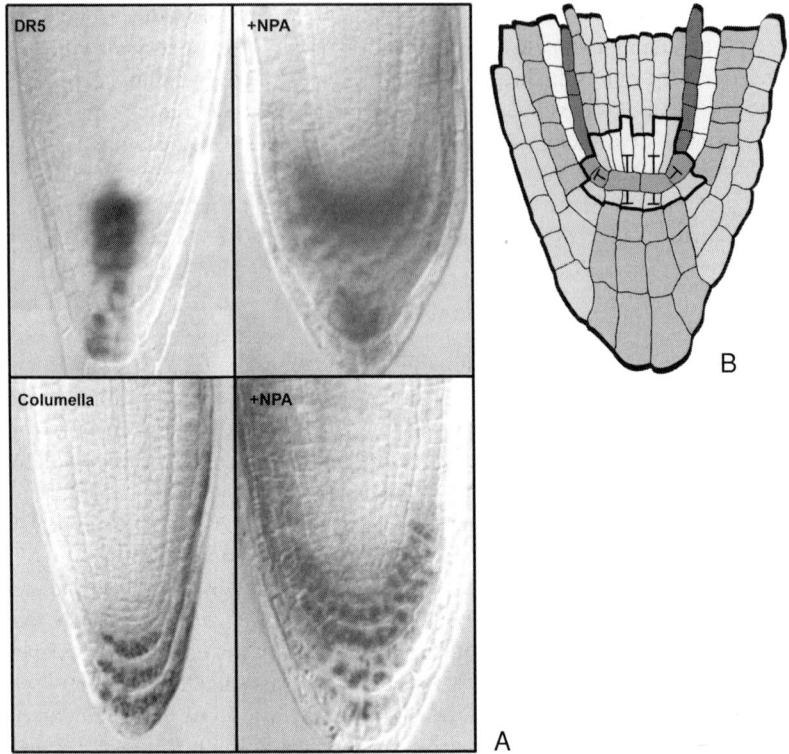

Figure 8.3 The auxin response element promoter DR5, fused with the GUS gene marks the auxin maximum in the distal most cells of the root tip. (A) GUS expression is severely changed after treatment with polar auxin transport inhibitor naphtylphtalamic acid (NPA). NPA treatment also alters the size and shape of the columella root cap domain, indicated here by staining of the starch granules. (B) Stem cell maintenance model. The cells that surround the QC are kept in an undifferentiated state by short-range signals that suppress progression of cell differentiation. Adapted from Scheres et al. (2002).

The octant stage (when the hypophysis cell is formed) is the first time when the apical–basal polarity can be observed. In *gnom* mutants the hypophysis cell does not develop and in severe mutants the apical–basal polarity is completely missing (Mayer et al., 1991, 1993). *GNOM* encodes a brefeldin A (BFA)-sensitive ARF GDP/GTP exchange factor that is required for the proper polar localization of auxin transporter PIN1. Treatment of embryos with BFA resulted in mislocalization of PIN1 (Geldner et al., 2001), which could be avoided with engineered BFA-resistant GNOM (Geldner et al., 2003). This supports the idea of polar auxin transport being the mediator in establishing apical–basal polarity.

The embryo starts its own auxin production most likely at the globular stage, which was shown by the accumulation of auxin when embryos at this stage were treated with NPA or in *gnom* mutants. During this stage, PIN1 is polar localized to the basal membranes of the provascular cells while PIN7 localization is reversed, directing auxin out of the embryo (Friml *et al.*, 2003). The auxin flow in the suspensor is reversed, leading to a new auxin maximum in the topmost suspensor cell, which becomes the hypophysis. Since *pin7* mutants start to recover the apical–basal axis at this stage, it could be concluded that PIN7 function becomes redundant at this stage.

At the late globular stage, *PIN4* expression is first detected along the surface of the hypophysis and at the basal end of the adjacent suspensor cell. It is thought to support PIN1 and PIN7 functions. During the later stages of embryogenesis, its expression pattern shifts slightly, disappearing from the suspensor cell, and appearing in the provascular cells. Following the expression pattern of *PIN4*, the *DR5* marker was visualized from the early heart stage onwards in the basal part of the embryo. In *pin4* mutants this expression pattern was altered significantly. The *DR5* expression domain was enlarged towards all sub-epidermal cells of globular stage embryos, and in later stages, the expression domain was mostly located in the presumptive vascular tissue, suggesting that PIN4 is necessary for the regulation of both positioning and levels of the auxin response maximum during embryogenesis (Friml *et al.*, 2002).

Since there seems to be some functional redundancy between the different embryonically expressed PIN efflux carriers, Friml *et al.* (2003) constructed a set of double, triple and quadruple loss of function mutants of combinations of *pin7* with *pin1*, *pin3* and *pin4*. These mutants showed a series of defects that became increasingly severe according to the number of knocked out *PIN* genes. Quadruple mutants completely failed to recover the embryonic axis and showed strong *gnom*-like defects. In the most severe cases the mutants were ball shaped, completely lacking apical–basal polarity. Similar defects could be achieved in wild-type embryos by interfering with the auxin homeostasis by chemical inhibition of auxin transport. These results suggest that the recovery of the apical–basal axis at the globular stage, which can be seen in the *pin7* mutant, is also caused by a PIN-dependent auxin distribution. This mechanism seems to be non-redundant.

In mature roots, PIN4 is localized in a polar manner at the basal ends of the proximal meristem, suggesting an auxin flow through the vascular cylinder which is directed towards the QC and columella initials. Both inhibition of auxin transport with NPA and via *pin4* mutants disrupted the *DR5* expression pattern and shifted its maximum to the cells in the vascular tissue that normally display polar localization of PIN4 (Sabatini *et al.*, 1999; Friml *et al.*, 2002). This suggests that PIN4 drives the formation of the auxin maximum in the root tip and that in the root tip auxin is conducted almost exclusively through polar transport. Besides the change of the location of the auxin maximum, cell fate within the root apical meristem (RAM) was also changed. Several markers for QC were misexpressed or not expressed at all in *pin4* mutants. The cells at the location of the QC, columella initials and endodermis also showed irregular cell divisions in the mutants. In addition, cells at the location of and slightly above the new maximum acquired features of columella initials and QC respectively, linking the auxin maximum to cell fate (Friml *et al.*, 2002).

Sterols have also been implicated in playing a role in embryo development. The first evidence for their involvement was a mutation in the *FACKEL* (*FK*) gene, which encodes a sterol C-14 reductase that acts in the brassinosteroid (BS) biosynthesis pathway (Jang et al., 2000; Schrick et al., 2000). Recently, two more mutants in the BS pathway have been isolated, *cephalopod (cph)* and *hydra1 (hyd1)*. *CPH* encodes a C-24 sterol methyltransferase and *HYD1* a sterol C-8,7 isomerase. CPH catalyses the first step in the BS pathway, and FK and HYD1 act sequentially a bit further in the pathway. All three mutants have a similar phenotype in which cell division is disturbed and/or polarity decisions are randomized. This results both in embryos where the early decisions fail and in very severe seedling phenotypes (Schrick et al., 2002). Souter et al. (2002) showed that both *hyd* and *fk* mutants are defective in ethylene- and auxin-mediated gene expression and cell differentiation, while the auxin transporter PIN is normally localized. The auxin influx component AUX1 can be blocked with NPA, yet the mutants showed auxin responses after two, 4-D treatment. *axr1* and *axr3* mutations, which block some auxin responses, were partially able to rescue the *hyd* mutant phenotype. This all suggests that the *hyd/fk* mutants are defective in membrane-bound protein activity, and thus indirectly affect polar auxin transport. This can be interpreted as further support for auxin being the key regulator in embryonic axis formation/root meristem formation.

The analysis of various mutations has suggested how auxin action might be executed during the patterning of the root meristem. Auxin is thought to initiate a specific signal transduction pathway. Although the receptor for auxin remains elusive, the transcriptional control related to auxin action is conceptually understood (Gray et al., 2001). It involves auxin response factors (ARFs) that activate the primary response genes. These genes are repressed by auxin-inducible, short-lived AUX/IAA proteins, which are selectively targeted for degradation based on an auxin-regulated proteolytic activity (Hellmann et al., 2003). During auxin signalling the conserved degradation domain II of IAA proteins interacts with the F-box protein TIR1 of the E3 ubiquitin ligase complex SCF_{TIR1}, which then results in the degradation of IAA by the 26S proteasome (Hellmann & Estelle, 2002). The *MONOPTEROS (MP)* locus was first defined by recessive mutations that disrupted the apical–basal organization of the embryo and the continuity of the vascular strands (Berleth & Jurgens, 1993). Molecular cloning revealed *MP* as one of the first ARFs to be genetically characterized. *MP* encodes ARF5 (Hardtke & Berleth, 1998) that can bind to conserved 'auxin response' promoter elements of downstream genes (Ulmasov et al., 1997; Guilfoyle et al., 1998). *MP* is first expressed in a broad domain during embryogenesis and its expression subsequently becomes restricted to vascular tissues and the basal domain of the embryo, indicating its importance as a key factor controlling auxin-mediated specification of vascular and basal fate. A semi-dominant mutation in the *BODENLOS* locus results in an *mp*-like phenotype (Hamann et al., 1999). *BODENLOS* encodes an AUX/IAA repressor (IAA12) and the mutation (located in domain II) that prevents interaction with SCF_{TIR1} is likely to extend its stability. Since *MP* and *BDL* gene expression domains overlap and their products interact, it is likely that they constitute an activator/repressor module regulating vascular bundle formation and basal fate (Hamann et al., 2002). Analogously, the *axr6* mutation, located in a gene that codes for the SCF subunit CUL1 and which results in an *mp*-like phenotype, also causes stabilization of AUX/IAA proteins (Hellmann et al.,

2003). The *hobbit* (*hbt*) mutation, which has a slightly later but similar phenotype to the *mp*, *bdl* and *axr6* mutants, is located in a subunit of the anaphase promoting complex (APC), another ubiquitin protein ligase class. *hbt* mutants accumulate IAA17, which suggests that the APC regulates stability of the AUX/IAA proteins (Blilou *et al.*, 2002).

On the basis of the studies above, auxin may be considered as a positional cue to specify the position of the QC and columella initials. In embryos the cells of the QC and columella are descendants of the hypophyseal cell. In the *mp*, *bdl* and *axr6* mutants (in which auxin signalling is disturbed) this cell fails to take the hypophyseal identity (Friml *et al.*, 2003). A similar defect can be seen in the *hbt* mutant, which shows loss of QC and root meristem (Blilou *et al.*, 2002).

Currently it is not known whether this failure to differentiate could be the result of not receiving the auxin signal, not having normal auxin signal transduction in the uppermost suspensor cell or due to a combination of these reasons. Also, it is currently unknown whether the same pathways are active both during embryogenesis and after germination.

In addition to auxin, other types of signals have been implicated in regulating the integrity of the root meristem. As described above, in post-embryonic roots it has been established that the QC plays an important role in the maintenance of the root meristem. Ablation of a QC cell leads to differentiation of the initial cells previously attached to it (van den Berg *et al.*, 1997). Observations on mutants missing post-embryonic root cell divisions further supported the view that the QC functions by inhibiting differentiation of the surrounding initial cells (van den Berg *et al.*, 1997). Umeda *et al.* (2000) came to a similar conclusion after examination of plants with altered CAK (cyclin-dependent kinase–activating kinase) expression, which showed differentiation of the columella initials. The role of QC as a regulatory unit to keep the surrounding cells undifferentiated can be compared to the function of certain cells in the shoot apical meristem that express the *WUSCHEL* (*WUS*) gene (Laux *et al.*, 1996). Together with the CLAVATA receptor complex (Clark *et al.*, 1993) WUS is responsible for the maintenance of meristem function. The current model of the CLAVATA signalling pathway is as follows. CLV3 protein is secreted by the shoot meristem stem cells and some of it moves to the surrounding cells. In both the stem cells and the cells that receive CLV3, it is bound by the CLV1 receptor and as a result CLV3 represses the *WUS* promoter. In the daughter cells this means that they cannot become stem cells themselves and they initiate differentiation. In the organizing centre (OC) downward movement of CLV3 is restricted by strong *CLV1* expression, which sequesters the ligand and prevents it from moving further. This allows for *WUS* expression in the OC, ensuring that the stem cell population and activity is maintained (Lenhard & Laux, 2003; see also Chapter 6, this volume).

In roots, however, there have only recently been some hints of a CLAVATA-like pathway, which influences meristem maintenance (Casamitjana-Martinez *et al.*, 2003). Over-expression of the CLV3 homologue CLE19 in the root meristematic zone leads to a loss of meristematic cells and a reduced cell elongation zone. While QC and initial cell identity is maintained, their activity is also lessened. A similar loss of the meristematic zone was achieved by over-expressing CLV3 in the root. So far the CLAVATA-like pathway controlling meristematic activity during root development remains to be identified, though a mutation in a putative Zn^{2+}-carboxypeptidase can suppress the over-expressor phenotype, suggesting that it is involved in processing the ligand (Casamitjana-Martinez et al., 2003). Furthermore, a *WUSCHEL*-like gene has been recently shown to be expressed in the QC cells in rice (Kamiya et al., 2003). How exactly the QC keeps the surrounding initials undifferentiated remains as yet unknown. On the basis of the analysis by Casamitjana-Martinez et al. (2003), the CLAVATA pathway in the root meristem does not seem to act directly to maintain the undifferentiated state of the initials (as in shoot apical meristem). Clearly more analysis has to be carried out and yet other factors need to be identified. Based on a genetic analysis (Sabatini et al., 2003), another molecule required for the integrity of the QC is SCARECROW, a transcription factor that has a role also in radial patterning of the ground tissue (as described below).

8.4.2 Genetic control of initiation of secondary roots

Over the years auxin has been linked to the development of secondary roots by several mutants that have auxin-related problems (for a recent review see Casimiro et al., 2003). Indeed, recently several genetic and physiological studies have provided evidence that auxin is required to facilitate lateral root formation at several specific developmental stages. By using auxin-responsive reporter genes and sophisticated mass spectrometry techniques, the auxin concentrations in *Arabidopsis* root have been mapped in detail. Furthermore, the changes in auxin concentration and localization have been correlated with the function of various proteins (Casimiro et al., 2003). For example, the *aux1* mutation alters auxin content and distribution in the root, leading to reduced number of lateral root primordia. Thus, the auxin influx carrier AUX1, which is expressed in lateral root primordia before the first periclinal cell division, has been implicated in regulation of lateral root development by facilitating auxin transport between auxin source and sink tissues (Marchant et al., 2002). Auxin transport between adjacent cells seems to be the most important factor initiating lateral root formation.

Recently, the auxin reporter DR5 was used to analyse the presence of auxin in the developing lateral root in *Arabidopsis* (Benkova et al., 2003). They showed that already at stage 0 of lateral root initiation, before any cell divisions have taken place, there is an auxin maximum in the pericycle cells that will divide. This maximum was present in all cells after the anticlinal cell divisions that form the short initial cells in stage I. During the next stage (stage II) the maximum was confined to the central cells of the two layers of lateral root initials. During the following stages a gradient of *DR5* expression was established with its maximum in the columella initials of a newly formed lateral root primordium, forming a pattern identical to that of the primary root (Sabatini et al., 1999). This gradient was shown to be dependent on polar auxin transport by differentially expressed, functionally redundant PIN proteins, whose sub-cellular localization was rearranged during lateral root initiation

(Benkova et al., 2003). On the basis of this study, Benkova et al. (2003) suggested a model of primordium development in which the pericycle cells at the site of future initiation start accumulating auxin, possibly due to PIN-dependent auxin transport.

In the context of establishment of the vascular network, Sachs (1988, 1991) proposed a model in which vascular tissue is established through a positive feedback loop where auxin is a limiting factor that induces auxin transport capability and polarity. On the basis of the studies with various *gnom* alleles, Geldner et al. (2004) proposed a model in which a similar mechanism results in the establishment of the embryonic axis and the initiation of lateral roots. In embryos corresponding to strong *gnom* alleles, as a result of grossly misaligned auxin transporters the basal part of the embryo lacks sufficient auxin to be able to initiate a primary root meristem. Instead, auxin accumulates in the apical domain at the presumptive site of synthesis, leading to cotyledon fusion (Fig. 8.4A). The absence of auxin transport also results in a randomization of cell division and expansion, which increases axis diameter. During lateral root development some pericycle cells start to accumulate auxin and divide and they thus deplete auxin from the surrounding cells, which in turn inhibits their proliferation. The observed simultaneous peak of auxin within the primordium and the auxin-mediated lateral inhibition may be explained by the assumption that in a stage I primordium PIN1 is localized in a bipolar manner (in both apical and basal sides of the cell). When PIN1 fails to take on the bipolar localization, auxin peaks would lead to homogeneous proliferation of the pericycle, as in *gnom*^{R5} (Geldner et al., 2004). Since PIN1 localizes to the newly formed cell plates after a division (Geldner et al., 2001), in *gnom*^{R5} mutants this would lead to daughter cells having opposing PIN1 polarities (Fig. 8.4B). Geldner et al. (2004) propose that after a division a weak bias in the auxin signal results in one of the cells changing polarity, and that GNOM is essential at this step. Subsequent cell divisions would be polarized in the direction set by the first two cells, leading to the 90° degree shift in polarity from the primary root axis (Fig. 8.4C). The retargeting of PIN proteins would cause the auxin flow to be redirected and allow auxin from the primary root vasculature to flow through the centre of the new lateral root primordium. At the tip of the new lateral root the flow is directed away from the meristem through the lateral rootcap, creating a 'fountain' model of auxin flow (Benkova et al., 2003).

8.4.3 *Molecular genetics of epidermal patterning*

The epidermis, the outermost layer of cells in the mature root, is composed of alternating files of trichoplasts, which give rise to hair cells (H cells), and atrichoplasts, which give rise to non-hair cells (N cells) (Dolan et al., 1994; Galway et al., 1994). In *Arabidopsis*, several mutants impaired in epidermal cell patterning have been identified. Three of these mutants, *werewolf* (*wer*), *transparent testa glabra* (*ttg*) and *glabra2* (*gl2*), show a hairy root phenotype due to ectopic root hair cell production. This implies that the normal role of the WER, TTG and GL2 is either to promote N cell differentiation or to repress root hair cell differentiation (DiCristina et al., 1996; Galway et al., 1994; Masucci et al., 1996; Lee & Schiefelbein, 1999). A fourth

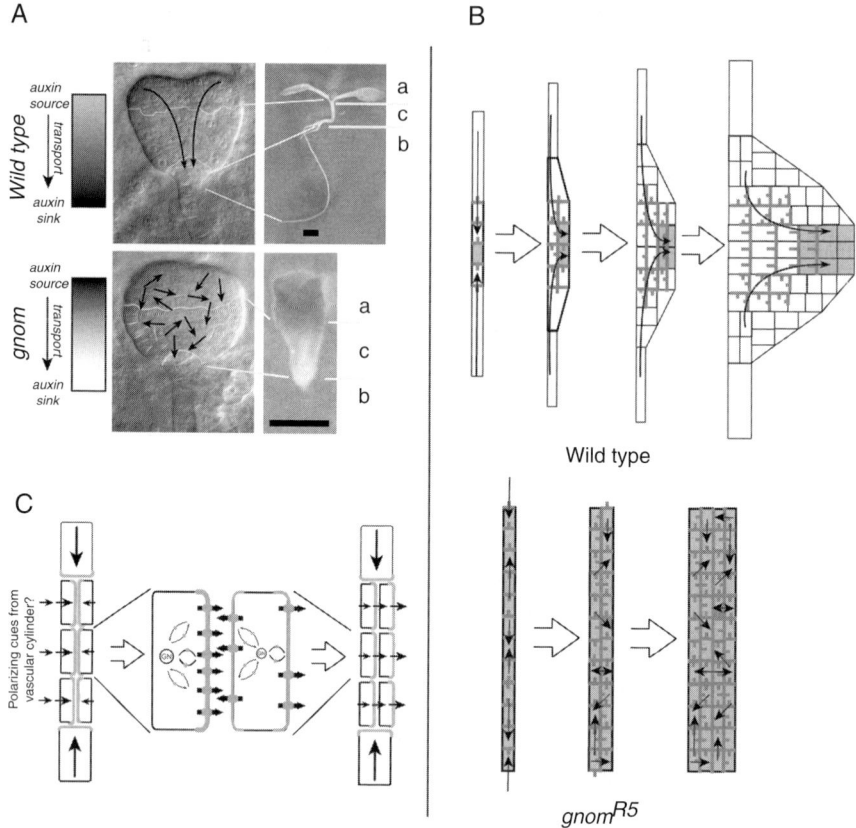

Figure 8.4 Speculative model for GNOM-mediated auxin canalization. (A) Wild type (top) and strong *gnom* allele (bottom) heart stage embryos. Apical (a), central (c) and basal (b) regions of the embryos and the successive relating areas in the mature seedling are indicated. Auxin flow from source to sink is indicated by black arrows. (B) Relation between PIN1 localization (grey stubs) and auxin-response gradient (grey filling) in lateral root primordium development. Gradual reorientation of the individual transport polarities of cells by auxin canalization are indicated by arrows. Grey stubs indicate the cell walls to which the respective PIN1 label is thought to belong. (C) Presumptive critical step during lateral root formation for the canalization of auxin flow. Stage II primordium immediately after division (left) with two daughter cells displaying opposite polarities. Gradual, GNOM-dependent, relocalization of the efflux carriers are possibly guided by weak polarizing cues from adjacent cells; this supplies more auxin to the inner layer, which as a result imposes its auxin transport on the outer layer. Arrows indicate the direction of the auxin flux; auxin efflux carrier PIN1 in grey; GN, GNOM-positive endosomes involved in recycling auxin carriers. Adapted from Geldner *et al.* (2004).

mutation, *caprice* (*cpc*), affects root epidermis cell specification in a different manner. Rather than causing ectopic root hair cells, the *cpc* mutant produces fewer hairs than normal root cells. This implies that CPC is a positive regulator of the root hair cell fate (Wada *et al.*, 1997).

All these genes encode putative transcriptional regulators. *WER* codes for a MYB-related putative transcriptional regulator that is mainly expressed in the N cells. Yeast two-hybrid assays showed that WER is able to interact with a bHLH-type protein to control epidermal cell patterning during *Arabidopsis* root development (Lee & Schiefelbein, 1999).

The *GL2* gene encodes a homeodomain-containing protein (Rerie *et al.*, 1994). It is preferentially expressed in the N cells (Masucci *et al.*, 1996) already early during embryogenesis (Lin & Schiefelbein, 2001; Costa & Dolan, 2003). It is first expressed in the future epidermis in the heart stage embryo and its expression is progressively restricted to those cells that will acquire an N identity at the transition between torpedo and mature stage (Costa & Dolan, 2003).

CPC encodes a small protein with a MYB-like DNA-binding domain, but it does not have a typical transcriptional activation domain (Wada *et al.*, 1997). It is preferentially expressed in the differentiating N cells (Wada *et al.*, 1997, 2002). CPC has been suggested either to work together with the *TTG* gene product or in an independent pathway that controls the number of root hairs upstream of GL2 in the developmental pathway of root hair formation. (Wada *et al.*, 1997). Interestingly, CPC protein is able to move to the hair-forming cells and repress gene expression (Wada *et al.*, 2002). These studies indicate that transcriptional feedback loops between the *WER*, *CPC* and *GL2* genes help to establish position-dependent epidermal patterning (Lee & Schiefelbein, 2002; Costa & Dolan, 2003).

Various gene expression studies have shown that WER positively regulates *GL2* expression in the N cells (Lee & Schiefelbein, 1999) and is thus required to specify the positional information for *GL2* expression (Hung *et al.*, 1998; Lin & Schiefelbein, 2001). The *WER* gene is also required for positive regulation of *CPC* transcription in the developing N cells, and CPC acts as a negative regulator of *GL2* (Wada *et al.*, 1997; Schellmann *et al.*, 2002; Wada *et al.*, 2002), *WER* and itself in the developing H cells (Lee & Schiefelbein, 2002).

Based on *GL2* promoter–reporter gene expression studies, it was proposed that the epidermal cell patterning mechanism in the root initiates during the early heart stage and that it occurs before the establishment of a functional meristem (Lin & Schiefelbein, 2001). Lee and Schiefelbein (2002) have proposed a simple model for the origin of the epidermal cell pattern. In this model, the specification of an H or N cells relies on the relative activity of two competing transcription factors, WER and CPC. In heart stage embryos, all epidermal cells express *WER* and *GL2* and, in the absence of positional cues, use CPC to inhibit their neighbours from expressing these genes. In the growing root the pattern is established by positional cues from the underlying tissue that break the symmetry of the inhibition and cause greater *WER* transcription in the cells overlying a single cortical cell (Fig. 8.5). This leads to a high level of *GL2* and *CPC* expression and to the N cell fate. In the alternate cell position, the CPC protein produced by the developing N cell is proposed to accumulate by virtue of its cell-to-cell trafficking and it represses *GL2*, *WER* and *CPC* expression, permitting H cell differentiation to proceed (Lee & Schiefelbein, 2002).

Recently, a different model for the establishment of the root epidermal pattern has been proposed based on more detailed examination of *GL2*, *WER* and *CPC* gene expression during embryogenesis (Costa & Dolan, 2003). According to this model the development of cell patterning in the root epidermis is also initiated

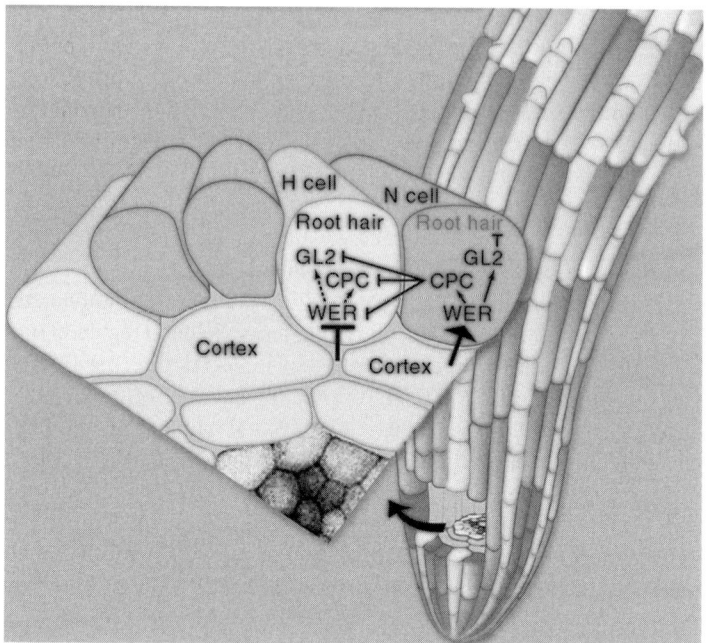

Figure 8.5 Model of root epidermal patterning in the growing root. Positional cues from underlying cells generate a bias in WER expression either because of inhibition of WER in H cells or because of activation of WER in N cells. This bias is enhanced by increased CPC levels, which carry an inhibitory signal from N cells to H cells. Dark grey tint corresponds to cell files that will produce non-hair cells (N cells). Light grey tint corresponds to cell files that will produce hair cells (H cells). Adapted from Larkin *et al.* (2003).

in the heart stage embryo, but it is not completed until the embryo reaches the mature stage. The model is based on the complex regulatory interactions between WER, CPC and GL2 that occur during the formation of epidermal pattern in the embryo. *GL2* is first expressed in the heart stage embryo in a subset of cells in the protoderm and *WER* positively regulates its expression. Then, by the torpedo stage, *GL2* expression has spread to all cells in the future epidermis. *WER* and *CPC* expression is then detectable and *WER* promotes *GL2* expression throughout the epidermis. *CPC* is in turn required for the preferential accumulation of *GL2* transcript in future N cells, perhaps by negatively regulating *GL2* transcription in H cells position. In the mature embryo, GL2 negatively regulates *WER* transcription whereas WER positively regulates *CPC* expression from the torpedo to mature stages. These events result in *GL2* being expressed at high levels in the future N cells and being absent from the future H cells in the mature embryo. The pattern of *GL2* expression is then maintained in the root of the seedling and accounts for the pattern of H cells and N cells in the root epidermis, where GL2 negatively regulates hair formation in cells located in the N position (Costa & Dolan, 2003).

In addition to *WER*, *GL2* and *CPC*, yet other genetic loci have been implicated in the root hair patterning. The *TTG* gene encodes a small WD40 repeat protein (Walker et al., 1999). It has been proposed to mediate protein–protein interactions (Galway et al., 1994) and to be involved in signal transduction to downstream transcription factors (Walker et al., 1999). Expression of the maize *R* gene (a transcription factor involved in anthocyanin biosynthetic pathway) under the CaMV 35S promoter complemented the *ttg1* mutant phenotype, suggesting that TTG1 might regulate an *R*-like gene in *Arabidopsis* to adopt a hairless cell fate (Lloyd et al., 1992; Galway et al., 1994). Indeed, yeast two-hybrid assays indicate that GLABRA3 (GL3), an R homologous bHLH protein expressed in shoot epidermal cells in *Arabidopsis*, interacts with TTG (Payne et al., 2000). Thus, TTG may act as a general regulator of epidermal cell patterning. However, its precise role in root epidermis remains to be elucidated.

In addition, the *TRY* (*TRIPTHYCON*) gene encodes a CPC-homologous MYB-related transcription factor that lacks a recognizable activation domain. TRY has been shown to function as a negative regulator of trichome development in the shoot but it is also expressed in roots, suggesting that it has a role in root hair patterning. Indeed, over-expression of *TRY* results in formation of extra root hairs. Furthermore, various expression studies with *TRY* have shown that *TRY* levels are controlled by TRY and CPC, and that TRY is likely to be expressed in N cells (Schellmann et al., 2002).

Although TTG and TRY involvement in root epidermal pattern formation has not been rigorously examined, several studies indicate that they have a role in this process. The emerging picture is that TTG and an as yet unknown bHLH-related transcription factor, together with WER, begin to act an early stage in embryonic development to positively regulate the expression of GL2 (and perhaps other as yet unidentified genes) in a cell position-dependent manner to specify the N cell type (Hung et al., 1998; Lin & Schiefelbein, 2001). Furthermore, it is likely that TRY, in addition to CPC, is also involved in the lateral suppression of GL2 expression in root hair files (Schellmann et al., 2002). The control of epidermal patterning in aerial tissue and its overlap with the system described in the root is explored further in Chapter 9, this volume.

8.4.4 Patterning of ground tissue

As discussed above, pattern formation in the *Arabidopsis* root is the result of highly asymmetric cell divisions and subsequent cell specification events. One of the best characterized examples is the radial pattern formation of ground tissue that originates from a set of stem cells ('initials') that undergo asymmetric cell divisions to give rise to the youngest cells of the endodermis and cortex cell lineages (Figs. 8.7A and 8.7B). Subsequently, cells in both lineages undergo differential expansion resulting in cell types with distinct morphologies and differentiated features. Two loci, *SHORT-ROOT* (*SHR*) and *SCARECROW* (*SCR*), controlling ground tissue patterning during root development have been identified in *Arabidopsis* (Benfey et al., 1993; Scheres et al., 1995). Mutations in these loci result in a single ground tissue layer, indicating

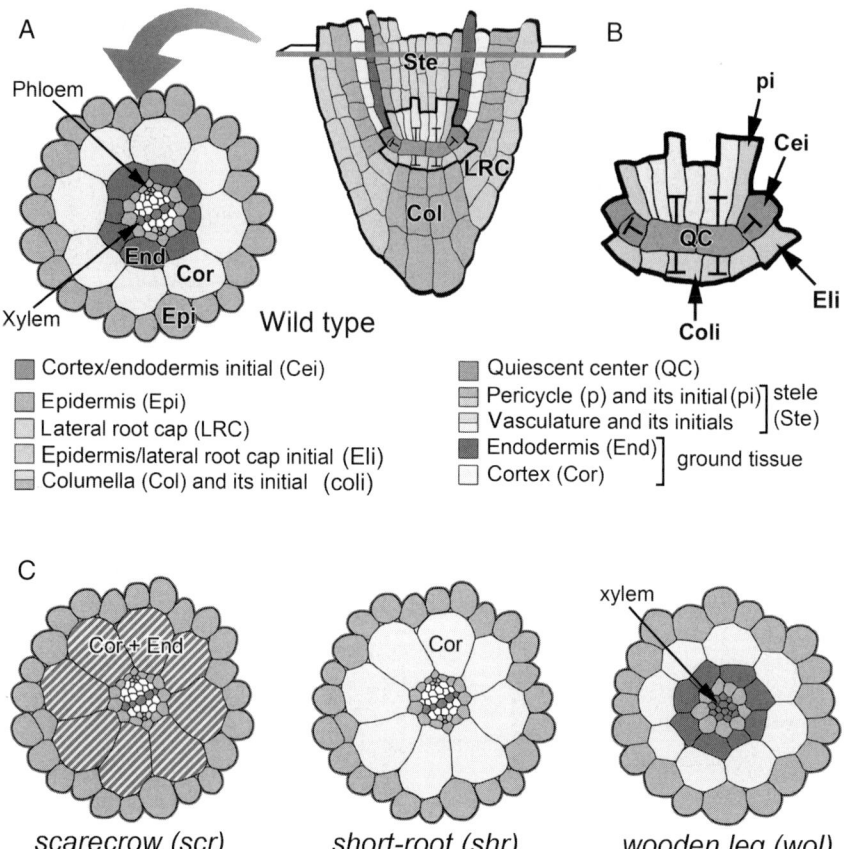

Figure 8.6 Schematics of wild-type and mutant *Arabidopsis* root structures. (A) Wild-type longitudinal and transverse schematic; (B) the QC keeps surrounding cells in an undifferentiated state; (C) radial patterns of three *Arabidopsis* mutants, *scr*, *shr* and *wol*. Adapted from Nakajima *et al.* (2001).

that both loci are essential for the periclinal asymmetric cell division that gives rise to the two distinct cell files (Fig. 8.6C). In *scr* mutants, the remaining layer has differentiated attributes of both endodermis and cortex, indicating that *SCR* is required for cell division but not differentiation of the ground tissue (Scheres *et al.*, 1995; Di Laurenzio *et al.*, 1996). In contrast to *scr*, the single layer of ground tissue in *shr* mutants is missing endodermal differentiation markers, indicating that SHR is essential for both cell division and cell specification (Benfey *et al.*, 1993).

Both *SCR* and *SHR* genes have been identified at the molecular level (Di Laurenzio *et al.*, 1996; Helariutta *et al.*, 2000). They encode members of the GRAS family of putative transcription factors, indicating that they regulate the asymmetric cell division (and subsequent endodermal differentiation in the case of SHR) at a transcriptional level (Pysh *et al.*, 1999). The *SCR* gene is expressed in the initial daughter cell before its asymmetric division and remains expressed in the endodermal cell layer after the division (Fig. 8.7A) (Di Laurenzio *et al.*, 1996; Wysocka-Diller *et al.*,

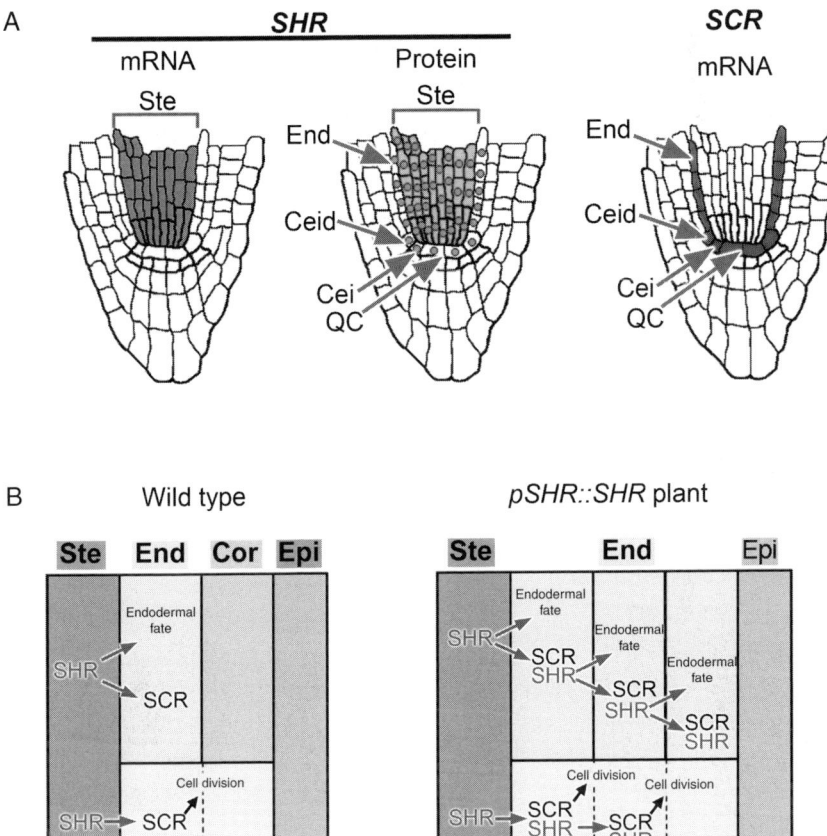

Figure 8.7 Radial patterning of the ground tissue in the root meristem. (A) *SHR* mRNA localization (left), SHR protein (middle) and *SCR* mRNA (right). (B) Intercellular signalling mechanism in radial pattern formation of wild type (left) and *pSCR::SHR* transgenic plants (right). In wild type, SHR protein moves from the stele to a single adjacent layer. In the adjacent layer it activates *SCR* transcription, which is essential for the asymmetric cell divisions that form the cortex and endodermis. In the mature region SHR presence in the adjacent layer confers endodermal cell fate. How SHR protein movement is restricted to a single cell layer is as yet unknown. In *pSCR::SHR* transgenic plants (right) the SCR promoter is repeatedly activated in adjacent layers by the production of SHR protein from the *pSCR::SHR* transgene, resulting in supernumerary layers due to cortex-endodermis initial daughter cell (Ceid) divisions and the acquisition of endodermal fate in these new layers. Abbreviations for the cell types are given in Figure 8.4A. Adapted from Nakajima *et al.* (2001).

2000). *SCR* expression is down-regulated in the *shr* background, indicating that SHR is upstream of *SCR*. On the other hand, ectopic SHR expression under the 35S promoter results in supernumerary cell divisions and altered cell specification including ectopic SCR expression, indicating that SHR is both necessary and sufficient to regulate cell division and cell specification in the root meristem.

Surprisingly, *SHR* RNA is not expressed in the ground tissue cell lineage but in the stele (pericycle and vascular cylinder) located immediately adjacent to it, suggesting a non-cell autonomous mode of action of SHR (Fig. 8.7A) (Helariutta *et al.*, 2000). In order to approach this non-cell autonomous mechanism, Nakajima *et al.* (2001) compared the protein accumulation to the RNA accumulation pattern and found that the SHR protein is transported from the stele to the adjacent endodermal layer, probably through plasmodesmata. This indicates that SHR acts as a mobile signal in exerting its functions as a transcriptional regulator. Although movement of transcription factors had been observed before (Lucas *et al.*, 1995), this was the first time a functional significance for such movement was demonstrated. Nakajima *et al.* (2001) also explored the outcome of introducing SHR ectopically to the endodermal layer under the SCR promoter. In this case a highly specific increase of ground tissue layers having endodermal characteristics was observed (Fig. 8.7B). Based on the morphological analysis of the *scr* and *shr* mutants as well as on a detailed spatio-temporal analysis of *SCR* and *SHR* gene expression, it is evident that the patterning of ground tissue is established already during early embryogenesis.

8.4.5 Vascular patterning

The formation of the vascular network of a plant takes place continuously at the meristematic regions of a plant during both shoot and root development. Even though under normal conditions vascular development is highly predictable, when the situation arises it can also react and adapt to either localized or environmental stimuli. As described in the context of distal patterning of the root, the establishment of the vascular network has also been shown to involve auxin transport and auxin signalling (Aloni, 1987; Sachs, 1991; Ye, 2002).

The vascular network in plants consists of transport tissues: xylem, which transports water and nutrients; and phloem, which transports photosynthates. Between them there is a third vascular tissue type, the (pro)cambium, that consists of the stem cells from which the xylem and phloem elements originate. There are several distinct patterns in which these three tissues are organized in different plant species (Ye, 2002).

Although xylem and phloem both are formed from cambial initial cells, their fate is quite different. The phloem is a system of several cell types: sieve elements (SE), companion cells (CC), phloem fibres and phloem parenchyma cells. The SE are the actual transport cells. They undergo a partial autolysis (involving disintegration of the nucleus) and in some regions of the plant deposit callose on its cell walls. The CC support SE with macromolecules (Esau, 1977; Kuhn *et al.*, 1997; Oparka and Turgeon, 1999). Xylem consists of tracheary elements (TE), xylem parenchyma cells and xylem fibres. The differentiation of TE involves deposition of elaborate cell wall thickenings (containing cellulose and lignin) and programmed cell death.

In contrast to other tissue types, the vascular pattern is established only during the last stages of embryonic development after the vascular initials have divided to first form a pattern of four near-identical poles, followed by a set of tangential and periclinal cell divisions associated with the formation of phloem poles (Plate 8.1) (Bonke *et al.*, 2003). Recently, a couple of informative *Arabidopsis* pattern formation mutants have been able to shed some light on the underlying mechanisms of vascular patterning in roots. In the *wooden leg* (*wol*) mutant cell divisions in the vascular bundle are reduced, leading to a phenotype in which only approximately 9–11 cells occupy the vascular cylinder and after germination all vascular cells differentiate into protoxylem. This phenotype is first seen during the torpedo stage of embryogenesis (Scheres *et al.*, 1995; Mähönen *et al.*, 2000). When *wol* was introduced into the *fass* background, phloem was restored. Thus, WOL does not appear to directly influence cell fate within the vascular bundle but has a more indirect influence by controlling the number of cells in the vascular cylinder.

WOL encodes a two-component hybrid-type receptor molecule (Mähönen *et al.*, 2000) and is identical to *CRE1/AHK4* (Inoue *et al.*, 2001; Suzuki *et al.*, 2001), a cytokinin receptor. It is expressed in the vascular initials already during the globular stage of embryogenesis, linking cytokinin signalling to vascular embryonic development.

Recently, Bonke *et al.* (2003) identified the gene *ALTERED PHLOEM DEVELOPMENT* (*APL*) as a MYB-CC transcription factor that is required for phloem development throughout the plant. In the *apl* mutant, phloem patterning is affected; the phloem-specific cell divisions occur less frequently and the cells in the phloem pole area take on a xylem identity.

APL has a complex gene expression pattern mirroring the dynamic nature of phloem development. It is expressed in developing protophloem and also further up in the companion cells and metaphloem. *APL* expression can first be detected during embryogenesis. Even though the phloem-specific asymmetric divisions are delayed in the *apl* mutant, *APL* expression is initiated only after these have occurred. This could indicate that APL acts as a cell non-autonomous factor to control these divisions. However, a GFP–APL protein fusion appears to be expressed in a spatially similar manner as *APL* RNA, indicating that APL itself probably does not act as the cell non-autonomous factor. It is also possible that metabolic defects resulting from loss of functional phloem may be the cause for the delay in the phloem-specific cell divisions (Bonke *et al.*, 2003).

APL expression driven by the procambium-specific *WOL* promoter showed that ectopic *APL* expression is able to fully suppress xylem differentiation in the protoxylem pole position and to some extent also in the metaxylem position. Importantly, the affected protoxylem cells retained their nucleus, which indicates that they did not change fate to phloem identity, meaning that *APL* is necessary, but not sufficient, for phloem identity. It has recently been established that in the aerial part of the plant, class III HD-ZIP and KANADI family transcription factors regulate the distribution pattern of xylem and phloem in stems and leaves (Bowman *et al.*, 2002; Emery *et al.*, 2003). These genes however do not lead to phloem defects. It remains to be studied if the class *III HD-ZIP* and *KANADI* genes regulate APL.

The exact mechanisms of vascular patterning are still relatively poorly understood, but the results so far seem to point to a temporal model where auxin is required for establishment of vascular tissue, cytokinin signalling is necessary for cell proliferation and the 'prepatterning' of four poles on which the vascular tissue identity (consisting of the two xylem poles and two phloem poles) is established. APL is necessary for this last phase of the vascular development.

8.5 Future prospects

Microarray technology provides a powerful tool for functional genomics and is being widely used to shed light on global changes in gene expression. By clustering genes according to their expression profile, it is possible to assign functions to genes with an unknown function and at the same time to assign new functions to known genes. Recently, a global gene expression map of the *Arabidopsis* root was made based on microarray studies of various root cell types and tissues at progressive developmental stages (Birnbaum *et al.*, 2003). The roots of plants expressing GFP in specific cell types (in stele, endodermis, endodermis plus cortex, epidermal atrichoblast cells and lateral root cap) were dissociated into single cells by enzymatic digestion of their cell walls, the GFP-expressing cells were then isolated with the use of a fluorescence-activated cell sorter and their mRNA was analysed with the use of microarrays.

Analysis of expression patterns revealed various clusters of genes with coordinated expression in various spatial and temporal domains in root. For example, three of these localized expression domains showed an aggregation of genes with known or putative roles in auxin, gibberellic acid or jasmonic acid pathways, suggesting the possibility of localized signalling centres which could mediate primary developmental cues (Birnbaum *et al.*, 2003). As described above, auxin is a major determinant of patterning in root meristem. Increasing number of studies have shown that additional factors are required during root development and growth, among them also other hormones. As increasing numbers of microarray experiments are completed, the collections of data retrieved from different analyses will contribute to the resolution of the complex relationships between the numerous signalling pathways that underlie cell communication determining pattern formation and cell differentiation during root development.

Acknowledgements

We thank Philip Benfey, Thomas Berleth, Gerd Jürgens, John C. Larkin, Ben Scheres and John Schiefelbein for their permission to use their figures.

References

Aloni, R. (1987) Differentiation of vascular tissues. *Annu. Rev. Plant Physiol.*, **38**, 179–204.
Baker, D.A. (2000) Long-distance vascular transport of endogenous hormones in plants and their role in source: sink regulation. *Isr. J. Plant Sci.*, **48**, 199–203.

Baum, S.F., Dubrovsky, J.G. & Rost, T.L. (2002) Apical organization and maturation of the cortex and vascular cylinder in *Arabidopsis thaliana* (Brassicaceae) roots. *Am. J. Bot.*, **89**, 908–920.

Benfey, P.N., Linstead, P.J., Roberts, K., Schiefelbein, J.W., Hauser, M.T. & Aeschbacher, R.A. (1993) Root development in *Arabidopsis*: four mutants with dramatically altered root morphogenesis. *Development*, **119**, 57–70.

Benkova, E., Michniewicz, M., Sauer, M., Teichmann, T., Seifertova, D., Jurgens, G. & Friml, J. (2003) Local, efflux-dependent auxin gradients as a common module for plant organ formation. *Cell*, **115**, 591–602.

Berger, F., Haseloff, J., Schiefelbein, J. & Dolan, L. (1998) Positional information in root epidermis is defined during embryogenesis and acts in domains with strict boundaries. *Curr. Biol.*, **8**, 421–430.

Berleth, T. & Chatfield, S. (2002) Embryogenesis: pattern formation from a single cell. In: *The Arabidopsis Book* (eds C.R. Somerville & E.M. Meyerowitz). American Society of Plant Biologists, Rockville, MD. Available at: http://www.aspb.org/publications/arabidopsis/. Access date Sept 30, 2002.

Berleth, T. & Jurgens, G. (1993) The role of the monopteros gene in organising the basal body region of the *Arabidopsis* embryo. *Development*, **118**, 575–587.

Birnbaum, K., Shasha, D.E., Wang, J.Y., Jung, J.W., Lambert, G.M., Galbraith, D.W. & Benfey, P.N. (2003) A gene expression map of the *Arabidopsis* root. *Science*, **302**, 1956–1960.

Blilou, I., Frugier, F., Folmer, S., Serralbo, O., Willemsen, V., Wolkenfelt, H., Eloy, N.B., Ferreira, P.C., Weisbeek, P. & Scheres, B. (2002) The *Arabidopsis* HOBBIT gene encodes a CDC27 homolog that links the plant cell cycle to progression of cell differentiation. *Genes Dev.*, **16**, 2566–2575.

Bonke, M., Thitamadee, S., Mähönen, A.P., Hauser, M.T. & Helariutta, Y. (2003) APL regulates vascular tissue identity in *Arabidopsis*. *Nature*, **426**, 181–186.

Bowman, J.L., Eshed, Y. & Baum, S.F. (2002) Establishment of polarity in angiosperm lateral organs. *Trends Genet.*, **18**, 134–141.

Casamitjana-Martinez, E., Hofhuis, H.F., Xu, J., Liu, C.M., Heidstra, R. & Scheres, B. (2003) Root-specific CLE19 overexpression and the sol1/2 suressors implicate a CLV-like pathway in the control of *Arabidopsis* root meristem maintenance. *Curr. Biol.*, **13**, 1435–1441.

Casimiro, I., Beeckman, T., Graham, N., Bhalerao, R., Zhang, H., Casero, P., Sandberg, G. & Bennett, M.J. (2003) Dissecting *Arabidopsis* lateral root development. *Trends Plant Sci.*, **8**, 165–171.

Clark, S.E., Running, M.P. & Meyerowitz, E.M. (1993) CLAVATA1, a regulator of meristem and flower development in *Arabidopsis*. *Development*, **119**, 397–418.

Clowes, F.A.l. (1956) Nucleic acids in root apical meristem of *Zea*. *New Phytol.*, **55**, 29–34.

Costa, S. & Dolan, L. (2003) Epidermal patterning genes are active during embryogenesis in *Arabidopsis*. *Development*, **130**, 2893–2901.

Di Cristina, M., Sessa, G., Dolan, L., Linstead, P., Baima, S., Ruberti, I. & Morelli, G. (1996) The *Arabidopsis* Athb-10 (GLABRA2) is an HD-Zip protein required for regulation of root hair development. *Plant J.*, **10**, 393–402.

Di Laurenzio, L., Wysocka-Diller, J., Malamy, J.E., Pysh, L., Helariutta, Y., Freshour, G., Hahn, M.G., Feldmann, K.A. & Benfey, P.N. (1996) The SCARECROW gene regulates an asymmetric cell division that is essential for generating the radial organization of the *Arabidopsis* root. *Cell*, **86**, 423–433.

Dolan, L., Duckett, C.M., Grierson, C., Lindstead, P., Schneider, K., Lawson, E., Dean, C., Poethig, R.S. & Roberts, K. (1994) Clonal relationships and cell patterning in the root epidermis of *Arabidopsis*. *Development*, **120**, 2465–2475.

Dolan, L., Janmaat, K., Willemsen, V., Linstead, P., Poethig, S., Roberts, K. & Scheres, B. (1993) Cellular organisation of the *Arabidopsis thaliana* root. *Development*, **119**, 71–84.

Dubrovsky, J.G., Doerner, P.W., Colon-Carmona, A. & Rost, T.L. (2000) Pericycle cell proliferation and lateral root initiation in *Arabidopsis*. *Plant Physiol.*, **124**, 1648–1657.

Emery, J.F., Floyd, S.K., Alvarez, J., Eshed, Y., Hawker, N.P., Izhaki, A., Baum, S.F. & Bowman, J.L. (2003) Radial patterning of *Arabidopsis* shoots by class III HD-ZIP and KANADI genes. *Curr. Biol.*, **13**, 1768–1774.

Esau, K. (1977) *Anatomy of Seed Plant*. Wiley, New York.

Friml, J., Benkova, E., Blilou, I., Wisniewska, J., Hamann, T., Ljung, K., Woody, S., Sandberg, G., Scheres, B., Jurgens, G. & Palme, K. (2002) AtPIN4 mediates sink-driven auxin gradients and root patterning in *Arabidopsis. Cell*, **108**, 661–673.

Friml, J., Vieten, A., Sauer, M., Weijers, D., Schwarz, H., Hamann, T., Offringa, R. & Jurgens, G. (2003) Efflux-dependent auxin gradients establish the apical–basal axis of *Arabidopsis. Nature*, **426**, 147–153.

Galway, M.E., Masucci, J.D., Lloyd, A.M., Walbot, V., Davis, R.W. & Schiefelbein, J.W. (1994) The TTG gene is required to specify epidermal cell fate and cell patterning in the *Arabidopsis* root. *Dev. Biol.*, **166**, 740–754.

Geldner, N., Anders, N., Wolters, H., Keicher, J., Kornberger, W., Muller, P., Delbarre, A., Ueda, T., Nakano, A. & Jurgens, G. (2003) The *Arabidopsis* GNOM ARF-GEF mediates endosomal recycling, auxin transport, and auxin-dependent plant growth. *Cell*, **112**, 219–230.

Geldner, N., Friml, J., Stierhof, Y.D., Jurgens, G. & Palme, K. (2001) Auxin transport inhibitors block PIN1 cycling and vesicle trafficking. *Nature*, **413**, 425–428.

Geldner, N., Richter, S., Vieten, A., Marquardt, S., Torres-Ruiz, R.A., Mayer, U. & Jurgens, G. (2004) Partial loss-of-function alleles reveal a role for GNOM in auxin transport-related, post-embryonic development of *Arabidopsis. Development*, **131**, 389–400.

Gray, W.M., Kepinski, S., Rouse, D., Leyser, O. & Estelle, M. (2001) Auxin regulates SCF(TIR1)-dependent degradation of AUX/IAA proteins. *Nature*, **414**, 271–276.

Guilfoyle, T., Hagen, G., Ulmasov, T. & Murfett, J. (1998) How does auxin turn on genes? *Plant Physiol.*, **118**, 341–347.

Hamann, T., Benkova, E., Baurle, I., Kientz, M. & Jurgens, G. (2002) The *Arabidopsis* BODENLOS gene encodes an auxin response protein inhibiting MONOPTEROS-mediated embryo patterning. *Genes Dev.*, **16**, 1610–1615.

Hamann, T., Mayer, U. & Jurgens, G. (1999) The auxin-insensitive bodenlos mutation affects primary root formation and apical–basal patterning in the *Arabidopsis* embryo. *Development*, **126**, 1387–1395.

Hardtke, C.S. & Berleth, T. (1998) The *Arabidopsis* gene MONOPTEROS encodes a transcription factor mediating embryo axis formation and vascular development. *EMBO J.*, **17**, 1405–1411.

Helariutta, Y., Fukaki, H., Wysocka-Diller, J., Nakajima, K., Jung, J., Sena, G., Hauser, M.T. & Benfey, P.N. (2000) The SHORT-ROOT gene controls radial patterning of the *Arabidopsis* root through radial signalling. *Cell*, **101**, 555–567.

Hellmann, H. & Estelle, M. (2002) Plant development: regulation by protein degradation. *Science*, **297**, 793–797.

Hellmann, H., Hobbie, L., Chapman, A., Dharmasiri, S., Dharmasiri, N., del Pozo, C., Reinhardt, D. & Estelle, M. (2003) *Arabidopsis* AXR6 encodes CUL1 implicating SCF E3 ligases in auxin regulation of embryogenesis. *EMBO J.*, **22**, 3314–3325.

Hung, C.Y., Lin, Y., Zhang, M., Pollock, S., Marks, M.D. & Schiefelbein, J. (1998) A common position-dependent mechanism controls cell-type patterning and GLABRA2 regulation in the root and hypocotyl epidermis of *Arabidopsis. Plant Physiol.*, **117**, 73–84.

Inoue, T., Higuchi, M., Hashimoto, Y., Seki, M., Kobayashi, M., Kato, T., Tabata, S., Shinozaki, K. & Kakimoto, T. (2001) Identification of CRE1 as a cytokinin receptor from *Arabidopsis. Nature*, **409**, 1060–1063.

Jang, J.C., Fujioka, S., Tasaka, M., Seto, H., Takatsuto, S., Ishii, A., Aida, M., Yoshida, S. & Sheen, J. (2000) A critical role of sterols in embryonic patterning and meristem programming revealed by the fackel mutants of *Arabidopsis thaliana. Genes Dev.*, **14**, 1485–1497.

Kamiya, N., Nagasaki, H., Morikami, A., Sato, Y. & Matsuoka, M. (2003) Isolation and characterization of a rice WUSCHEL-type homeobox gene that is specifically expressed in the central cells of a quiescent center in the root apical meristem. *Plant J.*, **35**, 429–441.

Kidner, C., Sundaresan, V., Roberts, K. & Dolan, L. (2000) Clonal analysis of the *Arabidopsis* root confirms that position, not lineage, determines cell fate. *Planta*, **211**, 191–199.

Kuhn, C., Franceschi, V.R., Schulz, A., Lemoine, R. & Frommer, W.B. (1997) Macromolecular trafficking indicated by localization and turnover of sucrose transporters in enucleate sieve elements. *Science*, **275**, 1298–1300.

Larkin, J.C., Brown, M.L. & Schiefelbein, J. (2003) How do cells know what they want to be when they grow up? Lessons from epidermal patterning in *Arabidopsis*. *Annu. Rev. Plant Biol.*, **54**, 403–430.

Laskowski, M.J., Williams, M.E., Nusbaum, H.C. & Sussex, I.M. (1995) Formation of lateral root meristems is a two-stage process. *Development*, **121**, 3303–3310.

Laux, T., Mayer, K.F., Berger, J. & Jurgens, G. (1996) The WUSCHEL gene is required for shoot and floral meristem integrity in *Arabidopsis*. *Development*, **122**, 87–96.

Lee, M.M. & Schiefelbein, J. (1999) WEREWOLF, a MYB-related protein in *Arabidopsis*, is a position-dependent regulator of epidermal cell patterning. *Cell*, **99**, 473–483.

Lee, M.M. & Schiefelbein, J. (2002) Cell pattern in the *Arabidopsis* root epidermis determined by lateral inhibition with feedback. *Plant Cell*, **14**, 611–618.

Lenhard, M. & Laux, T. (2003) Stem cell homeostasis in the *Arabidopsis* shoot meristem is regulated by intercellular movement of CLAVATA3 and its sequestration by CLAVATA1. *Development*, **130**, 3163–3173.

Lin, Y. & Schiefelbein, J. (2001) Embryonic control of epidermal cell patterning in the root and hypocotyl of *Arabidopsis*. *Development*, **128**, 3697–3705.

Lloyd, A.M., Walbot, V. & Davis, R.W. (1992) *Arabidopsis* and *Nicotiana* anthocyanin production activated by maize regulators R and C1. *Science*, **258**, 1773–1775.

Lucas, W.J., Bouche-Pillon, S., Jackson, D.P., Nguyen, L., Baker, L., Ding, B. & Hake, S. (1995) Selective trafficking of KNOTTED1 homeodomain protein and its mRNA through plasmodesmata. *Science*, **270**, 1980–1983.

Mähönen, A.P., Bonke, M., Kauinen, L., Riikonen, M., Benfey, P.N. & Helariutta, Y. (2000) A novel two-component hybrid molecule regulates vascular morphogenesis of the *Arabidopsis* root. *Genes Dev.*, **14**, 2938–2943.

Malamy, J.E. & Benfey, P.N. (1997) Organization and cell differentiation in lateral roots of *Arabidopsis thaliana*. *Development*, **124**, 33–44.

Mansfield, S.G. & Briarty, L.G. (1991) Early embryogenesis in *Arabidopsis thaliana*. *Can. J. Bot.*, **69**, 461–476.

Marchant, A., Bhalerao, R., Casimiro, I., Eklof, J., Casero, P.J., Bennett, M. & Sandberg, G. (2002) AUX1 promotes lateral root formation by facilitating indole-3-acetic acid distribution between sink and source tissues in the *Arabidopsis* seedling. *Plant Cell*, **14**, 589–597.

Masucci, J.D., Rerie, W.G., Foreman, D.R., Zhang, M., Galway, M.E., Marks, M.D. & Schiefelbein, J.W. (1996) The homeobox gene GLABRA2 is required for position-dependent cell differentiation in the root epidermis of *Arabidopsis thaliana*. *Development*, **122**, 1253–1260.

Mayer, U., Buttner, G. & Jurgens, G. (1993) Apical–basal pattern formation in the *Arabidopsis* embryo: studies on the role of the gnom gene. *Development*, **117**, 149–162.

Mayer, U., Torres Ruiz, R. A., Berleth, T., Miséra, S. & Jürgens, G. (1991) Mutations affecting body organization in the *Arabidopsis* embryo. *Nature*, **353**, 402–407.

Nakajima, K., Sena, G., Nawy, T. & Benfey, P.N. (2001) Intercellular movement of the putative transcription factor SHR in root patterning. *Nature*, **413**, 307–311.

Oparka, K.J. & Turgeon, R. (1999) Sieve elements and companion cells-traffic control centers of the phloem. *Plant Cell*, **11**, 739–750.

Payne, C.T., Zhang, F. & Lloyd, A.M. (2000) GL3 encodes a bHLH protein that regulates trichome development in *Arabidopsis* through interaction with GL1 and TTG1. *Genetics*, **156**, 1349–1362.

Pysh, L.D., Wysocka-Diller, J.W., Camilleri, C., Bouchez, D. & Benfey, P.N. (1999) The GRAS gene family in *Arabidopsis*: sequence characterization and basic expression analysis of the SCARECROW-LIKE genes. *Plant J.*, **18**, 111–119.

Rerie, W.G., Feldmann, K.A. & Marks, M.D. (1994) The GLABRA2 gene encodes a homeodomain protein required for normal trichome development in *Arabidopsis*. *Genes Dev.*, **8**, 1388–1399.

Sabatini, S., Beis, D., Wolkenfelt, H., Murfett, J., Guilfoyle, T., Malamy, J., Benfey, P., Leyser, O., Bechtold, N., Weisbeek, P. & Scheres, B. (1999) An auxin-dependent distal organizer of pattern and polarity in the *Arabidopsis* root. *Cell*, **99**, 463–472.

Sabatini, S., Heidstra, R., Wildwater, M. & Scheres, B. (2003) SCARECROW is involved in positioning the stem cell niche in the *Arabidopsis* root meristem, *Genes Dev.*, **17**, 354–358.

Sachs, T. (1988) Epigenetic selection: an alternative mechanism of pattern formation. *J. Theor. Biol.*, **134**, 547–559.

Sachs, T. (1991) Cell polarity and tissue patterning in plants. *Development*, **S1**, 83–93.

Schellmann, S., Schnittger, A., Kirik, V., Wada, T., Okada, K., Beermann, A., Thumfahrt, J., Jurgens, G. & Hulskamp, M. (2002) TRIPTYCHON and CAPRICE mediate lateral inhibition during trichome and root hair patterning in *Arabidopsis*. *EMBO J.*, **21**, 5036–5046.

Scheres, B., Benfey, P. & Dolan, L. (2002) Root development. In: *The Arabidopsis Book* (eds C.R. Somerville & E.M. Meyerowitz). American Society of Plant Biologists, Rockville, MD, pp. 1–18. Available at: http://www.aspb.org/publications/arabidopsis/. Access date Sept 30, 2002.

Scheres, B., Di Laurenzio, L., Willemsen, V., Hauser, M.T., Janmaat, K., Weisbeek, P. & Benfey, P.N. (1995) Mutations affecting the radial organisation of the *Arabidopsis* root display specific defects throughout the embryonic axis. *Development*, **121**, 53–62.

Scheres, B., Wolkenfelt, H., Willemsen, V., Terlouw, M., Lawson, E., Dean, C. & Weisbeek, P. (1994) Embryonic origin of the *Arabidopsis* primary root and root meristem initials. *Development*, **120**, 2475–2487.

Schrick, K., Mayer, U., Horrichs, A., Kuhnt, C., Bellini, C., Dangl, J., Schmidt, J. & Jurgens, G. (2000) FACKEL is a sterol C-14 reductase required for organized cell division and expansion in *Arabidopsis* embryogenesis. *Genes Dev.*, **14**, 1471–1484.

Schrick, K., Mayer, U., Martin, G., Bellini, C., Kuhnt, C., Schmidt, J. & Jurgens, G. (2002) Interactions between sterol biosynthesis genes in embryonic development of *Arabidopsis*. *Plant J.*, **31**, 61–73.

Souter, M., Toing, J., Pullen, M., Friml, J., Palme, K., Hackett, R., Grierson, D. & Lindsey, K. (2002) Hydra mutants of *Arabidopsis* are defective in sterol profiles and auxin and ethylene signalling. *Plant Cell*, **14**, 1017–1031.

Suzuki, T., Miwa, K., Ishikawa, K., Yamada, H., Aiba, H. & Mizuno, T. (2001) The *Arabidopsis* sensor His-kinase, AHk4, can respond to cytokinins. *Plant Cell Physiol.*, **42**, 107–113.

Swarup, R., Friml, J., Marchant, A., Ljung, K., Sandberg, G., Palme, K. & Bennett, M. (2001) Localization of the auxin permease AUX1 suggests two functionally distinct hormone transport pathways operate in the *Arabidopsis* root apex. *Genes Dev.*, **15**, 2648–2653.

Ulmasov, T., Hagen, G. & Guilfoyle, T.J. (1997) ARF1, a transcription factor that binds to auxin response elements. *Science*, **276**, 1865–1868.

Umeda, M., Umeda-Hara, C. & Uchimiya, H. (2000) A cyclin-dependent kinase-activating kinase regulates differentiation of root initial cells in *Arabidopsis*. *Proc. Natl. Acad. Sci. U.S.A.*, **97**, 13396–13400.

van den Berg, C., Willemsen, V., Hage, W., Weisbeek, P. & Scheres, B. (1995) Cell fate in the *Arabidopsis* root meristem determined by directional signalling. *Nature*, **378**, 62–65.

van den Berg, C., Willemsen, V., Hendriks, G., Weisbeek, P. & Scheres, B. (1997) Short-range control of cell differentiation in the *Arabidopsis* root meristem. *Nature*, **390**, 287–289.

Wada, T., Kurata, T., Tominaga, R., Koshino-Kimura, Y., Tachibana, T., Goto, K., Marks, M.D., Shimura, Y. & Okada, K. (2002) Role of a positive regulator of root hair development, CAPRICE, in *Arabidopsis* root epidermal cell differentiation. *Development*, **129**, 5409–5419.

Wada, T., Tachibana, T., Shimura, Y. & Okada, K. (1997) Epidermal cell differentiation in *Arabidopsis* determined by a Myb homolog CPC. *Science*, **277**, 1113–1116.

Walker, A.R., Davison, P.A., Bolognesi-Winfield, A.C., James, C.M., Srinivasan, N., Blundell, T.L., Esch, J.J., Marks, M.D. & Gray, J.C. (1999) The TRANSPARENT TESTA GLABRA1 locus, which regulates trichome differentiation and anthocyanin biosynthesis in *Arabidopsis*, encodes a WD40 repeat protein. *Plant Cell*, **11**, 1337–1350.

Wysocka-Diller, J.W., Helariutta, Y., Fukaki, H., Malamy, J.E. & Benfey, P.N. (2000) Molecular analysis of SCARECROW function reveals a radial patterning mechanism common to root and shoot. *Development*, **127**, 595–603.

Ye, Z.H. (2002) Vascular tissue differentiation and pattern formation in plants. *Annu. Rev. Plant. Biol.*, **53**, 183–202.

9 Lessons from leaf epidermal patterning in plants
Bhylahalli Purushottam Srinivas and Martin Hülskamp

9.1 Overview

The epidermis of mature leaves consists of three different types of cells. Most cells differentiate into normal epidermal pavement cells. Additionally, trichomes and stomata are found non-randomly distributed on the surface. The establishment of such distribution patterns involves different kinds of cell-to-cell communication. While trichome patterning appears to be primarily mediated by the movement of transcription factors through plasmodesmata, stomatal patterning seems to involve receptor-ligand-based cellular communication. This review compares the mechanisms of the two patterning systems.

9.2 Introduction

During plant development, all cells become successively restricted in their developmental potential and eventually adopt a distinct cell fate and differentiate accordingly. The spatial coordination of cell fate determination is mostly based on cell-to-cell communication, which may occur either over long distances (for example by hormones) or locally between neighbouring cells.

Most of our current knowledge on the cellular communication occurring during epidermal differentiation is derived from genetic analysis in the model plant *Arabidopsis*. Here, one of the best-studied models for analysing the mechanisms of cellular communication in the context of developmental processes is the leaf epidermis. Several aspects facilitate its analysis. On the one hand, the epidermis is readily accessible for visual inspection and experimental manipulations. On the other hand, the epidermis consists of only a few cell types that show a certain spacing or distribution pattern. Both trichomes as well as stomata are distributed non-randomly between epidermal pavement cells, suggesting the existence of mechanisms that regulate this pattern. The current data suggest that these mechanisms are fundamentally different for trichomes and stomata. The goal of this review is to compare these two patterning systems, with emphasis on the communication between epidermal cells.

9.3 Mechanisms of trichome patterning

Trichomes are found on the aerial surface of most higher plants and are important for the protection of the plant against insects and UV radiation and the reduction of water loss (Johnson, 1975). Trichomes in different plants vary dramatically in form, size and density, and may be unicellular or multicellular, secretory glandular or non-glandular hairs (Uphof, 1962; Esau, 1977; Theobald et al., 1979). The mechanisms underlying their development are best understood in the genetic model plant *Arabidopsis* and most of the data discussed in this review will focus on trichomes in *Arabidopsis*.

9.3.1 Trichome differentiation

Leaf trichomes in *Arabidopsis* are unicellular and branched cells that cover the surface of rosette leaves. Ecological studies have shown that they play a role in the defence against insect herbivores (Mauricio & Rausher, 1997). *Arabidopsis* trichomes develop from single protodermal cells. When they become recognizable as trichomes, they are separated from each other by an average of about three protodermal cells. On a given leaf the most mature trichomes are found at the tip, and concomitant with further leaf growth new trichomes are initiated at the leaf base (Larkin et al., 1996). The first morphological change of an incipient trichome cell is an increased nuclear size, which indicates that trichomes switch from mitotic cell divisions to endoreduplication cycles (Hülskamp et al., 1994). The surrounding epidermal pavement cells continue to divide, thereby increasing the distance between the already formed trichomes. The trichome cell undergoes on average four endoreduplication cycles, resulting in a mature trichome with a DNA content of about 32C (Melaragno et al., 1993; Hülskamp et al., 1994). The increase in nuclear size is accompanied by an enormous growth. Initially the cell expands out of the plane of the epidermis. Then two successive branching events occur with a stereotypical arrangement of the branches (Hülskamp et al., 1994; Folkers et al., 1997). The mature trichome reaches a size of up to 0.5 mm and is separated from the neighbouring trichomes by about 0.5–1 mm.

9.3.2 Why is a mechanism postulated to explain the trichome spacing pattern and what kind of underlying principles are operating?

In principle, the distribution of trichomes could simply be random and, in this case, no mechanism would be necessary to control the spatial arrangement. This, however, is not the case. If the observed distribution of trichomes were random, one would expect that trichomes would be initiated next to each other with a certain probability that depends on their density. A careful statistical analysis of trichome distribution revealed that clustered trichomes are much less frequently observed than expected for a random distribution (Larkin et al., 1996). Thus, some kind of patterning mechanism must exist that controls trichome spacing. As no correlation between the arrangement of underlying cells and trichome pattern was found, the pattern is expected to be created *de novo*. In principle, two types of mechanisms could account for the observed distribution pattern:

1. A cell lineage mechanism by which a stereotypical series of divisions segregates different cell fates;

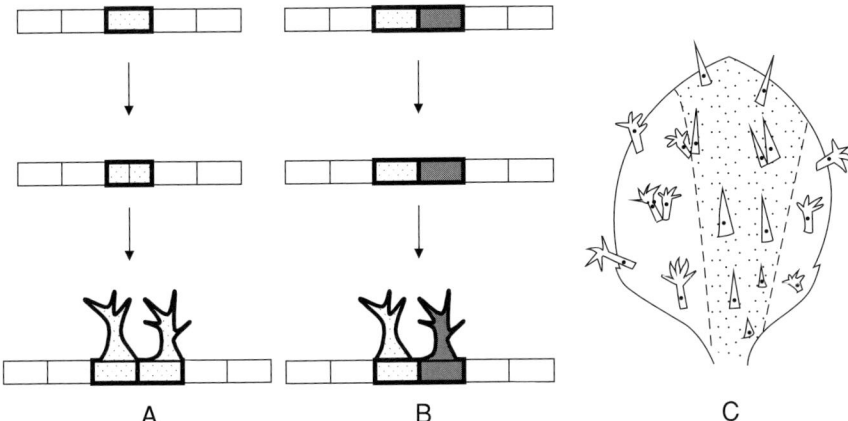

Figure 9.1 Cell lineage or mutual inhibition: clonal analysis as a tool. Two scenarios could account for the observed distribution pattern. In a cell lineage scenario, stereotypic division pattern segregating cell fates would create a pattern. In the second scenario, all cells are equivalent and competition between cells creates a pattern. The *try* mutant was used as a tool to distinguish between these two possibilities. (A) If the *TRY* gene were involved in a cell lineage scenario, the observed clusters in a *try* mutant should always contain clonally related cells. For example if a cell that is already determined to become a trichome cell (dotted box) divides again, a trichome cluster of cells with the same origin is formed. The function of *TRY* would be to inhibit such divisions. (B) If *TRY* were involved in the second scenario, clusters would be formed because *TRY*- mediated communication is disabled and this failure results in the formation of trichomes next to each other. These cells can have different ancestor cells (indicated as a dotted box and a grey box). (C) Clonal analysis showed that the second scenario is true. In a *try* homozygous and *stichel* heterozygous background, *stichel* clones were generated by EMS mutagenesis of seeds. If the second *stichel* allele is mutated in an individual cell, the descendants form a clonal sector of *stichel* homozygous cells. Such a sector on a leaf will show unbranched trichomes (marked by a dotted area). At the border of such clones *try* clusters were found that contained branched and unbranched trichomes (arrow), indicating that a cluster can consist of clonally unrelated cells.

2. A competition mechanism by which initially equivalent cells communicate with each other to select trichome cells separated from each other by an average minimum distance.

In order to distinguish between these two possibilities, the *triptychon* (*try*) mutant, which displays trichome clusters, was used (Fig. 9.1; Schnittger *et al.*, 1999). If the first scenario were true, all trichomes within a cluster should be derived from the same ancestor cell (clonally related). If the second scenario were true, trichomes in a cluster could have different ancestor cells. Clonal analysis of the *try* mutant showed that the latter outcome is observed, ruling out a cell-lineage-based mechanism to explain trichome spacing pattern.

9.3.3 Analysis of trichome initiation mutants

Several mutants have been identified that show aberrant trichome patterning. The corresponding genes appear to act as either positive or negative regulators of trichome initiation; the absence of positive factors reduces or abolishes trichome formation and the absence of negative regulators results in an increase in trichomes number.

9.3.3.1 Positive regulators of trichome initiation

Mutations in two genes, *GLABRA1* (*GL1*) and *TRANSPARENT TESTA GLABRA1* (*TTG1*), lead to the complete absence of trichomes. In *gl1* mutants, all trichomes on the leaf surface are absent. Only a few trichomes at the leaf margin are not affected. The *GL1* gene encodes an R2-R3 MYB related transcription factor (Oppenheimer et al., 1991). GL1 protein is found in the nucleus, indicating that GL1 acts as a transcription factor (Szymanski et al., 1998). GL1 is expressed initially ubiquitously in all epidermal cells of the young developing leaf. Then trichome precursor cells show increasingly higher expression levels and, eventually, expression is found only in developing trichomes but not in the surrounding epidermis cells (Larkin et al., 1993). In addition to the trichome phenotype, mutations in the *TTG1* gene result in a number of additional defects, including the initiation of ectopic root hairs, the lack of seed coat mucilage and reduced anthocyanin production (Koornneef, 1981; Galway et al., 1994). The control of these different processes is mediated by a set of MYB-related and MYC-related transcription factors such that a distinct combination of them controls one specific process (Zhang et al., 2003). *TTG1* encodes a protein containing WD40 repeats, which are known to function as protein–protein interaction domains in a variety of processes (Walker et al., 1999).

A less severe phenotype is found in *GLABRA3* (*GL3*) mutants. In *gl3* mutants, trichome number and size is drastically reduced. The *GL3* gene encodes a basic helix-loop-helix (bHLH)-like transcription factor (Payne et al., 2000) and, as the protein is found in the nucleus (Esch et al., 2003), it is likely to function as a transcription factor. The low importance of *GL3* in trichome patterning, as suggested by the weak phenotype of the *gl3* mutation, turned out to be misleading because of gene redundancy. A close homologue of *GL3*, the *ENHANCER OF GLABRA3* (*EGL3*) gene, turned out to act redundantly with *GL3*. While the single e*gl3* mutant exhibits no trichome phenotype, the e*gl3 gl3* double mutant is completely glabrous (Zhang et al., 2003).

In addition to the above-described genes, the *GLABRA2* (*GL2*) gene appears to be involved in trichome patterning. The phenotype of *GL2* mutants is pleiotropic and includes the reduction of seed coat mucilage, the overproduction of root hairs and a reduction of trichome growth (Rerie et al., 1994; Cristina et al., 1996). The reduced trichome phenotype suggested that *GL2* plays a role in the differentiation and morphogenesis of trichomes but not in patterning. Misexpression studies suggested an additional role in trichome patterning (Ohashi et al., 2002). Ubiquitous expression of GL2 appeared to be lethal but the introduction of additional copies of *GL2* under its own promoter resulted in an increased number of trichomes and the formation of trichome clusters. These findings suggest that *GL2* can also promote trichome formation and can, therefore, be considered to be a positive regulator of trichome initiation.

Several other mutants that were also shown to affect trichome initiation, such as *reduced trichome number, transparent testa glabra2, fiddlehead* and *increased chalcone synthase expression*, will not be considered here further, because no mechanistic data are available for these genes in the context of trichome patterning (Larkin *et al.*, 1996; Yephremov *et al.*, 1999; Wade *et al.*, 2001; Johnson *et al.*, 2002).

9.3.3.2 Negative regulators of trichome initiation

Two genes, *TRIPTYCHON* (*TRY*) and *CAPRICE* (*CPC*), were identified that appear to act as negative regulators of trichome initiation. In *try* mutants, trichomes are frequently arranged in clusters of up to three or four trichomes, suggesting that *TRY* locally suppresses adjacent trichome development (Hulskamp *et al.*, 1994). *TRY* encodes a small single-repeat MYB protein that lacks any recognizable activation domain, which suggests that TRY inhibits trichome initiation by competing with the activating factors at the level of DNA binding (Schellmann *et al.*, 2002). Such a negative role in trichome initiation is supported by the finding that over-expression of *TRY* causes a suppression of trichome formation (Schellmann *et al.*, 2002). *TRY* is initially expressed in all cells of the young leaf primordia. Surprisingly, *TRY* expression becomes stronger in trichome precursor cells and eventually is expressed exclusively in trichomes (Schellmann *et al.*, 2002). The *CPC* gene was initially identified as a root hair promoting factor since *cpc* mutants show a drastic reduction of root hairs (Wada *et al.*, 1997; see also Chapter 8, this volume). A more detailed analysis revealed that *cpc* mutants also have an increased density of leaf trichomes but no trichome clusters. *CPC* encodes a small single-repeat MYB protein with high sequence similarity to *TRY* and also shows a similar expression pattern (Wada *et al.*, 1997; Schellmann *et al.*, 2002). Double mutant analysis revealed that *TRY* and *CPC* function redundantly. The *try cpc* double mutant exhibits large clusters of up to 40 trichomes. These trichomes are initiated successively such that initially small clusters of two or three trichomes are found and at later stages also, neighbouring cells initiate trichome development (Schellmann *et al.*, 2002).

9.3.4 Interactions between the trichome initiation genes

The genetic analysis of the trichome initiation genes has revealed some insights into how the positive regulators depend on each other and how they control and are controlled by the negative factors.

Genetic evidence suggests that the positive regulators of trichome initiation act together, with *TTG1* having a promoting but not an essential role. This was inferred from over-expression studies. *GL3* and *GL1* appear to be essential, as the corresponding mutants cannot be rescued by any other factor (Larkin *et al.*, 1994; Lloyd *et al.*, 1994). Their combined over-expression results in a phenotype that is more than just additive (synergistic), which indicates that the two proteins interact with each other (Payne *et al.*, 2000). *TTG1* acts upstream of *GL3*, *EGL3* and *GL1*, which is evident from the finding that over-expression of *GL3* and *EGL3* or *GL3* and *GL1* can rescue the *TTG1* mutant trichome phenotype (Payne *et al.*, 2000; Zhang *et al.*,

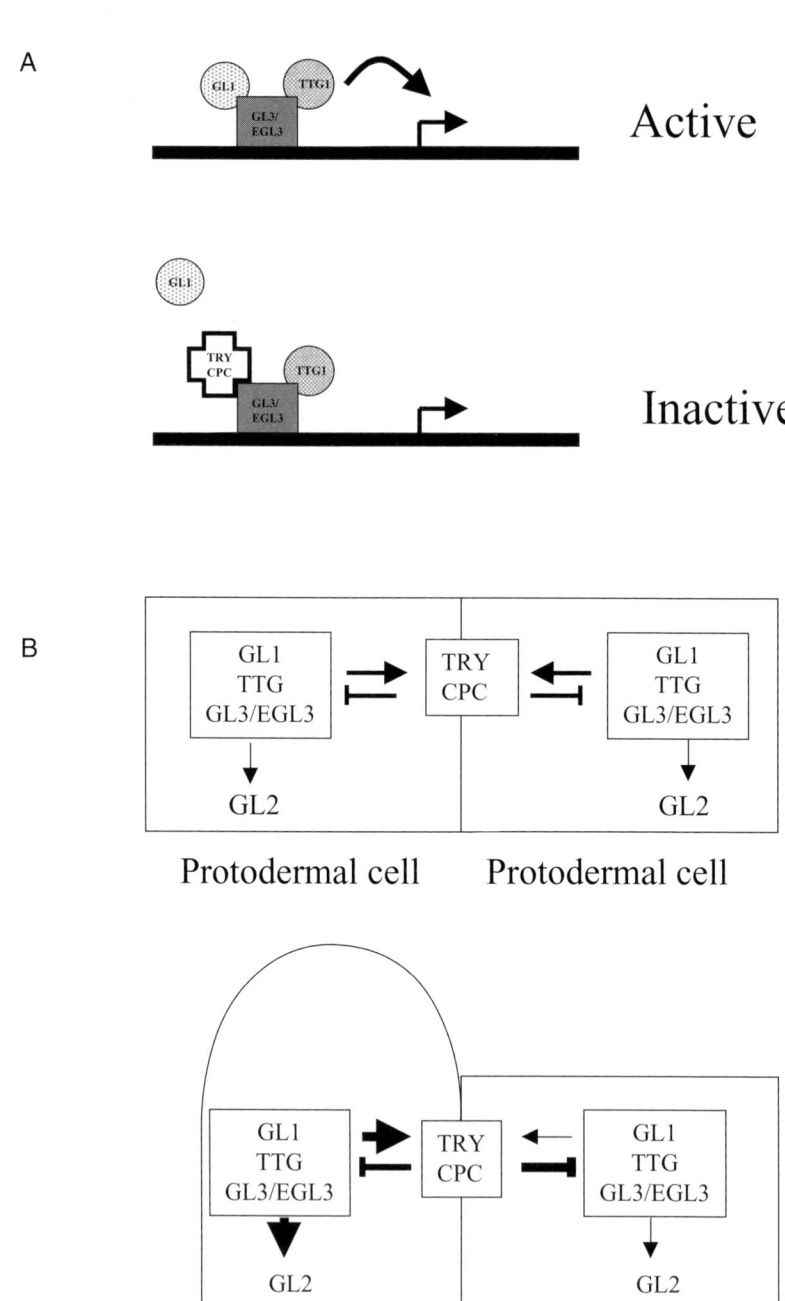

Figure 9.2

2003). That the positive regulators interact also with the negative regulators was shown by two types of experiments: Firstly, several mutant combinations in which the gene dosage of one of the positive regulators and of *TRY* was reduced to 50% (trans-heterozygous) exhibited patterning defects. Secondly, plants over-expressing *GL1* or *GL3* showed a much stronger phenotype in the absence of *TRY*, indicating that *TRY* represses them (Schnittger *et al.*, 1998; Szymanski & Marks, 1998).

The physical interactions between the transcription factors as revealed by yeast two-hybrid studies provided a more detailed picture. Strong interactions were found between *GL1* and *GL3* (Payne *et al.*, 2000). Using protein fragments, the interaction sites were mapped to the N-terminal MYB repeats of GL1 and the first 96 amino acids of GL3. GL3 can also homodimerize by its C-terminal region that contains the bHLH domains. TTG1 also binds to GL3 but not to GL1 (Payne *et al.*, 2000). The regions of GL3 responsible for GL1 and TTG1 binding are different. Given that TTG1 can be substituted by over-expression of GL1 and GL3, it is likely that TTG1 has a more accessory function, for example in the stabilization of the complex. The negative regulator TRY was shown to interact with GL3 (Esch *et al.*, 2003). Interestingly, TRY binds to the same domain of GL3 as GL1, and competition studies demonstrated that TRY can compete with GL1 for binding with GL3. This finding strongly suggests that two alternative complexes can be formed, an active complex consisting of GL1, TTG1 and GL3 and an inactive one containing TRY, TTG1 and GL3 (Fig. 9.2A).

A potential target gene of the putative transcriptional complex consisting of TTG1, GL3 and GL1 is the *GL2* gene. On the one hand, it is possible to turn on *GL2* expression ectopically by over-expressing other trichome promoting factors (Szymanski *et al.*, 1998). On the other hand, the *GL2* phenotype combines several aspects of other trichome morphogenesis mutants, suggesting that it acts as a master gene coordinating the proper differentiation of the trichome (Rerie *et al.*, 1994). Further studies are needed to clarify the above-mentioned additional role of *GL2* in trichome patterning.

9.3.5 Local cell–cell interactions leading to cell fate decisions: a model

As the position of trichome cells is determined neither by their relative position to other cells nor by cell lineage, the trichome pattern appears to be created *de novo*.

Figure 9.2 A model for the selection of trichomes from the protodermal cell pool. (A) GL1, TTG1 and GL3/EGL3 represent the active complex which turns on downstream target genes to specify trichome cell identity. TRY/CPC compete with GL1 for binding with GL3/EGL3, forming an inactive complex which fails to specify trichomes. (B) Initially, all epidermal cells are equivalent expressing the activators GL1, GL3/EGL3 and TTG1 and begin to communicate with each other via TRY/CPC that are believed to move from cell to cell (top diagram). A bias in the balance of the activator concentration is postulated to increase the activity of the activators in one cell. The increased levels of the activator leads to trichome cell fate determination and causes increased levels of the inhibitor, which in turn laterally suppresses the neighbouring cells from gaining trichome fate (bottom diagram).

In this scenario, all epidermal cells are initially equivalent and the position of individual trichome cells is determined by intercellular communication. How could the trichome patterning genes described above explain such a patterning mechanism? In particular, how is it possible that not only the positively acting patterning genes but also the negatively acting genes are expressed in trichomes?

A model that explains *de novo* patterning and accommodates the above-mentioned observations is the reaction–diffusion model by Meinhardt and Gierer (Meinhardt & Gierer, 1974; Meinhardt, 1982, 1994). It is a theoretical model that attempts to define the minimal requirements to create a spacing pattern in a two-dimensional field of cells (Fig. 9.3). According to this model, the activator component activates its own inhibitor whereas the inhibitor inhibits the activator. Both the activator and the inhibitor may move, but the movement of the inhibitor

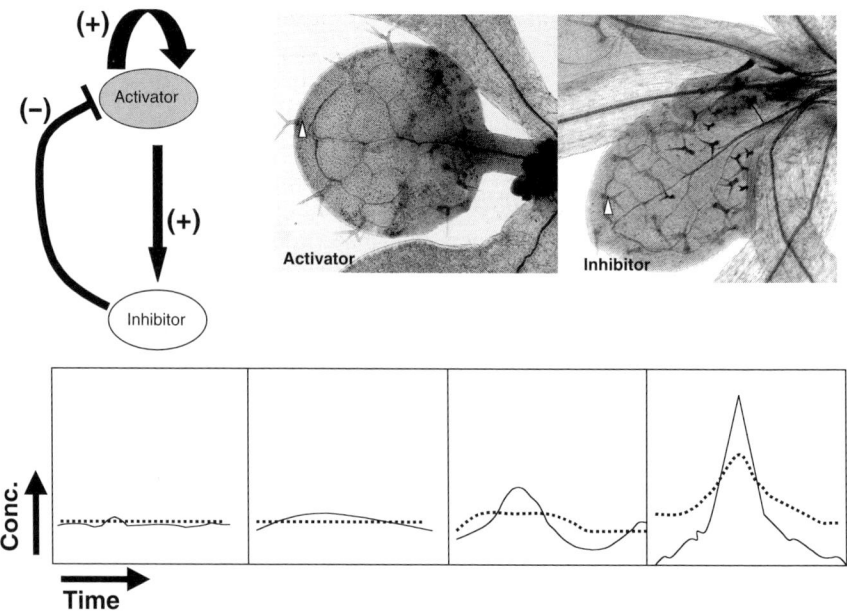

Figure 9.3 Models and expression patterns. The Meinhardt and Gierer model can explain the creation of a two-dimensional pattern by a two-component system (top-left scheme). The activator activates the production of an inhibitor, which in turn can suppress the production of the activator. In addition, a self-activation of the activator is postulated to theoretically allow the rapid amplification of small differences. Simulations are shown below. Small fluctuations result in the local amplification of the activator (solid line) and as a consequence also of that of the inhibitor (dotted line). In the end, a common peak of the activator and the inhibitor is formed. The expression of the inhibitors and the activators in the trichome system show this predicted co-expression (top right). As an example for a trichome activator the expression of *GL1* as revealed by a GL1:GUS construct is shown. As an example for an inhibitor, the expression of a CPC:GUS construct is shown. Note that in both lines the leaf base shows general expression (arrows) and that in more mature parts of the leaf only trichome cells show expression (arrow heads).

needs to be faster. In addition, it is necessary that the activator can activate its own synthesis because this positive feedback loop allows the rapid amplification of small fluctuations. In such a scenario, small fluctuations create a local enhancement of the activator and also of the inhibitor, which in turn leads to the suppression of the activator around this local peak. At a distance from already existing peaks, new local enhancements can occur. This model predicts that the activator and inhibitor show maximal levels at the same position, which intuitively would seem to be paradoxical.

Currently it is speculated that trichome patterning is in principle based on this model. The positive patterning genes *GL1*, *TTG* and *GL3* are assumed to locally activate their own expression and that of *TRY* and *CPC*. The inhibitors can counteract their activity by a competition mechanism as described above. Cell–cell interactions are likely to be mediated by the movement of TRY and CPC through the plasmodesmata (Fig. 9.2B) (see also Chapter 5, this volume). This is supported by the finding that in the root system CPC can move from the cells in which it is expressed into neighbouring cells (Wada *et al.*, 2002; see also Chapter 8, this volume).

While this model is consistent with the current data, many aspects of the model still remain to be proven.

9.3.6 Long-range control of trichome initiation by hormones

Among the plant hormones, only gibberellins have been implicated in the control of trichome initiation. Plants deficient for gibberellin develop no trichomes and the application of exogenous gibberellin rescues the trichome phenotype (Chien & Sussex, 1996). Gibberellin appears to act upstream of the trichome patterning genes as the over-expression of *GL1* and the maize homologue of *GL3* (the R gene) rescues the trichome phenotype in gibberellin-deficient plants (Perazza *et al.*, 1998). This regulation of trichome initiation by gibberellins appears to be important to regulate trichome density on organs and to change the density accompanying the switch from vegetative to reproductive growth (Telfer *et al.*, 1997).

9.4 Stomatal development and patterning

Stomata are small openings in the plant epidermis that control gas exchange between the plant and its environment to balance the water household and the levels of carbon dioxide in relation to photosynthetic activity. The actual pore consists of two guard cells that can regulate the size of the opening by their turgor (Nadeau & Sack, 2002b). In dicots the distribution of stomata is not regular, suggesting a random distribution. However, some patterning mechanism must exist, as stomata are normally not found directly next to each other. Since new stomata can develop between already existing stomata, Bünning proposed that stomata create an inhibitory field that prevents new stomata from developing in their neighbourhood (Bünning & Sagromsky, 1948; Bünning, 1956). The finding that stomata development involved a stereotypic series of cell divisions that placed new stomata away from already existing ones suggested

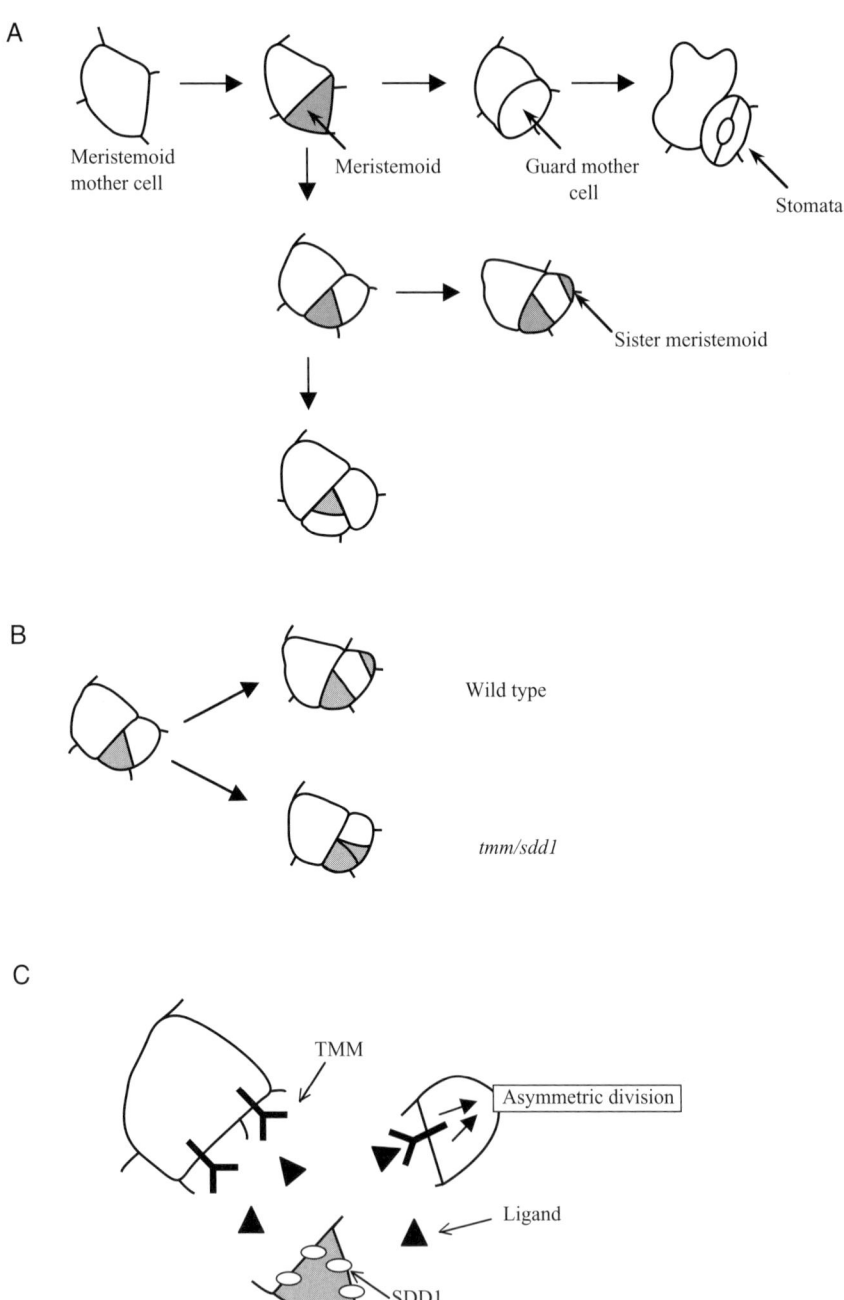

Figure 9.4

that cell lineage could be one or the sole mechanism responsible for the observed stomata pattern (Sachs, 1978).

9.4.1 Cell division pattern during stomata patterning

In *Arabidopsis*, stomata differentiation begins with the asymmetric division of the meristemoid mother cell (Geisler *et al.*, 2000). This division produces a small cell, called the meristemoid, and a large sister cell. The meristemoid cell may continue to divide up to three times asymmetrically to produce another meristemoid and a sister cell. In this case, division planes are arranged such that cells divide in an inward spiral, thereby placing the meristemoid in the middle of three ancestor cells (Fig. 9.4A) (Serna & Fenoll, 1997; Berger & Altmann, 2000; Serna, 2001). This division pattern would explain why stomata are not found next to each other. However, deviations from this pattern were also found because the meristemoid cells may differentiate into a guard mother cell any time. In this case the meristemoid cell adopts a round shape and divides symmetrically to produce the two guard cells of the mature stomate. If this occurs already after the first or second division, a stomate would not be surrounded by its own ancestor cells and another stomate could be positioned next to it. Also the larger sister cell resulting from the asymmetric divisions may adopt a stomatal pathway. Meristemoid cells resulting from such a sister cell are called satellite meristemoid cells.

These findings raise the question of what mechanism prevents two stomata from being placed next to each other. Cell lineage does not play a major role because a stomate is not always positioned in the middle of its own ancestor cells. Also inhibitory signals that prevent the formation of meristemoid cells in the neighbourhood of already existing ones can be excluded because meristemoid cells can form adjacent to another (Geisler *et al.*, 2000). Recent data show that the key patterning mechanism is the control of orientation of the asymmetric cell division plane (Geisler *et al.*, 2000). The plane of these divisions is always placed so that the new meristemoid develops on the side away from an adjacent guard mother cell or meristemoid. As this spatial control of the asymmetric divisions is not occurring in the context of cell lineage, interactions between the cells are necessary.

Figure 9.4 Stomatal patterning. (A) Overview of cell types during stomatal patterning. A meristemoid mother cell divides asymmetrically, producing a meristemoid and a sister cell. Meristemoids either divide again asymmetrically (up to three times) to produce another meristemoid and a sister cell or become a guard mother cell, which divides symmetrically to give rise to a stomate. (B) Comparison of cell division pattern in wild type and *tmm/sdd1* mutants during stomata formation. Unlike in the wild type, the *tmm/sdd1* mutants fail to orient the asymmetric cell division plane of the sister cell to produce the new meristemoid distal to the existing meristemoid. This leads to the formation of clustered stomata. (C) Model for the mechanism of TMM and SDD1 action in stomatal patterning. SDD1, which is highly expressed in meristemoids and guard mother cells, is thought to produce a ligand which relays the signal to the neighbouring cells. The signal is received by TMM, which is membrane localized, in the neighbour cells. As a result, the asymmetric division of the sister cell is oriented away from the pre-existing meristemoid cell.

9.4.2 Cell signalling and the control of asymmetric cell divisions during stomata development

The genetic, morphological and molecular characterization of two genes, *TOO MANY MOUTHS (TMM)* and *STOMATAL DENSITY AND DISTRIBUTION1 (SDD1)*, has revealed some insights into the underlying mechanism of stomata patterning.

In *tmm* mutants, stomata are found in clusters that can contain more than 20 individual stomata (Yang & Sack, 1995). The main defects in *tmm* mutants are the randomization of the orientation and the number of asymmetric divisions in cells adjacent to pre-existing stomata or meristemoids (Fig. 9.4B) (Geisler *et al.*, 2000). The molecular cloning of the *TMM* gene indicated that TMM serves to receive and relay signals coming from the neighbouring cells (Nadeau & Sack, 2002a). *TMM* encodes a protein with sequence similarity to leucine-rich-repeat-containing receptor-like proteins (LRR-RLP). The presence of a signal peptide and a transmembrane domain and *in vivo* localization studies with a TMM:GFP fusion protein suggest that *TMM* protein localizes to the plasma membrane. By analogy to other known LRR-RLP proteins, a likely molecular function of TMM is to perceive extracellular signals via its extracellular domain. Upon binding of a specific ligand the signal would be relayed into the cell. As TMM has no cytoplasmic kinase domain, it is likely that TMM interacts with additional factors for intracellular signal transduction. Consistent with its proposed function in controlling asymmetric cell divisions in meristemoids and neighbouring cells, TMM is expressed in exactly these cells.

The *sdd1* mutants show increased stomatal density and a very much higher frequency of adjacent stomata than wild-type plants. However, individual clusters contain far fewer stomata than *tmm* mutants. Also, the proportion of cells that enter the stomatal pathway is increased by about twofold compared to wild type. It has been found that all the stomata in these clusters are derived from satellite meristemoids (Fig. 9.4B). *SDD1* encodes a subtilisin-like serine protease and it is therefore likely that SDD1 acts to cleave or modify other proteins (Berger & Altmann, 2000; Von Groll *et al.*, 2002). The expression of SDD1 was found specifically in meristemoids and not in neighbouring cells. Although SDD1 has neither a predicted transmembrane domain nor post-translational membrane-association motifs, *in vivo* localization experiments using a SDD1:GFP fusion showed that the fusion protein is localized to the plasma membrane. It is therefore possible that SDD1 is involved in the production of the ligand that relays the signal to TMM (Fig. 9.4C). This view is supported by genetic experiments that show that *SDD1* and *TMM* act in the same pathway. Over-expression of *SDD1* causes a drastic reduction of stomata. This dominant effect of *SDD1* over-expression is rescued when *TMM* is absent, indicating that *SDD1* function is mediated by *TMM* (Von Groll *et al.*, 2002).

A likely scenario of how asymmetric cell divisions are controlled is that SDD1 is expressed in meristemoids or guard mother cells where it produces an extracellular signal. This is received and relayed into meristemoid as well as the sister cells by

TMM which serves, by controlling the orientation of cell division, to inhibit the formation of a new meristemoid adjacent to the pre-existing ones.

9.5 Perspective

Trichome and stomata patterning represent two formally very similar examples of *de novo* pattern formation in a two-dimensional field. Yet the underlying mechanisms by which the cells communicate and determine the pattern are different. Whereas cell-to-cell communication during trichome patterning is thought to be based on the movement of transcription factors through plasmodesmata, stomatal patterning appears to involve a receptor–ligand system. Whereas during trichome development, cells are likely to be selected by a reaction–diffusion mechanism, stomatal patterning is regulated by orienting the division plane properly during asymmetric cell divisions. Many of the discussed aspects are still very speculative and are based on better-studied animal models. It remains to be explored as to what extent these analogies and models hold true.

References

Berger, D. & Altmann, T. (2000) A subtilisin-like serine protease involved in the regulation of stomatal density and distribution in *Arabidopsis thaliana*. *Genes Dev.*, **14**, 1119–1131.

Bünning, E. (1956) General processes of differentiation. In: *The Growth of Leaves* (ed. F. Milthorpe), pp. 18–30. Butterworths, London.

Bünning, E. & Sagromsky, H. (1948) Die Bildung des Spaltöffnungsmusters in der Blattepidermis. *Z. Naturforsch. Teil B*, **3**, 203–216.

Chien, J.C. & Sussex, I.M. (1996) Differential regulation of trichome formation on the adaxial and abaxial leaf surfaces by gibberellins and photoperiod in *Arabidopsis thaliana* (L.) Heynh. *Plant Physiol.*, **111**, 1321–1328.

Cristina, M.D., Sessa, G., Dolan, L., Linstead, P., Ruberti, S. & Morelli, G. (1996) The *Arabidopsis* Athb-10 (GLABRA2) is an HD-Zip protein required for regulation of root hair development. *Plant J.*, **10**, 393–402.

Esau, K. (1977) *Anatomy of Seed Plants*. Wiley, New York.

Esch, J.J., Chen, M., Sanders, M., Hillestad, M., Ndkium, S., Idelkope, B., Neizer, J. & Marks, M.D. (2003) A contradictory GLABRA3 allele helps define gene interactions controlling trichome development in *Arabidopsis*. *Development*, **130**, 5885–5894.

Folkers, U., Berger, J. & Hulskamp, M. (1997) Cell morphogenesis of trichomes in *Arabidopsis*: differential control of primary and secondary branching by branch initiation regulators and cell growth. *Development*, **124**, 3779–3786.

Galway, M.E., Masucci, J.D., Lloyd, A.M., Walbot, V., Davis, R.W. & Schiefelbein, J.W. (1994) The *TTG* gene is required to specify epidermal cell fate and cell patterning in the *Arabidopsis* root. *Dev. Biol.*, **166**, 740–754.

Geisler, M., Nadeau, J. & Sack, F.D. (2000) Oriented asymmetric divisions that generate the stomatal spacing pattern in *Arabidopsis* are disrupted by the too many mouths mutation. *Plant Cell*, **12**, 2075–2086.

Hulskamp, M., Misera, S. & Jürgens, G. (1994) Genetic dissection of trichome cell development in *Arabidopsis*. *Cell*, **76**, 555–566.

Johnson, C.S., Kolevski, B. & Smyth, D.R. (2002) TRANSPARENT TESTA GLABRA2, a trichome and seed coat development gene of *Arabidopsis*, encodes a WRKY transcription factor. *Plant Cell*, **14**, 1359–1375.

Johnson, H.B. (1975) Plant pubescence: an ecological perspective. *Bot. Rev.*, **41**, 233–258.

Koornneef, M. (1981) The complex syndrome of *ttg* mutants. *Arabidopsis Inf. Serv.*, **18**, 45–51.

Larkin, J.C., Oppenheimer, D.G., Lloyd, A.M., Paparozzi, E.T. & Marks, M.D. (1994) The roles of the *GLABROUS1* and *TRANPARENT TESTA GLABRA* genes in *Arabidopsis* trichome development. *Plant Cell*, **6**, 1065–1076.

Larkin, J.C., Oppenheimer, D.G., Pollock, S. & Marks, M.D. (1993) *Arabidopsis GLABROUS1* gene requires downstream sequences for function. *Plant Cell*, **5**, 1739–1748.

Larkin, J.C., Young, N., Prigge, M. & Marks, M.D. (1996) The control of trichome spacing and number in *Arabidopsis*. *Development*, **122**, 997–1005.

Lloyd, A.M., Schena, M., Walbot, V. & Davis, R.W. (1994) Epidermal cell fate determination in *Arabidopsis*: patterns defined by steroid-inducible regulator. *Science*, **266**, 436–439.

Mauricio, R. & Rausher, M.D. (1997) Experimental manipulation of putative selective agents provides evidence for the role of natural enemies in the evolution of plant defense. *Evolution*, **51**, 1435–1444.

Meinhardt, H. (1982) *Models of Biological Pattern Formation*. Academic Press, London.

Meinhardt, H. (1994) Biological pattern formation: new observations provide support for theoretical predictions. *BioEssay*, **16**, 627–632.

Meinhardt, H. & Gierer, A. (1974) Applications of a theory of biological pattern formation based on lateral inhibition. *J. Cell Sci.*, **15**, 321–346.

Melaragno, J.E., Mehrotra, B. & Coleman, A.W. (1993) Relationship between endopolyploidy and cell size in epidermal tissue of *Arabidopsis*. *Plant Cell*, **5**, 1661–1668.

Nadeau, J.A. & Sack, F.D. (2002a) Control of stomatal distribution on the *Arabidopsis* leaf surface. *Science*, **296**, 1697–700.

Nadeau, J.A. & Sack, F.D. (2002b) Stomatal development in *Arabidopsis*. In: *The Arabidopsis Book* (eds C. Sommerville & E. Meyerowith). American Society of Plant Biology, Maryland. Available at: http://www.aspb.org.

Ohashi, Y., Ruberti, I., Morelli, G. & Aoyama, T. (2002) Entopically additive expression of GLABRA2 alters the frequency and spacing of trichome initiation. *Plant J.*, **21**, 5036–5046.

Oppenheimer, D.G., Herman, P.L., Sivakumaran, S., Esch, J. & Marks, M.D. (1991) A *myb* gene required for leaf trichome differentiation in *Arabidopsis* is expressed in stipules. *Cell*, **67**, 483–493.

Payne, C.T., Zhang, F. & Lloyd, A.M. (2000) GL3 encodes a bHLH protein that regulates trichome development in *Arabidopsis* through interaction with GL1 and TTG1. *Genetics*, **156**, 1349–1362.

Perazza, D., Vachon, G. & Herzog, M. (1998) Gibberellins promote trichome formation by up-regulating glabrous1 in *Arabidopsis*. *Plant Physiol.*, **117**, 375–383.

Rerie, W.G., Feldmann, K.A. & Marks, M.D. (1994) The glabra 2 gene encodes a homeodomain protein required for normal trichome development in *Arabidopsis*. *Genes Dev.*, **8**, 1388–1399.

Sachs, T. (1978) In: *The Clonal Basis of Development* (eds S. Subtelny & I.M. Sussex), pp. 161–183. Academic Press, New York.

Schellmann, S., Schnittger, A., Kirik, V., Wada, T., Okada, K., Beermann, A., Thumfahrt, J., Jurgens, G. & Hulskamp, M. (2002) TRIPTYCHON and CAPRICE mediate lateral inhibition during trichome and root hair patterning in *Arabidopsis*. *EMBO J.*, **21**, 5036–5046.

Schnittger, A., Folkers, U., Schwab, B., Jürgens, G. & Hulskamp, M. (1999) Generation of a spacing pattern: the role of *TRIPTYCHON* in trichome patterning in *Arabidopsis*. *Plant Cell*, **11**, 1105–1116.

Schnittger, A., Jurgens, G. & Hulskamp, M. (1998) Tissue layer and organ specificity of trichome formation are regulated by GLABRA1 and TRIPTYCHON in *Arabidopsis*. *Development*, **125**, 2283–2289.

Serna, L. (2001) Stomatal biology. *Trends Plant Sci.*, **6**, 554–555.

Serna, L. & Fenoll, C. (1997) Tracing the ontogeny of stomatal clusters in *Arabidopsis* with molecular markers. *Plant J.*, **12**, 747–755.

Szymanski, D.B. & Marks, M.D. (1998) *GLABROUS1* overexpression and *TRIPTYCHON* alter the cell cycle and trichome cell fate in *Arabidopsis*. *Plant Cell*, **10**, 2047–2062.

Szymanski, D.B., Jilk, R.A., Pollock, S.M. & Marks, M.D. (1998) Control of GL2 expression in *Arabidopsis* leaves and trichomes. *Development*, **125**, 1161–1171.

Telfer, A., Bollman, K.M. & Poethig, R.S. (1997) Phase change and the regulation of trichome distribution in*Arabidopsis thaliana*. *Development*, **124**, 645–654.

Theobald, W.L., Krahulik, J.L. & Rollins, R.C. (1979) Trichome description and classification. In: *Anatomy of the Dicotyledons*, Vol. 1 (eds C.R. Metcalfe & L. Chalk), pp. 41–53. Oxford Science, Oxford.

Uphof, J.C.T. (1962) *Plant Hairs*. Gebr. Bornträger, Berlin.

Von Groll, U., Berger, D. & Altmann, T. (2002) The subtilisin-like serine protease SDD1 mediates cell-to-cell signaling during *Arabidopsis* stomatal development. *Plant Cell*, **14**, 1527–1539.

Wada, T., Kurata, T., Tominaga, R., Koshino-Kimura, Y., Tachibana, T., Goto, K., Marks, M. D., Shimura, Y. & Okada, K. (2002) Role of a positive regulator of root hair development, CAPRICE, in *Arabidopsis* root epidermal cell differentiation. *Development*, **129**, 5409–5419.

Wada, T., Tachibana, T., Shimura, Y. & Okada, K. (1997) Epidermal cell differentiation in *Arabidopsis* determined by a *myb* homolog, *CPC*. *Science*, **277**, 1113–1116.

Wade, H.K., Bibikova, T.N., Valentine, W.J. & Jenkins, G.I. (2001) Interactions within a network of phytochrome, cryptochrome and UV-B phototransduction pathways regulate chalcone synthase gene expression in *Arabidopsis* leaf tissue. *Plant J.*, **25**, 675–685.

Walker, A.R., Davison, P.A., Bolognesi-Winfield, A.C., James, C.M., Srinivasan, N., Blundell, T.L., Esch, J.J., Marks, M.D. & Gray, J.C. (1999) The TTG1 (transparent testa, glabra1) locus which regulates trichome differentiation and anthocyanin biosynthesis in *Arabidopsis* encodes a WD40-repeat protein. *Plant Cell*, **11**, 1337–1350.

Yang, M. & Sack, F.D. (1995) The too many mouths and four lips mutations affect stomatal production in *Arabidopsis*. *Plant Cell*, **7**, 2227–2239.

Yephremov, A., Wisman, E., Huijser, P., Huijser, C., Wellesen, K. & Saedler, H. (1999) Characterization of the FIDDLEHEAD gene of *Arabidopsis* reveals a link between adhesion response and cell differentiation in the epidermis. *Plant Cell*, **11**, 2187–2201.

Zhang, F., Gonzalez, A., Zhao, M., Payne, C.T. & Lloyd, A. (2003) A network of redundant bHLH proteins functions in all TTG1-dependent pathways of *Arabidopsis*. *Development*, **130**, 4859–4869.

10 Lessons on signalling in plant self-incompatibility systems

Andrew G. McCubbin

10.1 Introduction

One of the key adaptations that have led to the success of the angiosperms is the flower, which functions to improve reproductive efficiency in a number of ways. One vital function performed by the flowers is screening of the genetic relatedness of gametes prior to fertilisation. Pollen from other species is generally prevented from germinating on the stigma or growing down the style, providing an effective interspecific, pre-zygotic breeding barrier. The majority of flowers are 'perfect', bearing both male and female reproductive organs in close proximity. Left unchecked, this results in a tendency to self-fertilise and inbreed. In response to this problem, a variety of reproductive strategies have arisen, one of the most widespread being termed self-incompatibility (SI). SI allows the pistil of a flower to discriminate between pollen from the same species that is genetically related and that which is unrelated. This self-/non-self-recognition results in the inhibition of self-pollen germination at the stigma surface or tube growth within the style, so promoting outbreeding. Whilst the effects of SI are consistent, i.e. rejection of genetically related pollen, both the phenomenology associated with rejection and molecular data suggest that a wide variety of signalling mechanisms have been recruited to achieve this goal.

SI appears to have evolved at least 21 different times during the evolution of flowering plants (Steinbachs & Holsinger, 2002) and historically has been split using two criteria. The first is whether mating types differ morphologically (heteromorphic SI) or are anatomically indistinguishable (homomorphic SI). The second factor is the genetic regulation of the system, the number of loci involved in encoding specificity and whether pollen phenotype is related to its own haploid genotype (termed gametophytic) or the diploid genotype of its parent plant (sporophytic) (see Fig. 10.1). For single-locus SI systems, the locus has in each case been termed S, but importantly it is evident that multiple genes reside within each S-locus, and the allelic complex of genes has been termed the S-haplotype.

Extensive molecular information is available for only three types of SI at present, two distinct forms of single-locus gametophytic SI and the single-locus sporophytic system of the Brassicaceae. In the most phylogenetically widespread form of single-locus gametophytic SI found in the Solanaceae, Rosaceae and Scrophulariaceae (Steinbachs & Holsinger, 2002), the pistil S-gene product is a glycoprotein (Kehyr-Pour & Pernes, 1985) with ribonuclease activity (McClure et al., 1989) and the

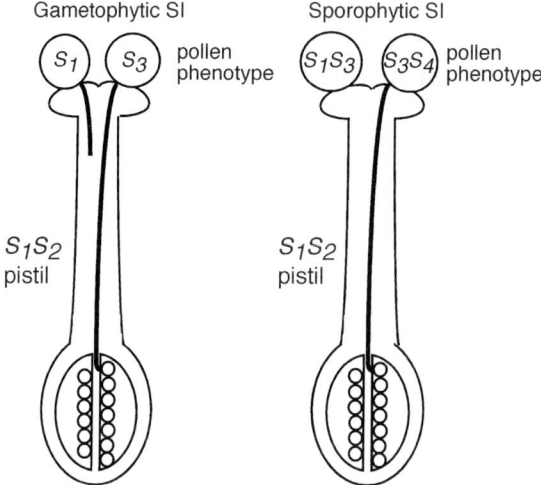

Figure 10.1 Genetics of pollen rejection in gametophytic and sporophytic SI. In gametophytic SI, pollen is rejected based on the matching of its haploid S-genotype with one of the two S-alleles of the pistil. In sporophytic SI the pollen behaves as diploid, and is rejected if either of the S-alleles matches either of the two alleles in the pistil.

pollen S-gene product has recently been identified as an F-box protein (Sijacic et al., 2004). In the Papaveraceae, SI single-locus gametophytic incompatability involves a complex series of events including changes in calcium ion concentration, phosphorylation of specific proteins, transcription of pollen genes and DNA fragmentation in nuclei (Jordan et al., 2000; Snowman et al., 2000). The pistil S-gene has been cloned but bears no significant homology to any gene of known function (Foote et al., 1994; Ride et al., 1999). In the single-locus sporophytic SI system of *Brassica*, both pollen and pistil S-genes have been identified, the pollen S-gene product encoding an extracellular ligand that mediates SI through activation of a protein receptor kinase – the stigmatic S-gene product (Schopfer et al., 1999; Takasaki et al., 2000). This chapter focuses on our current knowledge regarding the signalling events and mechanisms of action of these three different SI systems.

10.2 S-RNase-based single-locus gametophytic SI

S-allele products were first identified in members of the Solanaceae by electrophoretic analysis of pistil proteins (Bredemeijer & Blass, 1981). A polymorphic series of basic glycoproteins was identified which vary in molecular weight ($\sim$22–35 kDa) and isoelectric point ($\sim$8–10) and segregate with S-alleles. The location and developmental expression of these proteins also correlates with the SI response, they are largely confined to the extracellular space in the upper third of the stylar transmitting tract (the site of self-pollen tube inhibition), and are absent 1 day prior to anthesis (immature pistil are self-compatible) but rapidly accumulate to 1–10% of total protein at pollen release. These proteins were initially termed S-proteins and have been estimated to reach concentrations of 10–50 mg/mL in the extracellular matrix of the stylar transmitting tract of solanaceous species (Jahnen et al., 1989).

10.2.1 S-RNases encode S-specificity in the pistil

The first gene encoding an S-protein was cloned from *Nicotiana alata* (Anderson et al., 1986) and currently the sequences of more than 70 alleles from a variety of species belonging to the Solanaceae, Rosaceae and Scrophulariaceae have been reported. S-proteins are highly polymorphic with amino acid identity ranging from 38 to 98%. Nonetheless, sequence comparisons have identified five regions of conservation, named C1 to C5 (Ioerger et al., 1991; Tsai et al., 1992). Of these, two (C2 and C3) share significant sequence similarity with the corresponding regions of fungal ribonucleases (RNases), RNase T2 (Kawata et al., 1988) and RNase Rh (Horiuchi et al., 1988), a similarity that rapidly led to the discovery that S-proteins are themselves RNases (McClure et al., 1989) and resulted in their renaming as S-RNases.

Direct confirmation of the involvement of S-RNases in SI has been achieved through the application of transgenic methodologies in *Petunia*, *Nicotiana* and *Solanum* (Lee et al., 1994; Murfett et al., 1994; Matton et al., 1997). Expression of a novel S-RNase or repression of a native S-RNase in transgenic plants causes a gain or loss of *S*-specificity, respectively (Lee et al., 1994; Murfett et al., 1994). These studies also demonstrate that the extremely high levels of S-RNase found in wild-type pistils are necessary for pollen rejection to be complete. These data are sometimes interpreted as demonstrating the S-RNases that are necessary and sufficient for SI in the pistil. However, genes not residing at the *S*-locus have been shown to affect the SI response (McClure et al., 1999) and the presence or absence of active copies of these genes can vary between genetic backgrounds. As a result it can be concluded that S-RNases are necessary for SI and encode pistil *S*-specificity but are not always sufficient for SI.

10.2.2 S-RNase structure/function

The functional involvement of RNase activity in the action of S-RNases in SI was initially inferred from the study of a self-compatible *S*-allele in *Lycoperiscon peruvianum*. In this allele, one of two histidine residues essential for catalytic activity was found to be mutated, causing a loss of RNase activity (Royo et al., 1994). In the absence of a functional allele of the same specificity and knowledge of the functionality of the pollen component of the interaction, conclusive evidence that loss of SI was caused by the loss of RNase activity was lacking. This evidence has been obtained using site-directed mutagenesis to change a single codon for one of the two catalytically essential histidine residues to encode for asparagine, generating a mutant S-RNase lacking RNase activity, and transforming this gene into plants (Huang et al., 1994). This experiment showed that production of this mutant

S-RNase in transgenic plants did not confer a gain of S-function (unlike the wild-type protein); hence, intrinsic RNase activity is indeed an integral part of the function of S-RNases.

S-RNases are glycoproteins that contain one or more N-linked glycan chains, raising the question as to the function of these groups. One possibility, drawn by analogy with lectins, is that allelic specificity might be encoded by the sugar moieties in the glycan chains. Again this question has been addressed by site-directed mutagenesis and plant transformation. An S-RNase gene was engineered in which the asparagine codon essential for glycosylation of the sole N-glycosylation site of an S-RNase was replaced with a codon for aspartic acid. Analysis of transgenic plants expressing this non-glycosylated S-RNase demonstrated that the resultant protein was indistinguishable from wild-type S-RNase with regards to rejecting self-pollen (Karunanandaa et al., 1994). Thus, the encoding of S-specificity resides not in the glycan side chains but in the protein backbone of S-RNases. The function of the glycan side chains is unresolved but there has been speculation that they may be involved in protein stability.

As S-specificity is encoded in the primary structure of these proteins, efforts have been made to determine which amino acid residues are involved in this function. A sequence comparison by Ioerger et al. (1991) identified two hypervariable regions, termed HVa and HVb. In addition to being highly polymorphic, these regions are also highly hydrophilic and likely to be exposed on the surface of the molecules, making them prime candidates to be regions involved in the encoding of specificity (Ioerger et al., 1991; Tsai et al., 1992). Recent determination of the crystal structure of S_{F11}-RNase by X-ray diffraction (Ida et al., 2001) has confirmed that HVa and HVb regions are indeed juxtaposed on the surface of S-RNases.

Several groups have employed transgenic approaches to obtain direct evidence concerning regions and amino acid residues involved in the encoding of allelic specificity. In these experiments, chimeric S-RNase genes have been constructed and introduced into transgenic plants. In all cases, one allele of the S-RNase gene was used as a 'backbone' and sequences of a particular region were exchanged with the sequence of a corresponding region from another allele. The S-specificity displayed by hybrid S-RNases was then assayed by crossing to plants carrying the allelic progenitors of the chimaeras. The results of these experiments are superficially contradictory. In two separate experiments, exchanges between pairs of S-RNases with a high degree of sequence diversity led to the loss of the backbone S-allele specificity with all regions exchanged (which covered the whole gene) (Kao & McCubbin, 1996; Zurek et al., 1997). Further, no gain of the S-specificity of the donor allele was found, despite the fact that all hybrid S-RNases exhibited 'normal' levels of RNase activity (suggesting that overall protein structure was probably unaffected by the exchanges) (Kao & McCubbin, 1996; Zurek et al., 1997). An apparently different result was obtained in a separate experiment employing two very closely related S-RNases (S_{11} and S_{13}) of *Solanum chacoense* (Matton et al., 1997). These S-RNases differ by only 10 amino acids, three of these being in HVa and one in HVb. Exchange of both hypervariable regions of S_{11}-RNase with those of

S_{13}-RNase resulted in an S-RNase that exhibited S_{13}-specificity but not S_{11}-specificity in transgenic plants. These results have been interpreted as suggesting that HVa and HVb together are sufficient for S-haplotype specificity. Clearly, however, domain-swapping experiments assess only functions of amino acids that differ between the two proteins under study. Combined, all three experiments do suggest that HVa and HVb regions play a key role in encoding allelic specificity in S-RNases, but that in addition, amino acids outside HVa and HVb (conserved between S_{11}-RNase and S_{13}-RNase) are also likely to be involved in encoding allelic specificity in S-RNases (Verica et al., 1998).

10.2.3 The pollen S-gene

The nature of SI reactions dictates that there is recognition of pollen by the pistil; hence, recognition molecules are required to be present in both tissues. Early models of the mechanism of gametophytic SI were based upon a single gene. The product of this gene was envisaged to either act via a dimerisation event within the pollen tube, or alternatively different pistil and pollen products would be generated by differential processing of a single gene or operon (Lewis, 1954). Since that time, a considerable amount of data has been amassed which suggests that the pistil and pollen S-components are in fact separate genes, culminating with the recent identification of the pollen S-gene (Sijacic et al., 2004). Searches for pollen proteins that interact with S-RNases have identified a calcium-dependent protein-kinase-like activity, which phosphorylate S-RNases *in vitro* (Kunz et al., 1996), and, more recently, a protein (PhSBP1) containing a RING-HC domain, which potentially may be involved in the ubiquitin-ligase-mediated protein degradation pathway (Sims & Ordanic, 2001). Neither of these interactions is S-allele specific, and recent data suggest that although these molecules might participate in the SI response, neither is the pollen S-specificity-encoding gene.

Identification of the pollen S-gene in this system has resulted from extensive physical and genetic mapping of the S-locus. S-loci are inherently regions of low or no recombination, and thus they tend to accumulate repetitive elements. The extremely repetitive nature of the regions flanking the *S-RNase* gene dissuaded attempts at chromosome walking in this region until very recently (Coleman & Kao, 1992). Novel technologies, in particular the development of Bacterial Artificial Chromosome (BAC) libraries, greatly improved the feasibility of this task by increasing the upper size limit of genomic clones, thereby reducing the number of 'walking steps' required. BAC clones containing *S-RNase* genes have been identified in *Petunia inflata* (McCubbin et al., 2000) and *Antirrhinum hispanicum* (Lai et al., 2002).

A 63.7-kb BAC clone of the S_2-locus from *Antirrhinum*, which contains the *S-RNase*, has been fully sequenced (Lai et al., 2002). Subsequent analysis of this sequence revealed six putative genes with homology to previously reported sequences, and of these four encode retrotransposons. More interestingly, a gene termed *SLF* (*S*-locus F-box), which encodes an F-box-containing protein and is located approximately 9 kb downstream of the *S-RNase* gene, was identified. This gene is expressed in the tapetum and (importantly) in pollen, as predicted for the pollen S-gene (Lai et al., 2002).

Independent sequence analysis of a larger region of the S_2-locus of *P. inflata* (a 328-kb contig of three BAC clones containing *S-RNase*) has shown a similar abundance of retrotransposons and, interestingly, also contains a pollen-expressed gene similar to *SLF* though it lies approximately 167 kb from *S-RNase*. Sequence analysis of *SLF* from additional S-haplotypes of *P. inflata* revealed ~90% amino acid identity between three haplotypes and only ~30% identity to *Antirrhinum SLF-S_2* (Sijacic et al., 2004). This sequence polymorphism combined with the fact that the *SLF* gene appears to be the closest pollen expressed gene to *S-RNase* in both *Antirrhinum* and *Petunia*, made *SLF* a prime candidate for being the pollen S-component of this SI system.

Confirmation that *SLF* is indeed the pollen *S*-gene has recently been achieved using a transgenic approach (Sijacic et al., 2004). The design of these experiments required rather more thought than standard gain- or loss-of-function approaches. Results from X-ray mutagenesis studies had previously suggested that a loss of pollen S-function (i.e. self-compatibility) results from a phenomenon termed *competitive interaction* (Golz et al., 1999, 2001). Competitive interaction occurs when a plant carries two different S-haplotypes as a result of the S-locus or some part of it being duplicated, or as a result of tetraploidy. Of the pollen produced by such plants, the pollen that carry two different pollen S-alleles (heteroalleic pollen), but not those that carry two identical pollen S-alleles, fail to be rejected in the SI response. In a series of experiments that capitalise on this information, the SLF_2 allele of *P. inflata* has been transformed into plants of S_1S_1, S_1S_2 and S_2S_3 genotypes under regulation of its native promoter. It was found that S_1S_1/SLF_2 transgenic plants were self-compatible and that all progeny resulting from self-pollination carried the SLF_2 transgene. S_1S_2/SLF_2 and S_2S_3/SLF_2 transgenic plants were also self-compatible and analysis of progeny generated by selfing these plants again demonstrated that all progeny carried the transgene, but importantly no S_2S_2 progeny were found. Hence, the SLF_2 transgene causes a breakdown of SI only in heteroalleic pollen, precisely as the pollen S-gene product is predicted to behave by competitive interaction (Sijacic et al., 2004).

The possession of an F-box by SLF implicates a particular mode of action for these proteins. F-box-containing proteins are components of ubiquitin ligase complexes, which, together with ubiquitin-activating enzymes and ubiquitin-conjugating enzymes, mediate protein degradation by the 26S proteosome (Bai et al., 1996). Recent results suggest that SLF is indeed part of an SCF complex (Qiao et al., 2004). Hence it is envisaged (though not proven) that SLF acts by mediating the degradation of S-RNases in compatible pollinations, and that some form of specific interaction between S-RNase and SLF of the same haplotype blocks ubiquitination in a self-incompatible interaction. A question that remains unanswered with regards to SLF and 'competitive interaction' is whether or not 'knocking out' *SLF* is pollen lethal, as has been predicted. If SLF is indeed mediating degradation of S-RNase, this would be expected, as the inability to neutralise cross S-RNases would lead to universally incompatible pollen. A further possibility is that SLF is essential to pollen tube growth even in the absence of S-RNases. This could be assessed by generating plants in which *SLF* is 'knocked out' by transformation with an antisense transgene. These plants could then be used to pollinate both wild-type plants and plants in which both alleles of S-RNase are 'knocked out' by antisense or co-suppression, and determining whether the antisense SLF transgene is present in either of the resulting progenies. Such an experiment is currently in progress (T.-h. Kao, personal communication, 2004).

10.2.4 Non-S-linked components of S-RNase-based SI

Although S-RNase and SLF have been established as being the specificity-encoding components of this system, various studies, particularly of the breakdown of SI, provide evidence that additional factors are involved. As these are not directly involved in specificity there is no requirement for them to be located at the *S*-locus and, indeed, there is considerable genetic evidence that otherwise functional *S*-loci lose their ability to cause pollen rejection in certain genetic backgrounds (Martin, 1968; Ai et al., 1991; Bernatzky et al., 1995). The loci responsible for this breakdown have been termed 'modifier' loci and three separate classes have been designated (McClure et al., 2000).

Group 1 factors directly affect the expression of *S*-locus genes at the level of transcription, translation or post-translational modification. For example, in a particular background, self-fertility has been shown to result from the lack of transcription of the otherwise functional S_{13} allele (and only this allele) of *Petunia axillaris*, though the precise mechanism and the gene(s) responsible have not been identified (Tsukamoto et al., 1999, 2003).

Group 2 factors are required for pollen rejection, but do not directly affect the *S*-locus products and are not essential to pollination and/or fertilisation. Self-compatible cultivars of *Petunia hybrida* (Ai et al., 1991) and interspecies hybrids of *Lycopersicon hirsutum* × *esculentum* (Bernatzky et al., 1995) possess and express apparently functional S-RNases and apparently possess genes or alleles that represent Group 2 modifiers (Cruz-Garcia et al., 2003). One gene encoding a Group 2 modifier, designated *HT*, has been identified by differentially screening pistil cDNAs expressed in *N.* (SI) and *N. plumbaginifolia* (SC) (McClure et al., 1999). Confirmation that this gene is essential for the SI response was achieved in an antisense experiment, where down-regulation of *HT* expression was shown to cause self-compatibility without affecting S-RNase expression (McClure et al., 1999). The role of *HT* is unclear, as it does not have significant homology to any protein of known function. Homologues of this gene have also been identified and found to modify SI in *Lycopersicon* and *Solanum* (Kondo et al., 2002). A putative N-terminal secretion signal and an ND domain comprising a stretch of asparagine and aspartate residues near the C-terminus are conserved between the three homologues, but the overall amino acid identity excluding the secretion signal is only ∼32% (McClure et al., 1999; Kondo et al., 2002). Although the mode of action of *HT* is unknown, as the HT protein must be present for SI to act and also appears to be secreted into the extracellular matrix of the stylar transmitting tissue, it may somehow mediate the uptake of S-RNase into pollen tubes.

Group 3 factors are required for pollen rejection and also are essential for pollination and/or fertilisation. These characteristics make their identification problematic, as null mutations are unlikely to be heritable. Perhaps as a result of this, Group 3 factors have not been demonstrated genetically. There is, however, some biochemical evidence for the existence of such factors. A number of stylar proteins have been identified that bind to S-RNase columns (McClure *et al.*, 2000). Most significantly, the protein NaTTS, the tobacco homologue of TTS (Wu *et al.*, 1995), was identified (for discussion of all proteins identified, see Cruz-Garcia *et al.*, 2003). There is evidence that TTS stimulates pollen tube growth *in vitro*, and *in vivo* TTS is associated with the surface of pollen tubes and appears to be deglycosylated by growing tubes (Wu *et al.*, 1995). TTS appears to fuel pollen tube growth and the demonstration that it binds S-RNase suggests that it could conceivably fall into the Group 3 category of SI modifiers.

10.2.5 Model for the operation of S-RNase-based SI

Currently two separate gene products have been established as being essential to the process of signalling the pistil *S*-genotype to the pollen tubes. S-RNases encode allelic specificity and possess RNase activity that is required for pollen rejection. Self-pollen tubes are most likely inhibited through degradation of RNA (Huang *et al.*, 1994), but which form(s) of RNA is the primary target is less clear. S-RNases do not demonstrate detectable substrate specificity *in vitro* (Singh *et al.*, 1991), though whether this is the case *in vivo* is not certain. Ribosomal RNA (rRNA) is a prime potential target as rRNA genes are apparently not transcribed in pollen tubes (McClure *et al.*, 1990; Mascarenhas, 1993). As a result, degradation of rRNA would rapidly lead to an irreversible cessation of protein synthesis. Degradation of rRNA in pollen tubes specific to SI pollinations has been reported (McClure *et al.*, 1990) supporting this hypothesis, but as the tips of incompatible pollen tubes frequently burst, spilling out their contents into an extracellular matrix with high concentrations of S-RNase, it is not clear whether this is a cause or effect of SI. In addition, data has been reported that suggests that SI may be reversible, at least in some pollen tubes. By grafting the upper segment of a self-incompatibly pollinated pistil onto the lower segment of a compatible pistil, it has been demonstrated that at least some of the pollen tubes can recover from the effects of S-RNase (Lush & Clarke, 1997). This suggests that either another class of RNA is the target of S-RNases, or if rRNA is indeed the target, the general belief that rRNA genes cannot be transcribed in pollen tubes is false. The second pistil component that is required for SI is the HT protein (McClure *et al.*, 1999); the precise function of HT is not clear but it is envisaged to be involved in a complex with S-RNases and to be required for their entry into pollen tubes.

Recent confirmation that *SLF* is the gene that encodes pollen *S*-specificity, and that transgenics expressing this gene behave in accordance with the phenomenon of 'competitive interaction', is in agreement with the previously proposed 'inhibitor model' (McCubbin & Kao, 2000).

According to this model, pollen S functions to modulate RNase activity, such that RNase activity is active in self-pollinations, but inhibited in cross-pollinations. This model had been previously favoured over one based on selective entry of S-RNases into pollen tubes, as S-RNases appear to enter the cytoplasm of both cross- and self-pollen tubes (Luu et al., 2000). According to the inhibitor model, each pollen S-gene product is capable of inhibiting all S-RNases except self, which remains functional to inhibit self-pollen tube growth. The self-compatibility seen in 'competitive interaction' results from the possession of two different pollen S-alleles, each capable of inhibiting all S-RNases except self. In this situation, each of the two S-RNases present in the pistil are neutralised by the pollen S-inhibitor from the other S-haplotype.

The fact that SLF is an F-box-containing protein suggests a potential mechanism for the inhibition – or more correctly neutralisation of S-RNase activity. F-box proteins are components of SCF (Skip1/Cullin/F-box) complexes, which target proteins for degradation through the ubiquitin ligase pathway. Clearly, degradation S-RNases would be an effective method of neutralising them. In this scenario all cross S-RNases would be targeted for degradation, while self-S-RNase would not. An interesting implication of this is that SLF must interact with all S-RNases, presumably in a conserved region of these molecules (which must also be conserved in the self-S-RNase); hence, there must also be a specific interaction between SLF and its cognate self-S-RNase that somehow overrides the general interaction, so preventing ubiquitination.

It is tempting, though highly speculative, to propose that the other S-RNase interacting proteins that have been identified might also be involved in ubiquitin-mediated degradation of S-RNases. A RING-HC-containing protein, PhSBP1 (Sims & Ordanic, 2001), is a putative E3 ubiquitin ligase that interacts with S-RNase in a non-allele-specific manner, and may well be a component of the SCF complex involved in ubiquitinating S-RNases. In addition, there is also a potential role for the calcium-dependent protein kinase (CDPK) that has previously been shown to phosphorylate S-RNases (Kunz et al., 1996). The SCF substrates that have been characterised to date are recognised in a strictly phosphorylation-dependent manner, usually by phosphorylation of serine residues on the target protein (for review, see Craig & Tyers, 1999). CDPK has been shown to phosphorylate S-RNases on serine residues in a non-allele-specific manner, and this phosphorylation could potentially be the signal for SCF to bind to S-RNase.

The model shown in Fig. 10.2 draws together both the established data and speculation made above. S-RNase complexed with HT enters both cross- and self-pollen tubes. CDPK phosphorylates S-RNase, signalling the binding of an SCF complex, two components of which are SLF (which provides specificity) and PhSBP1 (which provides ubiquitin ligase activity). In a compatible interaction, S-RNase is ubiquitinated and subsequently degraded by the 26S proteosome. In an incompatible interaction some alternative 'high-specificity' interaction occurs between SLF and

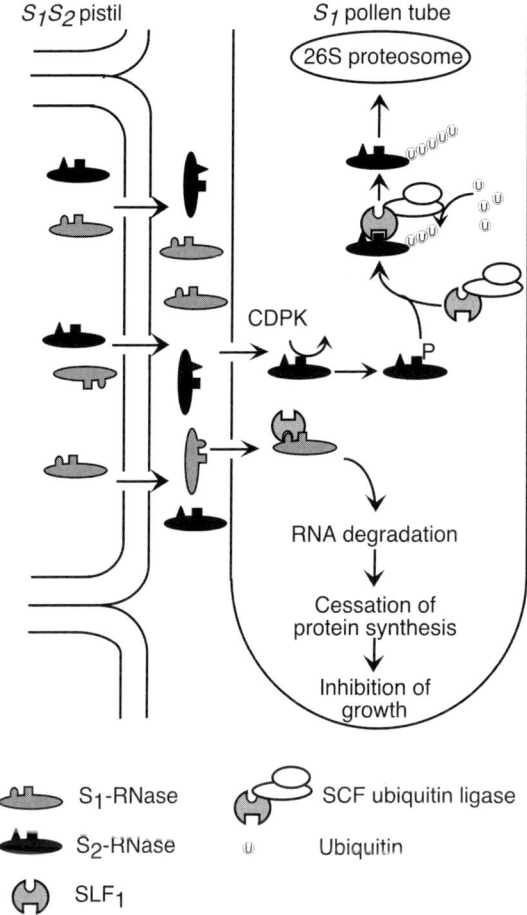

Figure 10.2 Model for the mechanism of S-RNase-based SI. Cross, S_2-RNase is targeted for degradation by the 26S proteosome via general interaction with SLF_1 through the (largely speculative) steps illustrated. Self, S_1-RNase interacts with SLF1 in an alternative *S*-specific manner that precludes ubiquitination and leaves RNase activity intact to degrade pollen RNA and inhibit tube growth.

S-RNase that prevents ubiquitination of S-RNase, so leaving RNase activity intact to inhibit pollen tube growth.

10.3 Gametophytic self-incompatability in the Papaveraceae

A second gametophytic SI system has been found in *Papaver rhoeas* L. (field poppy) and is mechanistically distinct from that found in the families described above. In this species, in contrast to the Solanaceae, inhibition of incompatible pollen occurs on the stigmatic papillae and initial arrest of pollen growth is rapid, occurring within minutes. As in the S-RNase-based system, SI is controlled by a single, multiallelic S-locus (Lawrence et al., 1978). A number of alleles of the stigmatic S-gene product have been cloned (Foote et al., 1994; Walker et al., 1996; Kurup et al., 1998). The proteins encoded by these S-genes are ~14 kDa in size and have no homology to S-RNases or any other proteins of known function. These S-proteins are developmentally expressed, appearing at maturity in the stigmatic papillae and are secreted to the stigma surface.

This species has proven to be recalcitrant to transformation, preventing transgenic approaches to examine gene function; however, a reliable and efficient *in vitro* bioassay for the SI response has been developed (Franklin-Tong et al., 1988). This assay has been used very productively to identify and study the stigma proteins involved in pollen rejection, and to characterise the biochemical events that occur on self-pollination. Both stigmatic extracts and recombinant S-proteins have been shown to have S-specific biological activity (Franklin-Tong *et al.*, 1988; Foote *et al.*, 1994). Current data suggest that binding of stigmatic S-proteins to an as yet unidentified pollen S-receptor activates a signal transduction cascade that ultimately leads to inhibition of pollen tube growth (Rudd & Franklin-Tong, 2003). A number of components of this signalling cascade have been identified. One of the earliest detected signalling events is an increase in the cytosolic-free Ca^{2+} concentration ($[Ca^{2+}]_i$), which acts as a second messenger (Franklin-Tong *et al.*, 1993). This elevation of $[Ca^{2+}]_i$ results in a rapid loss of the apical Ca^{2+} gradient associated with growing pollen tubes and accompanies the inhibition of pollen tube growth. Changes in the phosphorylation state of pollen phosphoproteins have been observed, presumably triggered by this $[Ca^{2+}]_i$ elevation (Rudd *et al.*, 1996) along with dramatic alterations in the actin cytoskeleton (Geitmann *et al.*, 2000; Snowman *et al.*, 2000, 2002; Staiger & Franklin-Tong, 2003). These initial events are thought to be associated with the cessation of pollen tube growth and are at least temporarily reversible. Later events including activation of a putative mitogen-activated protein kinase (MAPK) (Rudd *et al.*, 2003) are believed to irreversibly seal the fate of incompatible pollen, possibly through the triggering of programmed cell death (PCD) (Jordan *et al.*, 2000; Rudd & Franklin-Tong, 2003).

10.3.1 The S-gene controlling stigma function in P. rhoeas

Five alleles of the *Papaver* stigmatic S-gene have been cloned and sequenced (Foote *et al.*, 1994; Walker *et al.*, 1996; Kurup *et al.*, 1998). *Papaver* S-proteins, unlike S-RNases, are not abundant in pistils (nanograms per pistil as opposed to micrograms per pistil). Like S-RNases, however, they are highly polymorphic sharing between 51.3 and 63.7% amino acid sequence identity, and can be readily differentiated by isoelectric focusing. Significant sequence similarity has been found between S-proteins and a large family of open reading frames of the *Arabidopsis* genome (Ride *et al.*, 1999). There are approximately 100 S-protein homologues (SPHs) in *Arabidopsis*, but they are conspicuously absent in EST databases, suggesting that they may be expressed at very low levels, in narrow developmental windows, and/or transiently in response to certain environmental or biotic stimuli. As *Arabidopsis* is a self-compatible species and is also a member of the Brassicaceae, the role of these S-protein homologues is unclear, but the established role of S-proteins as signalling molecules in *Papaver*, combined with the number of members of the gene family in *Arabidopsis*, suggests that they may play significant roles in a variety of signalling pathways.

The pollen *S*-gene product in this system has not been identified, but is believed to be a plasma membrane receptor. One pollen protein that specifically binds S-proteins (S-protein binding protein, SBP) has been identified (Hearn *et al.*, 1996). SBP is a plasma membrane protein that is variably glycosylated, resulting in a size distribution from 70 to 120 kDa. *In vitro* studies suggest that though SBP specifically binds S-proteins, this interaction is not allele-specific. As a result, it has been proposed that SBP may be an accessory receptor involved in SI but is not likely to be the pollen S-specificity component.

10.3.2 Structure/function of S-proteins

Structural predictions suggest that despite extensive polymorphism between *S*-haplotypes, S-proteins share a virtually identical secondary structure. They are predicted to be composed of a series of six b strands followed by two helical regions at the C-terminus, all linked by seven hydrophilic loops (Walker *et al.*, 1996; Kurup *et al.*, 1998). A number of these S-proteins are glycosylated; the function of the glycan side chains is unclear as non-glycosylated S-protein expressed in *Escherichia coli* was found to be fully functional in the *in vitro* bioassay (Jordan *et al.*, 1999). As recombinant S-proteins are functional in the *in vitro* bioassay, it has been possible to investigate structure/function relationships using *E. coli* expressed S-proteins. Screening a number of mutant S-proteins for functionality in the SI response has led to the identification of amino acid residues in surface loop 6 as being critical to S-protein function (Kakeda *et al.*, 1998). Changing the only hypervariable amino acid residue in this loop resulted in a complete loss of the ability of S_1-protein to inhibit S_1 pollen, as did changing any of several highly conserved amino acids adjacent to this residue. Point mutations in other surface loops, however, were found not to affect S-protein function (Kakeda *et al.*, 1998). These results suggest that loop 6 may be directly involved in recognition events essential for the SI response.

10.3.3 Biochemical responses in pollen following self-recognition

A number of investigations into the biochemical events that occur in pollen during the SI response have also been carried out, including changes in pollen gene expression (Franklin-Tong *et al.*, 1990), protein phosphorylation (Franklin-Tong *et al.*, 1992) and cytosolic calcium levels (Franklin-Tong *et al.*, 1993).

Inhibition of transcription by actinomycin D does not affect pollen germination, presumably because mature pollen contains all the RNAs required for pollen tube growth. In contrast, actinomycin D does partially alleviate pollen inhibition by self-S-protein (Franklin-Tong *et al.*, 1990), suggesting that gene expression is required for a full SI response. Comparison of the transcripts produced by pollen treated with self- or non-self-S-protein using an *in vitro* translation system has led to the identification of several proteins specific to an incompatible reaction (Franklin-Tong *et al.*, 1990). Whether these genes are important mechanistically to the operation of SI or are expressed as a result of the incompatibility response is not known, but their induction suggests that signal transduction events in this system extend as far as the activation of transcription.

10.3.3.1 Ca^{2+} signalling in the SI response

$[Ca^{2+}]_i$ is well established as a second messenger in many signal transduction processes in plants (Hepler & Wayne, 1985; Trewavas & Gilroy, 1991). Signalling in pollen germination and tube growth is no exception and calcium imaging studies have amply demonstrated that an oscillating $[Ca^{2+}]_i$ gradient is not only present at the tip of growing pollen tubes but is also essential for tip extension and growth (Obermeyer & Weisensel, 1991; Miller *et al.*, 1992; Franklin-Tong, 1999). Imaging $[Ca^{2+}]_i$ within pollen tubes using Ca^{2+}-selective dyes and concomitantly challenging them with self-S-protein has demonstrated that the inhibition of pollen tube growth is preceded by a transient increase in $[Ca^{2+}]_i$ of up to ~ 1.5 mM in the shank of the pollen tube (Franklin-Tong *et al.*, 1993, 1995, 1997). This increase is seen within seconds of application of S-protein, peaks after 4–6 min and gradually declines to basal levels by ~ 10–12 min. Ca^{2+} imaging suggests that the elevation appears to originate from the nuclear complex and the endoplasmic reticulum associated with it, perhaps suggesting that $[Ca^{2+}]_i$ may be involved in regulating gene expression (Franklin-Tong *et al.*, 1993). These increases in $[Ca^{2+}]_i$ are thought to initiate the SI signalling cascade, which ultimately results in death of the incompatible pollen. Elevation of $[Ca^{2+}]_i$ in the shank of the tubes is likely in its own right to be sufficient to lead to the initial cessation of tube growth as it disrupts the apical $[Ca^{2+}]_i$ gradient. Low levels of apical $[Ca^{2+}]_i$ are associated with loss of growth, and it is likely that vesicle fusion at the tip requires high $[Ca^{2+}]_i$ and is rapidly inhibited in the SI response, leading to pollen tube inhibition. However, as $[Ca^{2+}]_i$ returns to basal levels after 10–12 min, it is clear that additional events must take place that prevent pollen tubes from recovering and resuming tip growth.

Normal growth of pollen tubes has also been shown to be inhibited by inositol trisphosphate (Ins[1,4,5]P_3) mediated Ca^{2+} release (Franklin-Tong *et al.*, 1996) and a recent study indicates that there are brief but reproducible decreases of PtdIns(4,5)P_2 in incompatible *P. rhoeas* pollen, which may indicate that InsP$_3$ is released (Straatman *et al.*, 2001). However, there is still no conclusive evidence that an inositide signalling pathway is involved in the SI response and further studies are required.

As mentioned above, imaging data suggest that the source of the ions that lead to the increase in $[Ca^{2+}]_i$ is intracellular; however, evidence is emerging that import of extracellular calcium may be involved. Using a calcium ion selective vibrating probe, extracellular fluxes have been measured around *P. rhoeas* pollen tubes and there is good evidence that Ca^{2+} influx occurs in the shanks of normally growing tubes (20–100 mm behind the pollen tube tip) (Kuhtreiber & Jaffe, 1990; Smith *et al.*, 1994). This is the region in which $[Ca^{2+}]_i$ transients are seen in the SI response and raises the possibility that the SI-related transient might be generated through uptake of extracellular calcium. Experimental evidence that this is indeed the case has been gained using the *in vitro* bioassay and measuring extracellular Ca^{2+} fluxes 50 mm behind the pollen tube tip. Challenging pollen tubes with incompatible S-proteins was found to stimulate influx of Ca^{2+} 13.6-fold over a time period comparable to the increases in $[Ca^{2+}]_i$ observed by imaging (Franklin-Tong *et al.*, 2002). A small increase in Ca^{2+} influx was also seen when compatible S-protein was used in the assay, but was not comparable to that seen in the incompatible challenge. These data provide convincing evidence that influx of extracellular Ca^{2+} plays a major role in generating the $[Ca^{2+}]_i$ transient in the SI response.

10.3.3.2 *Protein kinase activity and the SI response*
A number of studies have been carried out to determine what lies downstream of the $[Ca^{2+}]_i$ transient in this signalling pathway. One obvious avenue to explore is that of protein phosphorylation and a number of pollen proteins have been identified that exhibit either an increase or a decrease in phosphorylation when challenged with self-S-protein (Rudd *et al.*, 1996).

Two proteins that are specifically phosphorylated when pollen tubes are challenged with self-S-protein have been characterised in some detail. A 26-kDa pollen protein (with a pI of 6.2) termed p26 is phosphorylated within 90 s of challenging pollen with self-S-protein, with a further increase occurring at 400 s. This phosphorylation is Ca^{2+}-dependent and coincides with the transient increase in $[Ca^{2+}]_i$ on treatment with self-S-protein, suggesting that the $[Ca^{2+}]_i$ transient may directly stimulate phosphorylation of p26 through activation of a Ca^{2+}-dependent protein kinase (Rudd *et al.*, 1996). Recently, p26 was cloned and has been found to share 80–90% amino acid identity with plant-soluble inorganic pyrophosphatases (Rudd & Franklin-Tong, 2003). Biochemical assays using recombinant p26 protein indicate that it does indeed possess pyrophosphatase activity and that this activity is dependent on Mg^{2+} and inhibited by high $[Ca^{2+}]_i$, as expected of an enzyme of this class (Rudd & Franklin-Tong, 2003). Further, there is a large decrease in pyrophosphatase activity in crude pollen tube extracts at nanomolar Ca^{2+} concentrations that would allow p26 phosphorylation (Rudd & Franklin-Tong, 2003). Hence, the activity of p26 is almost certainly altered in the SI response. Soluble inorganic pyrophosphatases play an important role in cellular biosynthesis. They have been shown to be involved in the generation of both ATP, which drives cellular reactions and biopolymers required for the synthesis of membranes and cell walls (Cooperman *et al.*, 1992). Given these functions it is not difficult to envisage a pivotal role for p26 in the SI response and it has been proposed that the calcium-induced phosphorylation of p26 leads to a reduction in pyrophosphatase activity, causing a depletion of biopolymers (long-chain carbohydrates and proteins) and ultimately a lack of building materials available for pollen tube growth (Rudd & Franklin-Tong, 2003).

A second phosphoprotein, designated p68 (68 kDa, pI of 6.10–6.45), is also phosphorylated in response to self-S-protein, but in a Ca^{2+}-independent manner. Phosphorylation of p68 occurs somewhat later than that of p26, being barely detectable at 240 s but much increased at 400 s (Rudd et al., 1997). This suggests that p68 is likely to be downstream of p26 in the signal cascade. Inhibition of the pollen tube growth occurs within 1–2 min of challenge with self-S-protein (coinciding with the phosphorylation of p26), and is well in advance of p68 phosphorylation. This might suggest that p68 is not involved in the initial inhibition of tube growth, but perhaps in later events that cement the fate of incompatible pollen tubes.

A third pollen protein that is involved in phosphorylation events associated with the SI response is a putative MAPK, p56. This protein was identified using in-gel kinase assays employing the classic MAPK substrate myelin basic protein (MBP) (Rudd et al., 2003). p56 exhibits a basal level of protein kinase activity in growing pollen tubes, which is significantly enhanced in incompatible pollen tubes. p56 has not been cloned, but several lines of evidence suggest that it is a MAPK having the ability to phosphorylate MBP in the absence of Ca^{2+} ions, evidence that activated p56 is phosphorylated on a tyrosine residue in vivo, cross-reactivity to a phospho-MAPK-specific antibody and inhibition by apigenin (Rudd et al., 2003). p56 is not envisaged to be involved in the initial inhibition of tip growth as its SI-stimulated activation peaks several minutes after initial inhibition of pollen tube growth (Rudd et al., 2003; Rudd & Franklin-Tong, 2003). Inhibited pollen remains viable for some 40–60 min however, and p56 may be involved in later, downstream events involved in making the growth inhibition irreversible (Rudd et al., 2003). In regard to putative sites of action for p56, there are two plausible possibilities. The first is the actin cytoskeleton, which is critical for pollen tube growth (see below). In animal systems, MAPK activation has been linked to changes in the cytoskeleton and evidence that this may also occur in plants is emerging (Staiger, 2000; Samaj et al., 2002). A second possibility is that p56 feeds into a PCD signalling cascade as emerging data have demonstrated that MAPK activation plays an important role in this process in plants (Zhang et al., 2000; Yang et al., 2001; Zhang & Liu, 2001; Ren et al., 2002).

10.3.4 S-protein-binding proteins in pollen

A pollen protein, which interacts with the pistil S-protein, has been identified by Western ligand blotting. This protein, termed S-protein binding protein (SBP), binds to pistil S-proteins of all S-haplotypes examined, suggesting that it is unlikely to be the pollen S-specificity component (Hearn et al., 1996). SBP is, however, an integral plasma membrane protein, as one might expect of a cell surface receptor. SBP is 70–120 kDa in size, this wide variation apparently being due to variable glycosylation. The glycan moieties appear to be required for the interaction between SBP and S-proteins as the interaction is abolished by periodate treatment of the pollen protein blots prior to incubation with S-proteins (Hearn et al., 1996). Correlative evidence that SBP might be involved in the SI response has come from the use of mutant S-proteins in an *in vitro* bioassay combined with Western ligand blotting. In general, there is a good correlation between binding of mutant S-proteins to SBP and their functionality in the SI response (Jordan et al., 1999). Amino acid changes in the predicted loops 6 and 2 of S_1-protein simultaneously reduced the ability of S_1-protein to bind SBP and to effect the SI response *in vitro* (changes in loop 6 having the greatest affect). However, an N-terminal deletion of 16 amino acids of the S_1-protein completely abolished the ability of the protein to inhibit self-pollen yet did not affect the binding activity to SBP. One explanation for these results is that a second non-*S*-specific interaction involving the N-terminal amino acids and an as yet unidentified protein is essential for the SI response. Indeed, SBP has been proposed to function as an accessory receptor, which might modulate the interaction of S-proteins with the pollen S-receptor (Jordan et al., 1999). In this scenario, SBP and S-proteins would interact to form a recognition complex. Credence to this argument comes from analogy to systems found in the animal kingdom, such as the fibroblast growth factor (FGF) signalling system. Here a family of membrane proteins, the heparin sulfate proteoglycans (HSPG), function as accessory receptors and low-affinity, high-capacity, binding of FGF to glycan side chains of HSPG facilitates oligomerisation of the ligand and enables interaction with the high-affinity FGF receptor leading to signal transduction (Spivak-Kroizman et al., 1994).

10.3.5 Changes in the actin cytoskeleton

The actin cytoskeleton has been amply demonstrated to play a critical role in tip growth in general (Gibbon et al., 1999). In the poppy SI response, rapid and dramatic changes in the organisation of the F-actin cytoskeleton have been observed in incompatible pollen tubes. Punctate actin foci are formed and continue to form for several hours after stimulation (Geitmann et al., 2000; Snowman et al., 2002). In addition, the SI response rapidly elicits a large and sustained depolymerisation of F-actin. Concentrations of F-actin within pollen tubes decrease by 50% within 5 min of being challenged by self-S-protein and remain low for at least an hour (Snowman et al., 2002). Increases in $[Ca^{2+}]_i$ in pollen tubes have been shown stimulate actin depolymerisation; hence, it is envisaged that the SI-induced $[Ca^{2+}]_i$ increase discussed above is likely to be causing this effect, though the mechanism is unknown (Snowman et al., 2002).

10.3.6 PCD in the SI response

As mentioned above, SI in *Papaver* appears to be a two-step process, the initial step being a rapid but reversible inhibition of tip growth, followed by a second, more gradual but irreversible, termination of pollen tube viability. Preliminary evidence suggests that induction of PCD may be the cause of the irreversible step in SI. There is now good evidence for the involvement of PCD in a number of developmental processes in plants, including xylogenesis, aerenchyma formation and senescence, as well as in response to pathogens and various abiotic stresses (Greenberg et al., 1994). The phenotype of PCD in plant cells has a number of similarities with that of apoptosis in animal cells, including DNA laddering, and cytochrome c release from mitochondria (reviewed in Dannon et al., 2000). The pattern of depolymerisation of F-actin observed in self-incompatible *Papaver* pollen tubes is unusual, but is often associated with PCD in animal cells (Korichneva & Hämmerling, 1999; Rao et al., 1999). Indeed, F-actin depolymerisation has been found to actually stimulate apoptosis in some cell lines (Janmey, 1998). Hence, studies to investigate the potential role of PCD in *Papaver*-type SI have been initiated.

Caspase-induced cleavage of nuclear DNA at regular intervals to produce oligosomal fragments is considered to be a diagnostic characteristic of PCD. Evidence that nuclear DNA degradation occurs in plant cells undergoing the hypersensitive response has been gathered using the technique of Fragment End Labelling (FragEL) (Ryerson & Heath, 1996; Wang et al., 1996). Using this technique, it has been established that nuclear DNA fragmentation occurs in self-incompatible but not self-compatible pollen (Jordan et al., 2000). This nuclear DNA fragmentation is only detectable several hours after induction of SI, consistent with it being one of the last steps of PCD and after irreversible termination of pollen tube growth. DNA fragmentation is also observed after prolonged culture of pollen in which $[Ca^{2+}]_i$ is artificially elevated by treatment with mastoparan (Jordan et al., 2000). This suggests that the elevation of $[Ca^{2+}]_i$ observed in the SI response may be involved in initiating the PCD pathway. In agreement with this, calcium signalling has also been shown to be involved in the induction of PCD in the hypersensitive response in plant cells (Levine et al., 1997).

Although nuclear DNA fragmentation is associated with PCD, it is also possible that it results from necrosis, i.e. is a result of cell death rather than being a cause of it. To more conclusively affirm that PCD is occurring, it is necessary to identify upstream components of the PCD pathway, which are more easily separable from 'general death' phenomena. In plant systems, this process has been complicated by the apparent absence of homologues of a number of the key, highly conserved components of animal PCD pathways in plant genomes (Rudd & Franklin-Tong, 2003). Particularly notable is the absence of genes with homology to caspases, which are directly involved in DNA fragmentation, though a family of putative cysteine proteases that are distantly related to caspases has been recently identified in plants (termed metacaspases) and demonstrated to exhibit cysteine protease activity (Uren et al., 2000; Szallies et al., 2002). Inhibitors of caspase activity have been demonstrated to block pathogenesis in the plant hypersensitive response (Richael et al., 2001), providing further evidence for PCD and providing an alternative assay.

In the *Papaver* SI response, application of the tetrapeptide caspase-3 inhibitor Ac-DEVD-CHO has been found to block the DNA fragmentation normally observed in pollen challenged with self-S-protein (Rudd & Franklin-Tong, 2003). In addition, incompatible pollen tubes treated with Ac-DEVD-CHO grow to approximately the same length as compatible tubes, suggesting that caspase-like activity is likely to be involved early in the primary pathway of the signalling cascade, which leads to SI (Rudd & Franklin-Tong, 2003). This significantly bolsters the argument that the DNA fragmentation observed in SI is a result of PCD and not necrosis. Further, this result suggests that the initial reversible cessation of tube growth is not only reversible, but also temporary. This raises the interesting question as to whether the signalling events involved in the initial reversible inhibition of tube growth are truly functionally involved in the operation of this SI system or are an indirect effect that results from elevation of $[Ca^{2+}]_i$ within the pollen tube. Ca^{2+} is a promiscuous secondary messenger and while its elevation appears to be required for the activation of PCD, other $[Ca^{2+}]_i$ regulated pathways will also be affected, some of which are very likely to be involved in tip growth. The answer to this question lies in the timing of the various processes *in vivo*. Can activation of PCD alone at the stigma surface terminate the growth of all pollen tubes before they reach the ovules? PCD is not a particularly rapid process; in the hypersensitive response in tomato leaves, necrotic lesions appear after ~16 h (Hoeberichts *et al.*, 2003). Although it is unclear how long a pollen tube would continue to grow after the induction of PCD, it is entirely possible that some pollen tubes would have sufficient time to effect fertilisation. The function of the rapid reversible inhibition of pollen tube growth is, thus, likely to be to stop the pollen tubes while PCD comes into effect, and in so doing strengthen the SI response.

10.3.7 Model for the mechanism of self-incompatibility in P. rhoeas

The data described above suggest that SI in *P. rhoeas* is mediated by a number of signal transduction cascades that are activated within incompatible pollen grains/tubes. The stigmatic S-proteins are well established as encoding female *S*-specificity, and function to communicate the *S*-genotype of the stigma to incoming pollen grains. The pollen *S*-specificity component has not been conclusively identified, but is thought to be an S-protein receptor located at the surface of pollen tubes. SBP both binds S-proteins and is located in the plasma membrane, but S-protein binding appears to be irrespective of *S*-genotype, suggesting that this protein functions as an accessory receptor (Hearn *et al.*, 1996). It remains possible, however, that binding of S-protein to the pollen receptor occurs in both cross- and self-interactions but that the result differs, as is proposed in Solanaceous SI.

Figure 10.3 illustrates a working model for the mechanism of SI in *P. rhoeas*. Binding between an S-protein and its cognate pollen *S*-receptor is envisaged to trigger intracellular signalling cascade(s) involving Ca^{2+} acting as a secondary messenger (Franklin-Tong *et al.*, 1993). Within approximately 1 min of the initial signal, there is an influx of extracellular Ca^{2+}, which, together with release of Ca^{2+} from internal stores, leads to an increase in cytosolic $[Ca^{2+}]_i$ in the tube shank (Franklin-Tong *et al.*, 2002). Elevation of cytosolic $[Ca^{2+}]_i$, has multiple rapid effects. The initial series of events are fully capable of stopping pollen tube growth, but are reversible. First is the loss of the apical $[Ca^{2+}]_i$ gradient, which may well be responsible for the initial arrest of pollen tube growth. Second is the rapid phosphorylation of

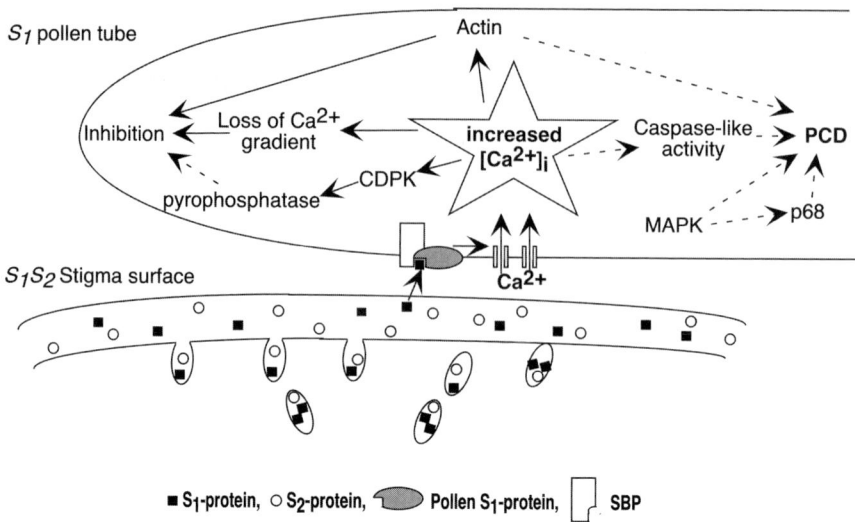

Figure 10.3 Model for the mechanism of gametophytic SI in *Papaver rhoeas*. S-protein secreted by the stigma binds to a pollen *S*-receptor complex of which SBP is a component. This causes the opening of ion channels in the shank of the tube, raising cytosollic Ca^{2+} levels. This leads to a loss of the Ca^{2+} gradient required for tip growth, restructuring of the actin cytoskeleton, and activation of CDPK (which then phosphorylates the pyrophosphatase p26), leading to rapid cessation of growth. In addition, raised Ca^{2+} levels ultimately lead to the activation of PCD and death of the pollen tube. Arrows with solid lines represent established steps, and those with dashed lines are speculative. Based on Rudd and Franklin-Tong (2003).

p26 by a calcium-regulated protein kinase (Rudd *et al.*, 1996). As p26 has inorganic pyrophosphatase activity, this is likely to contribute to growth arrest by stopping or reducing biosynthesis of membrane and wall material essential for growth. The third effect of $[Ca^{2+}]_i$ elevation is to cause dramatic alteration of the cytoskeleton, involving depolymerisation of F-actin (Geitmann *et al.*, 2000; Snowman *et al.*, 2002). Again it is established that tip growth is highly sensitive to the actin depolymerisation (Gibbon *et al.*, 1999). The depolymerisation of F-actin in the SI response is rapid and could also contribute to growth arrest. In addition, the size and longevity of depolymerisation response (several hours) suggest that it may play a role in downstream, longer term, inhibition (Rudd & Franklin-Tong, 2003).

A second series of events appears to be activated after the initial cessation of tip-growth, and it has been proposed that their function is to make growth inhibition irreversible (Rudd & Franklin-Tong, 2003). By removing S-proteins experiments at various time points after challenge in the bioassay, inhibition of pollen tube growth can be reversed up to 9 min after SI induction (Rudd & Franklin-Tong, 2003). Current data suggest that a PCD-like mechanism is responsible for cementing the fate of self-pollen and that it is triggered quite early in the response, but after the 9-min time point. One potential component of this cascade is an MAPK (p56), which is activated downstream of the Ca^{2+} signals after pollen tube growth arrest, suggesting that it probably is not involved in the initial cessation of tip growth. The precise nature of the signalling components that lead to the proposed activation of PCD requires further study. To date, the *Papaver* SI system appears to be the most complex and tiered yet studied, with multiple phenomena that each might easily lead to the initial reversible cessation of tube growth and ultimately a separate distinct mechanism through which pollen tubes are ultimately terminated.

10.4 Sporophytic self-incompatability

In sporophytic self-incompatibility (SSI), the pollen phenotype is determined by the genotype of the parent plant and hence the pollen whilst being haploid behaves as diploid. SSI has been most intensively studied in members of the Brassicacae, though it is not restricted to this family. The stigmas of plants in this group are made up of papillate cells that are covered by a cuticle and are classified as 'dry'. In a compatible pollination, there is a rapid series of events, which include adhesion to the stigma, hydration and germination of the pollen grains, and subsequent penetration of the stigma by the emerging pollen tube (Elleman *et al.*, 1992). The phenomena associated with incompatible pollination vary somewhat according to the strength of the alleles involved. In the strongest alleles, adhesion and even hydration are poor, whereas in weaker alleles pollen tubes may emerge and even penetrate the stigmatic cuticle, but fail to fully penetrate the stigma. This contrasts with the situation in GSI where incompatible pollen tube growth ceases approximately one third of the way down the style.

The fact that in sporophytic SI the pollen though haploid behaves genetically as diploid, and carries the specificity of both maternal *S*-haplotypes, suggests that recognition between the pollen and stigma is likely to involve the surface coating of pollen grains (pollen coat). Pollen coat is derived largely from the cells of the tapetum, which nourish developing microspores and, importantly, are diploid. Cytological studies using electron microscopy have revealed that on pollination the pollen coat rapidly flows into microchannels in the stigmatic cuticle and establishes molecular continuity with the stigma (Dickinson, 1995). Interaction of the SI determinants and inhibition of self-pollen occurs very rapidly after pollination and though variable between haplotypes, usually occurs within 45 min (Dickinson, 1995).

As with GSI, identification of the female component of SSI initially focused on the identification of stigmatic proteins that co-segregate with *S*-haplotypes. These initial efforts identified SLGs (*S*-locus glycoproteins), which are abundant in stigma extracts and polymorphic in their isoelectric points (demonstrated by isoelectric focusing) (Nishio & Hinata, 1977). Like S-RNases, they exhibit characteristics expected of proteins that would determine female *S*-haplotype specificity. They localise to the walls of stigmatic papillae, (the cells that receive pollen on pollination) and their temporal regulation coincides with the acquisition of SI, being barely detectable in immature self-compatible buds and increasing dramatically just prior to pollen release. Since the cloning of the first *SLG* cDNA (Nasrallah *et al.*, 1985), numerous alleles of *SLG* have been cloned, and these genes exhibit a high level of sequence polymorphism, pairwise sequence identities varying from 65 to 97.5% (Kusaba *et al.*, 1997). As a result, for a considerable time the *SLG* gene was believed to play a major role in the determination of S-haplotype specificity in the pistil.

Importantly, the process of screening cDNA libraries for *SLG* alleles uncovered additional homologous genes. These include two S-locus-related genes (*SLR1* and *SLR2*) that are homologous to *SLG* and but not linked to the S-locus, and the S-receptor kinase (*SRK*) gene that is located at the S-locus (Stein *et al.*, 1991), all of these genes show stigma-specific expression.

10.4.1 Brassica S-locus glycoproteins

Initial attempts to demonstrate the function of *SLG* by gain- and loss-of-function approaches did not yield conclusive results. An attempt to knock out *SLG* by introduction of an antisense *SLG* gene into transgenic plants led to the breakdown of SI, but the transcript levels of both *SLG* and *SRK* (S-locus receptor kinase) were reduced, complicating interpretation of this result (Shiba *et al.*, 1995). Similarly, analysis of a self-compatible mutant (*scf1*), whose stigma, but not pollen function was defective, showed that transcript and protein levels of both *SLG* and *SLR* (S-locus-related) were reduced (Nasrallah *et al.*, 1992). The difficulty in obtaining plants that had a reduction in *SLG* expression without simultaneously reducing expression of other related genes made it difficult to draw strong conclusions about the function of SLG. Similar problems plagued gain-of-function experiments. Introduction of *SLG* of a new S-haplotype did not confer the new S-haplotype specificity on the stigma of transgenic plants, but instead, caused breakdown of SI. This phenotype appeared to be a result of homology-dependent gene silencing, as the expression of the transgene, endogenous *SLG*, *SLRs* and *SRK* were all suppressed (Conner *et al.*, 1997). Several lines of evidence have cast doubt on the requirement of *SLG* in SSI. One line of self-incompatible *Brassica oleracea* has been found to produce lower levels of SLG than self-compatible immature buds of other self-incompatible lines (Gaude *et al.*, 1995). *SLG* is absent in a self-incompatible line of *B. oleracea* (Okazaki *et al.*, 1999). Introduction of SLG_{43} into $S_{52}S_{60}$ plants of *Brassica campestris* showed that transgenic plants producing SLG_{43} at a normal level fail to acquire the ability to reject S_{43} pollen (Takasaki *et al.*, 1999). Recently reported transformation experiments have clearly confirmed that SLG is not required for the S-haplotype specificity in *Brassica* stigmas; however, they do suggest that the SLG may still play a significant role in the operation of SSI (Takasaki *et al.*, 2000) (see below).

10.4.2 SRK encodes S-haplotype specificity in the stigma

SRK is the second highly polymorphic gene identified at the *Brassica* S-locus (Stein *et al.*, 1991). SRK is a classic receptor kinase and consists of an extracellular domain, a single transmembrane domain, and a cytoplasmic kinase domain that possesses serine/threonine kinase activity (Goring & Rothstein, 1992). The extracellular domain is called the *S*-domain and is 75–99% identical to SLG at the amino acid level. In addition to allelic polymorphism, other characteristics of SRK are also consistent with a key role in stigmatic haplotype specificity. SRK is predominantly expressed in stigmatic papillae, and localises to the plasma membrane (Delorme *et al.*, 1995) right at the site of pollination. Further like *SLG*, expression of *SRK* is low in immature buds that do not reject self-pollen and increases as SSI becomes activated.

Attempts to directly confirm SRK function *in vivo* by transformation were initially plagued by co-suppression effects that hindered expression of gain-of-function transgenes as well as endogenous *SRK* and *SLG*. This problem has been overcome by the use of genotypes with a high degree of sequence divergence. In a definitive set of experiments, SLG_{28} and SRK_{28} of *B. campestris* were introduced into $S_{52}S_{60}$ and $S_{60}S_{60}$ plants, respectively (Takasaki *et al.*, 2000). It was found that expression of SRK_{28} alone, but not of SLG_{28} alone, in the stigmas of the transgenic plants conferred the ability to reject S_{28} pollen. These results somewhat surprisingly demonstrate that *SRK* is the sole determinant of stigmatic *S*-haplotype specificity and rule out the involvement of *SLG* in this function.

If *SLG* is not required for stigmatic *S*-specificity, a question obviously arises as to what its function(s) might be. Interestingly, though *SLG* was not absolutely required for SI, the transgenic experiments did demonstrate that the strength of incompatibility may involve *SLG*. It was determined that the degree of self-pollen rejection by SRK_{28} varied according to the *S*-haplotype of the transformant. A good correlation was found between the efficiency of self-pollen rejection and the degree of amino acid sequence identity between the *S*-domain of SRK_{28} and the SLG produced in the same stigma: the higher the identity, the stronger the rejection (Takasaki *et al.*, 2000). As a result, it has been hypothesised that SLG interacts with the *S*-domain of SRK to facilitate the recognition reaction between SRK and the pollen *S*-specificity determinant. The strength of the association between SLG and SRK (which is akin to dimerisation) might then be affected by the level of sequence similarity between them. It is important to note, however, that the transgenic plants expressed SRK_{28} at only approximately 30% of the wild-type level and hence this phenomenon may not be significant in the natural situation. Another potential role for SLG is in pollen adhesion to the stigma surface. It has been demonstrated that treatment of stigmas with an anti-SLG antibody prior to pollination reduces the strength of pollen adhesion to the stigmatic papillae (Luu *et al.*, 1999).

10.4.3 SCR/SP11 encodes pollen S-haplotype specificity

Because of the genetically sporophytic nature of SSI, the determinant of pollen *S*-haplotype specificity has long been suspected to be a component of the pollen coat, which is derived from the diploid cells of the tapetum and also makes direct contact with the stigma surface (for a review of pollen coating, see Doughty *et al.*, 2000). A cyclohexane-based protocol has been developed for removing the highly lipidic pollen coat free from contamination by gametophytic proteins from within pollen grains. Experimental evidence that the pollen *S*-specificity component is indeed in the pollen coat was gained using a pollination bioassay (Stephenson *et al.*, 1997). In this assay, stigmatic papillae were treated with either cross- or self-pollen coat, pollinated with cross-pollen, and the ability of the pollen to hydrate was assessed. It was found that pre-treatment with self-pollen coat, but not cross pollen coat, significantly reduced hydration of cross pollen on the stigma (Stephenson *et al.*, 1997). These results strongly suggest that the molecule that encodes pollen *S*-haplotype specificity does indeed reside in the pollen coat. Subsequent fractionation of pollen coat proteins, followed by assessing their activity in this bioassay, determined that the pollen *S*-determinant is a member of a group of basic, cysteine-rich proteins of the PCP family with a molecular mass less than 10 kDa (Stephenson *et al.*, 1997).

The pollen *S*-determinant is predicted to have a number of other characteristics. It must exhibit sequence polymorphism between *S*-haplotypes and it has to be genetically and physically linked to the *S*-locus. Using these criteria, two separate groups undertook a comprehensive characterisation of the *Brassica S*-locus region surrounding *SLG* and *SRK* to identify potential candidates. In the first published report, a 76-kb region of the *S*-locus of the S_9-haplotype of *B. campestris* was cloned and completely sequenced. In addition to the *SLG* and *SRK* genes, it was found to contain 12 genes that are expressed in anthers and/or pistils (Suzuki *et al.*, 1999). One of these genes (*SP11*) attracted particular interest as a potential candidate for encoding the pollen *S*-determinant, as it encodes a protein with similar characteristics to those predicted using the above bioassy, being a small, basic cysteine-rich protein. Further, this gene is located in the immediate 3′ flanking region of SRK_9 and it is expressed predominantly in anthers (Suzuki *et al.*, 1999). Independently, a second allele of *SP11* was identified in a 13-kb region between *SRK* and *SLG* in the S_8-haplotype of *B. campestris*, though this gene was termed *SCR* (*S*-locus cysteine-rich) (Schopfer *et al.*, 1999). Additional alleles of *SP11/SCR* have been isolated and sequenced (Schopfer *et al.*, 1999; Takayama *et al.*, 2000) and found to be extremely polymorphic, with overall amino acid sequence identities of between 26 and 46%. These proteins are 74–77 amino acids in length and are hydrophilic, except for an N-terminal stretch of 19 amino acids. The hydrophobic N-terminus is highly conserved and is predicted to encode a signal peptide that leads to secretion of these proteins. The mature, secreted protein is predicted to be 8.4–8.6 kDa and a pI of 8.1–8.4, with only 12 amino acids conserved between haplotypes and 8 of these are cysteines hypothesised to form disulphide bonds involved in intramolecular structure. Moreover, *in situ* hybridisation of anther sections showed that SCR/SP11 is expressed in the tapetum of the anther (Takayama *et al.*, 2000). This sporophytic expression pattern explains why in *Brassica* the SI phenotype of pollen is determined by the genotype of the pollen parent.

Proof that *SP11/SCR* is indeed the pollen *S*-determinant has been attained using both the bioassay approach and, conclusively, transgenic experiments (Schopfer *et al.*, 1999; Takayama *et al.*, 2000). In a classic gain-of-function experiment, the

SCR_6 cDNA was introduced under control of the SCR_2 promoter into *B. oleraceae* $S_2 S_2$ plants. Pollen expressing the SCR_6 transgene was found to be fully incompatible when used to pollinate wild-type $S_6 S_6$ and fully compatible on wild-type $S_{22} S_{22}$ stigmas. In a pollination bioassay, recombinant $S_9 SCR/SP11$ was applied to $S_9 S_9$ and $S_8 S_8$ stigmas and the protein was found to elicit the SI response only in the former (self) stigmas, causing the inhibition of hydration of cross-pollen (Takayama *et al.*, 2000).

10.4.4 Regulation of SRK

Receptor kinase activity is generally tightly regulated to avoid their spontaneous activation. In the case of SRK, this is an important consideration as non-specific activation could easily lead to sterility (Kemp & Doughty, 2003). Multiple mechanisms have been identified and characterised in animal systems that act often in a concerted manner to down-regulate kinase activity (Cock *et al.*, 2002). *In vitro*, both the truncated kinase domain and the full-length SRK protein autophosphorylate in the absence of the cognate ligand (Goring & Rothstein, 1992; Giranton *et al.*, 2000). *In vivo*, however, SRK requires the addition of pollen coat for autophosphorylation (Cabrillac *et al.*, 2001), suggesting that some mechanism of repressing SRK is present in the stigma. Several SRK interacting proteins have been identified which may potentially regulate SRK activity. Two thioredoxin-h-like proteins (THL1 and THL2) were identified in yeast two-hybrid screens using the kinase domain of SRK as bait (Bower *et al.*, 1996). THL1 inhibits constitutive autophosphorylation of SRK in the absence of the SCR ligand (Cabrillac *et al.*, 2001) and this interaction requires the presence of a conserved cysteine residue on the cytoplasmic side of the SRK transmembrane domain (Mazzurco *et al.*, 2001). This inhibition can be overcome in a haplotype-specific manner by addition of SCR and suggests that THL1 (and possibly THL2) acts as basal state inhibitors of SRK. Ligand-mediated activation through binding of SCR to the extracellular domain of SRK presumably leads to a conformational change, causing derepression of kinase activity. In a superficially contradictory study, synthetic SCR_8 (but not SCR_9) was found to increase phosphorylation of SRK_8 extracted from microsomal membranes in the absence of thioredoxin or other regulators of SRK activity (Takayama *et al.*, 2001). One likely explanation for this is that the enzymatic properties of the SRK kinase domain may be substantially altered as a result of the conformational change induced by binding to SCR. The ligand-induced 'active state' might be expected to represent a more effective conformation of this protein with regards to kinase activity.

It is likely that multiple mechanisms regulate SRK activity. SRK also interacts *in vitro* with *Arabidopsis* kinase-associated protein phosphatase (KAPP) (Braun *et al.*, 1997) and recently a stigma-expressed *Brassica* homologue of KAPP has been shown to be phosphorylated by, and to dephosphorylate, the SRK kinase domain (Vanoosthuyse *et al.*, 2003). These results, together with the observation that *in vivo* SRK is phosphorylated approximately 1 h after incompatible pollination (Cabrillac *et al.*, 2001), suggest that KAPP may play a role in SRK down-regulation in a similar manner to that proposed for CLV1 and FLS2 (Williams *et al.*, 1997; Stone *et al.*, 1998; Gomez-Gomez *et al.*, 2001). There is also evidence that additional phosphatases may be involved in the regulation of SRK. SI in *Brassica* has been shown to breakdown on application of the phosphatase inhibitors okadeic acid and mycrostatin (Rundle *et al.*, 1993; Scutt *et al.*, 1993). These inhibitors are expected to act on type 1 and type 2A phosphatases, but not type 2C phosphatases, such as KAPP.

An additional physical regulation of SRK may also be present. An interaction between a sorting nexin (*Brassica* SNX1) and SRK has recently been demonstrated (Vanoosthuyse *et al.*, 2003). Sorting nexins have been implicated in the down-regulation of animal receptor kinases by endocytosis (Kurten *et al.*, 1996; Haft *et al.*, 1998; Chin *et al.*, 2001; Phillips *et al.*, 2001). Over-expression of sorting nexins increases the rate of internalisation and degradation of the epidermal growth factor receptor and the platelet-derived growth factor in mammalian cells (Kurten *et al.*, 1996; Phillips *et al.*, 2001). The SNX1 homologue (Vps5p/Grd2p) in yeast is part of a retromer complex that is essential for retrograde transport from the endosome to the Golgi apparatus (Seaman *et al.*, 1998). While such a protein trafficking pathway is yet to be established in plants, the *Arabidopsis* genome is predicted to encode orthologues of the retromer complex proteins involved, and hence, plants potentially possess all of the components of a sorting-nexin-based vesicle trafficking system (Vanoosthuyse *et al.*, 2003). The role of *Brassica* SNX1 seems likely to be in mediating endocytotic removal of SRK from the plasma membrane and in so doing to play an important role in SRK regulation.

10.4.5 SRK substrates

Binding of SP11/SCR by SRK is envisaged to lead to autophosphorylation of SRK and subsequent phosphorylation of intracellular substrate(s). The signal transduction cascade downstream of SRK has been partially characterised. Several substrates have been identified through yeast two-hybrid interaction screening and one has been confirmed to be involved in the mechanism of action of SI. *ARC1* (Armadillo repeat containing protein 1) is a gene that was identified as interacting with the kinase domain of SRK in a yeast two-hybrid library screen. ARC shows a phosphorylation-dependent interaction with the cytosolic domain of SRK *in vitro* (Gu *et al.*, 1998), and importantly, transgenic plants in which ARC1 production in the stigma was suppressed, exhibited a substantial breakdown of SI (Stone *et al.*, 1999). The function of ARC1 is not completely understood; however, in addition to the ARM (Armadillo) repeats, which are involved in protein–protein interactions, ARC1 also contains the recently identified U-box motif (Azevedo *et al.*, 2001). The U-box is involved in ubiquitination, specifically the transfer of ubiquitin to other proteins, in so doing targeting them for proteasome degradation (Koegl *et al.*, 1999). Hence, it is likely that ARC1 targets specific proteins for degradation; these targets are unknown but would presumably be either inhibitors of SI or stimulators of pollen growth. It should be noted, however, that ubiquitination can have roles other than targeting for degradation, including intercellular targeting, recruitment to molecular complexes (Kachroo *et al.*, 2002) and transcriptional regulation (Conaway *et al.*, 2002).

In addition to ARC1, the calcium-binding protein calmodulin has also been shown to interact with the SRK kinase domain (Vanoosthuyse *et al.*, 2003). Calmodulin is a known component of many signalling pathways in both plants and animals (Sanders *et al.*, 1999; Chin & Means, 2000) and calcium fluxes have been reported in both the cytoplasm and cell walls of stigmatic papillae after both compatible and incompatible pollination (Dearnaley *et al.*, 1997; Elleman & Dickinson, 1999). These fluxes occur directly under the point of contact with the pollen grains and small localised peaks in $[Ca^{2+}]_i$ are correlated with pollen hydration (Dearnaley *et al.*, 1997; Goring, 2000). As a result, calmodulin has been proposed to mediate cross-talk between calcium signalling in pollination and SI signal transduction (Vanoosthuyse *et al.*, 2003).

10.4.6 *Model for the action of SSI in Brassica*

As has been discussed above, SSI in *Brassica* is mediated in the stigma by a receptor kinase (SRK) and SI in this system is mediated by a recognition event that leads to a signal transduction cascade (Fig. 10.4). There appear to be multiple mechanisms to hold SRK in an inactive state, including thioredoxin (THL1) and protein phosphatase (KAPP). Upon pollination the pollen S-haplotype determinant, SCR/SP11, diffuses rapidly from the pollen coat and interacts in an *S*-haplotype-specific manner with the extracellular (*S*-) domain of SRK. This interaction causes derepression of the kinase domain of SRK within the cytoplasm of the papillar cell, and phosphorylation of specific substrates. Haplotype-specific interaction between SCR/SP11 and SRK has been demonstrated using two different biochemical approaches (Takayama *et al.*, 2001; Kachroo *et al.*, 2002). This interaction was shown to induce transphosphorylation ('autophosphorylation') of serine and threonine residues in the kinase domains of, presumably dimerised, SRKs (Schopfer & Nasrallah, 2001; Takayama *et al.*, 2001). The exact molecular make-up of the stigmatic receptor is a matter of some speculation. The majority of (though not all) animal receptor kinases are active as dimers. It is clear that at least one molecule of SRK must be present in the receptor complex and also that as the *S*-haplotypes function independently, dimerisation between SRKs of different haplotypes is unlikely. However, the presence of the SLGs of each haplotype complicates the situation. Given the high level of sequence identity between the S-domain of SRK and SLG within an *S*-haplotype, it is certainly possible that both SRK homodimers and SRK/SLG heterodimers are formed and that both could be active receptors for SCR/SP11. This scenario might explain the enhancement of the SRK-mediated SI response by SLG observed by Takasaki *et al.* (2000), as it would lead to an increase in the total number of receptors present at the stigma surface. One argument countering this is that if SLG can bind to SP11/SCR, presumably the presence of large amounts of SLG could titrate the ligand away from the receptor, causing breakdown of SI. As SLG is

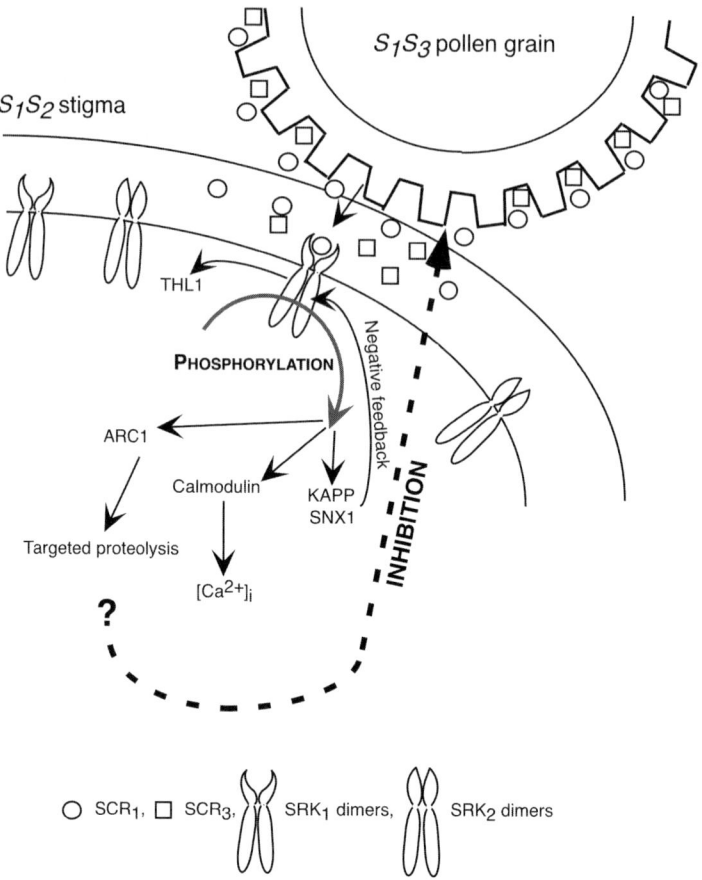

Figure 10.4 Model for the mechanism of sporophytic SI in Brassicaceae. Binding of SCR_1 to SRK_1 leads to derepression of SRK_1 by displacement of THL1. This leads to phosphorylation of both effectors (ARC1 and possibly calmodulin) and negative regulators of SRK1 (KAPP and SNX1). The dashed arrow represents unknown elements of the cascade, which ultimately lead to inhibition of pollen germination and/or tube growth. Based on McCubbin and Kao (2000).

more than 100-fold more abundant than SRK in the stigma, this is a significant consideration.

A key question is how does this sequence of signalling events ultimately lead to the inhibition of pollen germination or tube growth? ARC1 has been shown to be both phosphorylated by SRK and essential for the SI response. As a U-box containing protein ARC1 presumably targets various proteins for degradation – what these might be is currently unknown. Calmodulin also interacts with the SRK kinase domain and most likely serves to mediate cross-talk between calcium signalling in pollination *per se* and SI signal transduction (Vanoosthuyse *et al.*, 2003). Exactly how self-pollen is inhibited and or killed is still unclear. Pollen rejection can occur at different stages. In the strongest *S*-haplotypes self-pollen fail to even hydrate, and the decision to deny water must be extremely rapid to account for the speed of this rejection response (Dickinson, 1995). In weaker *S*-haplotypes rejection takes place after the initiation of pollen tube growth. It seems likely, therefore, that SI in *Brassica* can act at several levels as appears to be the case in *Papaver*, but in *Brassica* the primary molecular mechanism leading to rejection may vary between haplotypes.

Clearly, considerably more work is required to elucidate the precise mechanism of pollen rejection in this system. A recent development that is likely to speed this process has been the successful introduction of SSI into *Arabidopsis thaliana* (Nasrallah *et al.*, 2002). Importantly, the availability of SSI *A. thaliana* not only provides direct access to the genomic resources available for this species, but also makes it feasible to conduct large-scale mutant screens for loss of SI, and it is likely these tools will significantly speed the identification of downstream components of the SRK signalling cascade.

10.5 Summary

Research into SI has a long and distinguished history spanning well over 100 years. Our understanding of SI systems and how they function has understandably surged forward with each new wave of technological advances over that period. The molecular data now available implicate a wide variety of signalling processes and mechanisms. In terms of signalling complexity the most simple system appears to be S-RNase-based gametophytic SI, which is also the most phylogenetically widespread and is most likely the most ancient (Steinbachs & Holsinger, 2002). Both the gametophytic system found in *Papaver* and the sporophytic system in *Brassica* appear to involve extensive signal transduction cascades, which are capable of inhibiting pollen germination and/or tube growth through multiple mechanisms. While the precise signalling events appear to be specific to each system, some common mechanistic themes are emerging. Calcium signalling has been been strongly implicated in the SI mechanism of *Papaver* and may also be involved in both *Brassica* and S-RNase-based systems. Protein phosphorylation is involved in *Papaver* and *Brassica* and possibly the S-RNase-based system. Ubiquitin-mediated protein degradation has been implicated in *Brassica* and the S-RNase-based systems.

In the systems described, most of the *S*-locus-specific factors have now been identified. Many of the molecular components downstream of these factors may well be involved in other signalling processes in plant cells. Of particular note are the number of parallels that can be drawn between SI recognition systems and host–pathogen responses from SI systems (Hodgkin *et al.*, 1988). As work progresses over the next few years, it will be interesting to see how much overlap is found between the molecules involved in SI responses and signalling cascades involved in other plant responses.

References

Ai, Y., Kron, E. & Kao, T.-h. (1991) S-alleles are retained and expressed in a self-compatible cultivar of *Petunia hybrida*. *Mol. Gen. Genet.*, **230**, 353–358.

Anderson, M.A., Cornish, E.C., Mau, S.L., Williams, E.G., Hoggart, R.M., Atkinson, R.J., Roche, P.J., Haley, P., Penschow, J., Niall, H., Tregear, G., Coghlan, J., Crawford, R. & Clarke, A.E. (1986) Molecular cloning of a cDNA of a stylar glycoprotein associated with the expression of self-incompatibility in *Nicotiana alata* Link and Otto. *Nature*, **321**, 38–44.

Azevedo, C., Santos-Rosa, M.J. & Shirasu, K. (2001) The U-box protein family in plants. *Trends Plant Sci.*, **6**, 354–358.

Bai, C., Sen. P., Hofman, K., Ma, L., Goebel, M., Harper, W. & Elledge, S. (1996) *Skp1* connects cell cycle regulation to the ubiquitin proteolysis machinery through a novel motif, the F-box. *Cell*, **86**, 263–274.

Bernatzky, R., Glaven, R.H. & Rivers, B.A. (1995) S-related protein can be recombined with self-compatibility in interspecific derivatives of *Lycopersicon*. *Biochem. Genet.*, **33**, 215–225.

Bower, M.S., Matias, D.D., Fernandes-Carvalho, E., Mazzurco, M., Gu, T., Rothstein, S.J. & Goring, D.R. (1996) Two members of the thioredoxin-h family interact with the kinase domain of a *Brassica* S locus receptor kinase. *Plant Cell*, **8**, 1641–1650.

Braun D.M., Stone J.M. & Walker J.C. (1997) Interaction of the maize and *Arabidopsis* kinase interaction domains with a subset of receptor-like protein kinases: implications for transmembrane signaling in plants. *Plant J.*, **12**, 83–95.

Bredemeijer, G.M.M. & Blass, J. (1981) S-specific proteins in styles in styles of self-incompatible *Nicotiana alata*. *Theor. Appl. Genet.*, **59**, 185–190.

Cabrillac, D., Cock, J.M., Dumas, C. & Gaude, T. (2001) The *S*-locus receptor kinase is inhibited by thioredoxins and activated by pollen coat proteins. *Nature*, **410**, 220–223.

Chin, D. & Means, A.R. (2000) Calmodulin: a prototypical calcium sensor. *Trends Cell Biol.*, **10**, 322–328.

Chin, L.S., Raynor, M.C., Wei, X., Chen, H.Q. & Li, L. (2001) Hrs interacts with sorting nexin 1 and regulates degradation of epidermal growth factor receptor. *J. Biol. Chem.*, **276**, 7069–7078.

Cock, J.M., Vanoosthuyse, V. & Gaude, T. (2002) Receptor kinase signalling in plants and animals: distinct molecular systems with mechanistic similarities. *Curr. Opin. Cell Biol.*, **14**, 230–236.

Coleman, C.E. & Kao, T.-h. (1992) The flanking regions of two *Petunia inflata* S-alleles are heterogeneous and contain repetitive sequences. *Plant Mol. Biol.*, **18**, 725–737.

Conaway, R.C., Brower, C.S. & Conaway, J.W. (2002) Gene expression – emerging roles of ubiquitin in transcription regulation. *Science*, **296**, 1254–1258.

Conner, J.A., Tantikanjana, T., Stein, J.C., Kandasamy, M.K., Nasrallah, J.B. & Nasrallah, M.E. (1997) Transgene-induced silencing of *S*-locus genes and related genes in *Brassica*. *Plant J.*, **11**, 809–823.

Cooperman, B.S., Baykov, A.A. & Lahti R. (1992) Evolutionary conservation of the active site of soluble inorganic pyrophosphatase. *Trends Biochem. Sci.*, **17**, 262–266.

Craig, K.L. & Tyers, M. (1999) The F-box: a new motif for ubiquitin dependent proteolysis in cell cycle regulation and signal transduction. *Prog. Biophys. Mol. Biol.*, **72**, 299–328.

Cruz-Garcia, F., Hancock, C.N. & McClure B. (2003) S-RNase complexes and pollen rejection. *J. Exp. Bot.*, **54**, 123–130.

Dannon, A., Delorme, V., Mailhac, N. & Gallois, P. (2000) Plant programmed cell death: a common way to die. *Plant Physiol. Biochem.*, **38**, 647–655.

Dearnaley, J.D.W., Levina, N.N., Lew, R.R., Heath, I.B. & Goring, D.R. (1997) Inter-relationships between cytoplasmic Ca^{2+} peaks, pollen hydration and plasma membrane conductances during compatible and incompatible pollinations of *Brassica napus* papillae. *Plant Cell Physiol.*, **38**, 985–999.

Delorme, V., Giranton, J.-L., Hatzfeld, Y., Friry, A., Heizmann, P., Ariza, M.J., Dumas, C., Gaude, T. & Cock, J.M. (1995) Characterization of the S locus genes, *SLG* and *SRK*, of the *Brassica* S_3

haplotype: identification of a membrane-localized protein encoded by the *S* locus receptor kinase gene. *Plant J.*, **7**, 429–440.

Dickinson, H.G. (1995) Dry stigmas, water and self-incompatibility in *Brassica*. *Sex. Plant Reprod.*, **8**, 1–10.

Doughty, J., Wong, H.Y. & Dickinson, H.G. (2000) Cysteine-rich pollen coat proteins (PCPs) and their interactions with stigmatic *S* (incompatibility) and *S*-related proteins in *Brassica*: putative roles in SI and pollination. *Ann. Bot.*, **85**, 161–169.

Elleman, C. & Dickinson, H.G. (1999) Commonalities between pollen/stigma and host/pathogen interactions: calcium accumulation during stigmatic penetration by *Brassica oleracea* pollen. *Sex. Plant Reprod.*, **12**, 194–202.

Elleman, C., Franklin-Tong, V.E. & Dickinson, H.G. (1992) Pollination in species with dry stigmas: the nature of the early stigmatic response and the pathway taken by the pollen tubes. *New Phytol.*, **121**, 413–424.

Foote, H.C.C., Ride, J.P., Franklin-Tong, V.E., Walker, E.A., Lawrence, M.J. & Franklin, F.C.H. (1994) Cloning and expression of a distinctive class of self-incompatibility (*S*-) gene from *Papaver rhoeas* L. *Proc. Natl. Acad. Sci. U.S.A.*, **91**, 2265–2269.

Franklin-Tong, V.E. (1999) Signaling and the modulation of pollen tube growth. *Plant Cell*, **11**, 727–738.

Franklin-Tong, V.E., Drøbak, B.K., Allan, A.C. & Trewavas, A.J. (1996) Growth of pollen tubes of *Papaver rhoeas* is regulated by a slow moving calcium wave propagated by inositol (1,4,5) trisphosphate. *Plant Cell*, **8**, 1305–1321.

Franklin-Tong, V.E., Hackett, G. & Hepler, P.K. (1997) Ratioimaging of Ca^{2+} in the self-incompatibility response in pollen-tubes of *Papaver rhoeas*. *Plant J.*, **12**, 1375–1386.

Franklin-Tong, V.E., Holdaway-Clarke, T.L., Straatman, K.R., Kunkel, J.G. & Hepler, P.K. (2002) Involvement of-extracellular calcium influx in the self-incompatibility response of *Papaver rhoeas*. *Plant J.*, **29**, 333–345.

Franklin-Tong, V.E., Lawrence, M.J. & Franklin, F.C.H. (1988) An in vitro bioassay for the stigmatic product of the self incompatibility gene in *Papaver rhoeas* L. *New Phytol.*, **110**, 109–118.

Franklin-Tong, V.E., Lawrence, M.J. & Franklin, F.C.H. (1990) Self-incompatibility in *Papaver rhoeas* L. Inhibition of incompatible pollen is dependent of gene expression. *New Phytol.*, **166**, 310–324.

Franklin-Tong, V.E., Lawrence, M.J. Thorlby, G.J. & Franklin, F.C.H. (1992) Recognition, signals and pollen responses in the incompatibility reaction in *Papaver rhoeas*. In: *Angiosperm Pollen and Ovules* (eds E. Ottaviano, D.L. Mulcahy, G. Sari-Gorla & G. Bergamini-Mulcahy), pp. 83–93. Springer-Verlag, New York.

Franklin-Tong, V.E., Ride, J.P. & Franklin, F.C.H. (1995) Recombinant stigmatic self-incompatibility (*S*-) protein elicits a Ca^{2+} transient in pollen of *Papaver rhoeas*. *Plant J.*, **8**, 299–307.

Franklin-Tong, V.E., Ride, J.P., Read, N.D., Trewavas, A.J. & Franklin, F.C.H. (1993) The self-incompatibility response in *Papaver rhoeas* is mediated by cytosolic free calcium. *Plant J.*, **4**, 163–177.

Gaude, T., Rougier, M., Heizmann, P., Ockendon, D. & Dumas, C. (1995) Expression level of the *SLG* gene is not correlated with the self-incompatibility phenotype in the class II S haplotypes of *Brassica oleracea*. *Plant Mol. Biol.*, **27**, 1003–1114.

Geitmann, A., Snowman, B.N., Emons, A.M.C. & Franklin-Tong, V.E. (2000) Alterations to the actin cytoskeleton of pollen tubes are induced by the self-incompatibility reaction in *Papaver rhoeas*. *Plant Cell*, **12**, 1239–1252.

Gibbon, B.C., Kovar, D.R. & Staiger, C.J. (1999) Latrunculin B has different effects on pollen germination and tube growth. *Plant Cell*, **11**, 2349–2363.

Giranton, J.L., Dumas, C., Cock, J.M. & Gaude, T. (2000) The integral membrane *S*-locus receptor kinase of *Brassica* has serine/threonine kinase activity in a membranous environment and spontaneously forms oligomers in planta. *Proc. Natl. Acad. Sci. U.S.A.*, **97**, 3759–3764.

Golz, J.F., Oh, H.-Y., Su, V., Kusaba, M & Newbigin, E. (2001) Genetic analysis of *Nicotiana* pollen-part mutants is consistent with the presence of an S-ribonuclease inhibitor at the *S*-locus. *Proc. Natl. Acad. Sci. U.S.A.*, **98**, 15372–15376.

Golz, J.F., Su, V., Clarke, A.E. & Newbigin, E. (1999) A molecular description of mutations affecting the pollen component of the *Nicotiana alata* S locus. *Genetics*, **152**, 1123–1135.

Gomez-Gomez, L., Bauer, Z. & Boller, T. (2001) Both the extracellular leucinerich repeat domain and the kinase activity of FLS2 are required for flagellin binding and signaling in *Arabidopsis*. *Plant Cell*, **13**, 1155–1163.

Goring, D.R. (2000) The search for components of the self-incompatibility signaling pathway(s) in *Brassica napus*. *Ann. Bot.*, **85**, 147–153.

Goring, D.R. & Rothstein, S.J. (1992) The *S*-locus receptor kinase gene in a self-incompatible *Brassica napus* line encodes a functional serine/threonine kinase. *Plant Cell*, **4**, 1273–1281.

Greenberg, J.T., Guo, A., Klessig, D.F. & Ausubel, F.M. (1994) Programmed cell death in plants, a pathogen-triggered response activated co-ordinately with multiple defense functions. *Cell*, **77**, 551–563.

Gu, T., Mazzurco, M., Sulaman, W., Matias, D.D. & Goring, D.R. (1998) Binding of an arm repeat protein to the kinase domain of the *S*-locus receptor kinase. *Proc. Natl. Acad. Sci. U.S.A.*, **95**, 382–387.

Haft, C.R., de la Luz Sierra, M., Barr, V.A., Haft, D.H. & Taylor S.I. (1998) Identification of a family of sorting nexin molecules and characterization of their association with receptors. *Mol. Cell. Biol.*, **18**, 7278–7287.

Hearn, M.J., Franklin, F.C.H. & Ride, J.P. (1996) Identification of a membrane glycoprotein in pollen of *Papaver rhoeas* which binds stigmatic self-incompatibility (S-) proteins. *Plant J.*, **9**, 467–475.

Hepler, P.K. & Wayne, R.O (1985) Calcium and plant development. *Annu. Rev. Plant Physiol.*, **36**, 397–439.

Hodgkin, T., Lyon, G.D. & Dickinson, H.G. (1988) Recognition in flowering plants: a comparison of the *Brassica* self incompatibility system and plant–pathogen interactions. *New Phytol.*, **110**, 557–569.

Hoeberichts, F.A., ten Have, A. & Wolterung, E.J. (2003) A tomato metacaspase is upregulated during programmed cell death in *Botrytis cinerea* infected leaves. *Planta*, **217**, 517–522.

Horiuchi, H., Yanai, K., Takagai, M., Yano, K., Wakabayashi, E., Sanda, A., Mine, S., Ohgi, K. & Irie, M. (1988) Primary structure of a base non-specific ribonucleases from *Rhizopus niveus*. *J. Biochem.*, **103**, 408–418.

Huang, S., Lee, H.-S., Karunanandaa, B. & Kao T.-h. (1994) Ribonuclease activity of *Petunia inflata* S proteins is essential for rejection of self-pollen. *Plant Cell*, **6**, 1021–1028.

Ida, K., Norioka, S., Yamamoto, M., Kumasaka, T., Yamashita, E., Newbigin, E., Clarke, A.E., Sakiyama, F. & Sato, M. (2001) The 1.55 Å resolution structure of *Nicotiana alata* S(F11)-RNase associated with gametophytic self-incompatibility. *J. Mol. Biol.*, **314**, 103–112.

Ioerger, T.R., Gohlke, J.R., Xu, B. & Kao, T.-h. (1991) Primary structural features of the self-incompatibility protein in Solanaceae. *Sex. Plant Reprod.*, **4**, 81–87.

Jahnen, W., Lush, W.M. & Clarke, A.E. (1989) Inhibition of in vitro pollen tube growth by isolated *S*-glycoproteins in *Nicotiana alata*. *Plant Cell*, **1**, 501–510.

Janmey, P.A. (1998) The cytoskeleton and cell signalling: component localization and mechanical coupling. *Physiol. Rev.*, **78**, 763–781.

Jordan, N.D., Franklin, F.C.H. & Franklin-Tong, V.E. (2000). Evidence for DNA fragmentation triggered in the self-incompatibility response in pollen of *Papaver rhoeas*. *Plant J.*, **23**, 471–479.

Jordan. N.D., Kakeda, K., Conner, A., Ride, J.P., Franklin-Tong, V.E. & Franklin, F.C.H. (1999) S-protein mutants indicate a functional role for SBP in the self-incompatibility reaction of *Papaver*. *Plant J.*, **20**, 119–126.

Kachroo, A., Nasrallah, M.E. & Nasrallah, J.B. (2002) Self-incompatibility in the Brassicaceae: receptor-ligand signaling and cell–cell communication. *Plant Cell*, **14**, S227–S238.

Kakeda, K., Jordan, N.D., Conner, A., Ride, J.P., Franklin-Tong, V.E. & Franklin, F.C.H. (1998) Identification of residues in a hydrophilic loop of the *Papaver rhoeas* S protein that play a crucial role in recognition of incompatible pollen. *Plant Cell*, **10**, 1723–1731.

Kao, T.-h. & McCubbin, A.G. (1996) How flowering plants discriminate between self and non-self pollen to prevent inbreeding. *Proc. Natl. Acad. Sci. U.S.A.*, **93**, 12059–12065.

Karunanandaa, B., Huang, S. & Kao, T.-h. (1994) Carbohydrate moiety of the *Petunia inflata* S_3 protein is not required for self-incompatibility interactions between pollen and pistil. *Plant Cell*, **6**, 1933–1940.

Kawata, Y., Sakiyama, F. & Tamakoi, H. (1988) Amino-acid sequence of ribonucleases T2 from *Aspergillus oryzae*. *Eur. J. Biochem.*, **176**, 683–697.

Kehyr-Pour, A. & Pernes, J. (1985) A new *S*-allele and specific S-proteins associated with two *S*-alleles in *Nicotiana alata*. In: *Biotechnology and Ecology of Pollen. Proceedings of the International Conference on the Biotechnology and Ecology of Pollen*, 9–11 July, 1985, University of Massachusetts, Amherst (eds D.L. Mulcahy, G.B. Mulcahy & E. Ottaviano), pp. 191–196. Springer-Verlag, New York.

Kemp, B.P. & Doughty, J. (2003) Just how complex is the *Brassica S*-receptor complex? *J. Exp. Bot.*, **54**, 157–68.

Koegl, M., Hoppe, T., Schlenker, S., Ulrich, H.D., Mayer, T.U. & Jentsch, S. (1999) A novel ubiquitination factor, E4, is involved in multi-ubiquitin chain assembly. *Cell*, **96**, 635–644.

Kondo, K., Yamamoto, M., Itahashi, R., Sato, T., Egashira, H., Hattori, T. & Kowyama, Y. (2002) Insights into the evolution of self-compatibility in *Lycopersicon* from the study of stylar factors. *Plant J.*, **30**, 143–153.

Korichneva, I. & Hämmerling, U. (1999) F-actin as a functional target for retro-retinoids: a possible role in anhydroretinol triggered cell death. *J. Cell Sci.*, **112**, 2521–2528.

Kuhtreiber W.M. & Jaffe L.F. (1990) Detection of extracellular calcium gradients with a calcium-specific vibrating electrode. *J. Cell Biol.*, **110**, 1565–1573.

Kunz, C., Chang, A., Faure, J.-D. Clarke, A.E., Polya, G. & Anderson, M.A. (1996) Phosphorylation of style S-RNases by Ca^{2+}-dependent protein kinases from pollen tubes. *Sex. Plant Reprod.*, **9**, 25–34.

Kurten, R.C., Cadena, D.L. & Gill, G.N. (1996) Enhanced degradation of EGF receptors by a sorting nexin, SNX1. *Science*, **272**, 1008–1010.

Kurup, S., Ride, J.P., Jordan, N., Fletcher, G., Franklin-Tong, V.E. & Franklin, F.C.H. (1998) Identification and cloning of related self-incompatibility *S*-genes in *Papaver rhoeas* and *Papaver nudicaule*. *Sex. Plant Reprod.*, **11**, 192–198.

Kusaba, M., Nishio, T., Satta, Y., Hinata, K. & Ockendon, D. (1997) Striking sequence similarity in inter- and intra-specific comparisons of class I *SLG* alleles from *Brassica oleracea* and *Brassica campestris*: implications for the evolution and recognition mechanism. *Proc. Natl. Acad. Sci. U.S.A.*, **94**, 7673–7678.

Lai, Z., Ma, W., Han, B., Liang, L., Zhang, Y., Hong, G. & Xue, Y. (2002) An F-box gene linked to the self-incompatibility (*S*) locus of *Antirrhinum* is expressed specifically in pollen and tapetum. *Plant Mol. Biol.*, **50**, 29–42.

Lawrence, M.J., Afzal, M. & Kenrick, J. (1978) The genetical control of self-incompatibility in *Papaver rhoeas*. *Heredity*, **40**, 239–253.

Lee, H.-S., Huang, S. & Kao T.-h. (1994) S proteins control rejection of incompatible pollen in *Petunia inflata*. *Nature*, **367**, 560–563.

Levine, A., Pennell, R.I., Alverez, M.E., Palme, R. & Lamb, C. (1997) Calcium-mediated apoptosis in a plant hypersensitive disease resistance response. *Curr. Biol.*, **6**, 427–437.

Lewis, D. (1954) Comparative incompatibility in angiosperms and fungi. *Adv. Genet.*, **6**, 235–285.

Lush, W.M. & Clarke, A.E. (1997) Observations of pollen tube growth in *Nicotiana alata* and their implications for the mechanism of self-incompatibility. *Sex. Plant Reprod.*, **10**, 27–35.

Luu, D., Qin, K., Morse, D. & Cappadocia, M. (2000) S-RNase uptake by compatible pollen tubes in gametophytic self-incompatibility. *Nature*, **407**, 649–651.

Luu, D.T., Marty-Mazars, D., Trick, M., Dumas, C. & Heizmann, P. (1999) Pollen–stigma adhesion in *Brassica* spp. involves SLG and SLR1 glycoproteins. *Plant Cell*, **11**, 251–262.

Martin, F.W. (1968) The behavior of *Lycopersicon* incompatibility alleles in an alien genetic milieu. *Genetics*, **60**, 101–109.

Mascarenhas, J.P. (1993) Molecular mechanisms of pollen tube growth and differentiation. *Plant Cell*, **5**, 1303–1314.

Matton, D.P., Maes, O., Laublin, G., Xike, Q., Bertrand, C., Morse D. & Cappadocia, M. (1997) Hypervariable domains of self-incompatibility RNases mediate allele-specific pollen recognition. *Plant Cell*, **9**, 1757–1766.

Mazzurco, M., Sulaman, W., Elina, H., Cock, J.M. & Goring, D.R. (2001) Further analysis of the interactions between the *Brassica S* receptor kinase and three interacting proteins (ARC1, THL1 and THL2) in the yeast two-hybrid system. *Plant Mol. Biol.*, **45**, 365–376.

McClure, B.A., Cruz-Garcia, F., Beecher, B.S. & Sulaman, W. (2000) Factors affecting inter- and intraspecific pollen rejection in *Nicotiana*. *Ann. Bot.*, **85**, 113–123.

McClure, B.A., Gray, J.E., Anderson, M.A. & Clarke, A.E. (1990) Self-incompatibility in *Nicotiana alata* involves degradation of pollen rRNA. *Nature*, **347**, 757–760.

McClure, B.A., Haring, V., Ebert, P.R., Anderson, M.A., Simpson R.J., Sakiyama. F. & Clarke, A.E. (1989) Style self-incompatibility gene products of *Nicotiana alata* are ribonucleases. *Nature*, **342**, 955–957.

McClure, B.A., Mou, B., Canevascini, S. & Bernatzky, R. (1999) A small asparagine-rich protein required for *S*-allele-specific pollen rejection in *Nicotiana*. *Proc. Natl. Acad. Sci. U.S.A.*, **96**, 13548–13553.

McCubbin, A.G. & Kao, T.-h. (2000) Molecular recognition and response in pollen and pistil interactions. *Annu. Rev. Cell Dev. Biol.*, **16**, 333–364.

McCubbin, A.G, Zuniga, C. & Kao, T.-h. (2000) Construction of a binary bacterial artificial chromosome library of *Petunia inflata* and the isolation of large genomic fragments linked to the self-incompatibility (*S*) locus. *Genome*, **43**, 820–826.

Miller, D.D., Callaham, D.A., Gross, D.J. & Hepler, P.K. (1992) Free Ca^{2+} gradient in growing pollen tubes of *Lilium*. *J. Cell Sci.*, **101**, 7–12.

Murfett, J., Atherton, T.L., Mou, B., Gasser, C.S. & McClure, B.A. (1994) S-RNase expressed in transgenic *Nicotiana* causes *S*-allele-specific pollen rejection. *Nature*, **367**, 563–566.

Nasrallah, M.E., Kandasamy, M.K. & Nasrallah, J.B. (1992) A genetically defined trans-acting locus regulates *S*-locus function in *Brassica*. *Plant J.*, **2**, 497–506.

Nasrallah, J.B., Kao, T.-h., Goldberg, M.L. & Nasrallah, M.E. (1985) A cDNA clone encoding an *S*-locus specific glycoprotein from *Brassica oleracea*. *Nature*, **318**, 617–618.

Nasrallah, M.E., Liu, P. & Nasrallah, J.B. (2002) Generation of self-incompatible *Arabidopsis thaliana* by transfer of two *S* locus genes from *A. lyrata*. *Science*, **297**, 247–249.

Nishio, T. & Hinata, K. (1977). Analysis of *S*-specific proteins in stigma of *Brassica oleracea* by isoelectric focusing. *Heredity*, **38**, 391–396.

Obermeyer, G. & Weisensel, M.H. (1991) Calcium channel blocker and calmodulin antagonists affect the gradient of free calcium ions in lily pollen tubes. *Eur. J. Cell Biol.*, **56**, 319–327.

Okazaki, K., Kusaba, M., Ockendon, D.J. & Nishio, T. (1999) Characterization of *S* tester lines in *Brassica oleracea*: polymorphism of restriction fragment length of *SLG* homologues and isoelectric points of *S*-locus glycoproteins. *Theor. Appl. Genet.*, **98**, 1329–1334.

Phillips, S.A., Barr, V.A., Haft, D.H., Taylor, S.I. & Haft, C.R. (2001) Identification and characterization of SNX15, a novel sorting nexin involved in protein trafficking. *J. Biol. Chem.*, **276**, 5074–5084.

Qiao, H., Wang, H., Zhao, L., Zhou, J., Huang, J., Zhang, Y. & Xue, Y. (2004) The F-box protein AhSLF-S_2 physically interacts with S-RNases that may be inhibited by the ubiquitin/26S proteasome pathway of protein degradation during compatible pollination in *Antirrhinum*. *Plant Cell*, **16**, 582–595.

Rao, J.Y., Jin, Y.S., Zheng, Q.L., Cheng, J., Tai, J. & Hemstreet, G.P.I. (1999) Alterations of the actin polymerization status as an apoptotic morphological effector in HL-60 cells. *J. Cell. Biochem.*, **75**, 686–697.

Ren, D.T., Yang, H.P. & Zhang, S.Q. (2002) Cell death mediated by MAPK is associated with hydrogen peroxide production in *Arabidopsis*. *J. Biol. Chem.*, **277**, 559–565.

Richael, C., Lincoln, J.E., Bostock, R.M. & Gilchrist, D.G. (2001) Caspase inhibitors reduce symptom development and limit bacterial proliferation in susceptible plant tissues. *Physiol. Mol. Plant Pathol.*, **59**, 213–221.

Ride, J.P., Davies, E.M., Franklin, F.C.H. & Marshall, D.F. (1999) Analysis of *Arabidopsis* genome sequence reveals a large new gene family in plants. *Plant Mol. Biol.*, **39**, 927–932.

Royo, J., Kunz, C., Kowyama, Y., Anderson, M., Clarke, A.E. & Newbigin, E. (1994) Loss of a histidine residue at the active site of *S*-locus ribonuclease is associated with self-compatibility in *Lycopersicon peruvianum*. *Proc. Natl. Acad. Sci. U.S.A.*, **91**, 6511–6514.

Rudd, J.J. & Franklin-Tong, V.E. (2003) Signals and targets of the self-incompatibility response in pollen of *Papaver rhoeas*. *J. Exp. Bot.*, **54**, 141–148.

Rudd, J.J., Franklin, F.C.H. & Franklin-Tong, V.E. (1997) Ca^{2+}-independent phosphorylation of a 68 kDa pollen protein is stimulated by the self-incompatibility response in *Papaver rhoeas*. *Plant J.*, **12**, 507–514.

Rudd, J.J., Franklin, F.C.H., Lord, J.M. & Franklin-Tong, V.E. (1996) Increased phosphorylation of a 26 kD pollen protein is induced by the self-incompatibility response in *Papaver rhoeas*. *Plant Cell*, **8**, 713–724.

Rudd, J.J., Osman, K., Franklin, F.C.H. & Frankin-Tong, V.E. (2003) Activation of MAP kinase in pollen is stimulated by the self-incompatibility response. *FEBS Lett.*, **547**, 223–227.

Rundle, S.J., Nasrallah, M.E. & Nasrallah, J.B. (1993) Effects of inhibitors of protein serine/threonine phosphatases on pollination in *Brassica*. *Plant Physiol.*, **103**, 1165–1171.

Ryerson, D.E. & Heath, M.C. (1996) Cleavage of nuclear DNA into oligosomal fragments during cell death induced by fungal infection. *Plant Cell*, **8**, 393–402.

Samaj, J., Ovecka, M., Hlavacka, A., Lecourieux, F., Meskiene, I., Lichtscheidl, I., Lenart, P., Salaj, J., Volkmann, D., Bogre, L., Baluska, F. & Hirt, H. (2002) Involvement of the mitogen-activated protein kinase SIMK in regulation of root hair tip growth. *EMBO J.*, **21**, 3296–3306.

Sanders, D., Brownlee, C. & Harper, J.F. (1999) Communicating with calcium. *Plant Cell*, **11**, 691–706.

Schopfer, C.R. & Nasrallah, J.B. (2001) Self-incompatibility. Prospects for a novel putative peptide-signaling molecule. *Plant Physiol.*, **125**, 2203–2204.

Schopfer, C.R., Nasrallah, M.E. & Nasrallah, J.B. (1999) The male determinant of self-incompatibility in *Brassica*. *Science*, **286**, 1697–1700.

Scutt, C.P., Fordham-Skelton, A.P. & Croy, R.R.D. (1993) Okadaic acid causes breakdown of self-incompatibility in *Brassica oleracea*: evidence for the involvement of protein phosphatases in the incompatible response. *Sex. Plant Reprod.*, **6**, 282–285.

Seaman, M.N., McCaffery, J.M. & Emr, S.D. (1998) A membrane coat complex essential for endosome-to-Golgi retrograde transport in yeast. *J. Cell Biol.*, **142**, 665–681.

Shiba, H., Hinata, K., Suzuki, A. & Isogai, A. (1995) Breakdown of self-incompatibility in *Brassica* by the antisense RNA of the *SLG* gene. *Proc. Jpn. Acad.*, **71**, 81–83.

Sijacic, P., Wang, X., Skirpan, A.L., Wang, Y., Dowd, P.E., McCubbin, A.G., Huang, S. & Kao, T.-h. (2004) Identification of the pollen determinat of S-RNase-mediated self-incompatibility. *Nature*, **429**, 302–305.

Sims, T.L. & Ordanic, M. (2001) Identification of a S-ribonuclease-binding protein in *Petunia hybrida*. *Plant Mol. Biol.*, **47**, 771–783.

Singh, A., Ai, Y. & Kao, T.-h. (1991) Characterization of ribonuclease activity of three *S*-allele-associated proteins of *Petunia inflata*. *Plant Physiol.*, **96**, 61–68.

Smith, P.J.S., Sanger, R.H. & Jaffe, L.F. (1994) The vibrating Ca^{2+} electrode: a new technique for detecting plasma membrane regions of Ca^{2+} influx and efflux. *Methods Cell Biol.*, **40**, 115–134.

Snowman, B.N., Geitman, A., Clarke, S.R., Staiger, C.J., Franklin, F.C.H., Emons, A.M.C. & Franklin-Tong, V.E. (2000) Signaling and the cytoskeleton of pollen tubes of *Papaver rhoeas*. *Ann. Bot.*, **85**, 49–57.

Snowman, B.N., Kovar, D.R., Shevchenko, G., Franklin-Tong, V.E. & Staiger, C.J. (2002) Signal-mediated depolymerisation of actin in pollen during the self-incompatibility response. *Plant Cell*, **14**, 2613–2626.

Spivak-Kroizman, T., Lemmon, M.A., Dikic, I., Ladbury, J.E., Pinchasi, D., Huang, J., Jaye, M., Crumley, G., Schlessinger, J. & Lax, I. (1994) Heparin-induced oligomerization of FGF molecules is responsible for FGF receptor dimerization, activation, and cell proliferation. *Cell*, **16**, 1015–1024.

Staiger, C.J. (2000) Signalling to the actin cytoskeleton in plants. *Annu. Rev. Plant Physiol. Plant Mol. Biol.*, **51**, 257–288.

Staiger, C.J. & Franklin-Tong, V.E. (2003) The actin cytoskeleton is a target of the self-incompatibility reponse in *Papaver rhoeas*. *J. Exp. Bot.*, **54**, 103–113.

Stein, J.C., Howlett, B., Boyes, D.C., Nasrallah, M.E. & Nasrallah, J.B. (1991) Molecular cloning of a putative receptor kinase gene encoded by the self-incompatibility locus of *Brassica oleracea*. *Proc. Natl. Acad. Sci. U.S.A.*, **88**, 8816–8820.

Steinbachs, J.E. & Holsinger, K.E. (2002) S-RNase-mediated gametophytic self-incompatibility is ancestral in eudicots. *Mol. Biol. Evol.*, **19**, 825–829.

Stephenson, A.J., Doughty, J., Elleman, C.J., Hiscock, S.J. & Dickinson, H.G. (1997) The male determinant of self-incompatibility in *Brassica oleracea* is located in the pollen-coating. *Plant J.*, **12**, 1351–1359.

Stone, J.M., Trotochaud, A.E., Walker, J.C. & Clark, S.E. (1998) Control of meristem development by CLAVATA1 receptor kinase and kinase-associated protein phosphatase interactions. *Plant Physiol.*, **117**, 1217–1225.

Stone, S.L., Arnoldo, M. & Goring, D.R. (1999) A breakdown of *Brassica* self-incompatibility in ARC1 antisense transgenic plants. *Science*, **286**, 1729–1731.

Straatman, K.R., Dove, S.K., Holdaway-Clarke, T., Hepler, P.K., Kunkel, J.G. & Franklin-Tong, V.E. (2001) Calcium signalling in pollen of *Papaver rhoeas* undergoing the self incompatibility (SI) response. *Sex. Plant Reprod.*, **14**, 105–110.

Suzuki, G., Kai, N., Hirose, T., Fukui, K., Nishio, T., Takayama, S., Isogai, A., Watanabe, M. & Hinata, K. (1999) Genomic organization of the *S* locus: identification and characterization of genes in the SLG/SRK region of S_9 haplotype of *Brassica campestris* (syn. *rapa*). *Genetics*, **153**, 391–400.

Szallies, A., Kubata, B.K. & Duszenko, M. (2002) A metacaspase of *Trypanosoma brucei* causes loss of respiration competence and clonal death in the yeast *Saccharomyces cereviseae*. *FEBS Lett.*, **517**, 144–150.

Takasaki, T., Hatakeyama, K., Suzuki, G., Watanabe, M., Isogai, A. & Hinata, K. (2000) The *S* receptor kinase determines self-incompatibility in *Brassica* stigma. *Nature*, **403**, 913–916.

Takasaki, T., Hatakeyama, K., Watanabe, M., Toriyama, K., Isogai, A. & Hinata, K. (1999) Introduction of *SLG* (*S* locus glycoprotein) alters the phenotype of endogenous *S* haplotype, but confers no new *S* haplotype specificity in *Brassica rapa* L. *Plant Mol. Biol.*, **40**, 659–668.

Takayama, S., Shiba, H., Iwano, M., Shimosato, H., Che, F.S., Kai, N., Watanabe, M., Suzuki, G., Hinata, K. & Isogai, A. (2000) The pollen determinant of self-incompatibility in *Brassica campestris*. *Proc. Natl. Acad. Sci. U.S.A.*, **97**, 1920–1925.

Takayama, S., Shimosato, H., Shiba, H., Funato, M., Che, F.S., Watanabe, M., Iwano, M. & Isogai, A. (2001) Direct ligand–receptor complex interaction controls *Brassica* self-incompatibility. *Nature*, **413**, 534–538.

Trewavas, A.J. & Gilroy, S. (1991) Signal transduction in plant cells. *Trends Genet.*, **7**, 358–361.

Tsai, D.-S., Lee, H.-S., Post, L.C., Kreiling, K.M. & Kao, T.-h. (1992) Sequence of an S-protein of *Lycopersicon peruvianum* and comparison with other solanaceous S-proteins. *Sex. Plant Reprod.*, **5**, 256–263.

Tsukamoto, T., Ando, T., Kokubun, H., Watanabe, H., Masada, M., Zhu, X., Marchesi, E. & Kao T.-h. (1999) Breakdown of self-incompatibility in a natural population of *Petunia axillaris* (Solanaceae) in Uruguay containing both self-incompatible and self-compatible plants. *Sex. Plant Reprod.*, **12**, 6–13.

Tsukamoto, T., Ando, T., Kokubun, H., Watanabe, H., Sato, T., Masada, M., Marchesi, E. & Kao, T.-h. (2003) Breakdown of self-incompatibility in a natural population of *Petunia axillaris* caused by a modifier locus that suppresses the expression of an S-RNase gene. *Sex. Plant Reprod.*, **15**, 255–266.

Uren, A.G., O'Rourke, K., Aravind, L., Pisabarro, M.T., Seshagiri, S., Koonin, E.V. & Dixit, V.M. (2000) Identification of paracaspases and metacaspases. Two ancient families of caspase-like proteins, one of which plays a key role in MALT lymphoma. *Mol. Cell*, **6**, 961–967.

Vanoosthuyse, V., Tichtinsky, G., Dumas, C., Gaude, T. & Cock, J.M. (2003) Interaction of a sorting nexin and kinase-associated protein phosphatase with the *Brassica oleracea S* locus receptor kinase. *Plant Physiol.*, **133**, 919–929.

Verica, J.A., McCubbin, A.G. & Kao, T.-h. (1998) Are the hypervariable regions of *S* RNases sufficient for allele-specific recognition of pollen? *Plant Cell*, **10**, 314–316.

Walker, E.A., Ride, J.P., Kurup, S., Franklin-Tong, V.E., Lawrence, M.J. & Franklin, F.C.H. (1996) Molecular analysis of two functional homologues of the S_3 allele of the *Papaver rhoeas* self-incompatibility gene isolated from different populations. *Plant Mol. Biol.*, **30**, 983–994.

Wang, H., Li, J., Bostock, R.M. & Gilchrist, D.G. (1996) Apoptosis: a functional paradigm for pcd by a host-selective phytotoxin and invoked during development. *Plant Cell*, **8**, 375–391.

Williams, R.W., Wilson, J.M. & Meyerowitz, E.M. (1997) A possible role for kinase-associated protein phosphatase in the *Arabidopsis* CLAVATA1 signaling pathway. *Proc. Natl. Acad. Sci. U.S.A.*, **94**, 10467–10472.

Wu, H.-M., Wang, H. & Cheung, A.Y. (1995) A pollen tube growth stimulatory glycoprotein is deglycosylated by pollen tubes and displays a glycosylation gradient in the flower. *Cell*, **83**, 395–403.

Yang, K.-Y., Liu, Y. & Zhang, S. (2001) Activation of a mitogen-activated protein kinase pathway is involved in disease resistance in tobacco. *Proc. Natl. Acad. Sci. U.S.A.*, **98**, 373–378.

Zhang, S. & Liu, Y. (2001) Activation of salicylic acid-induced protein kinase, a mitogen-activated protein kinase, induces multiple defense responses in tobacco. *Plant Cell*, **13**, 1877–1889.

Zhang, S., Liu, Y. & Klessig, D.F. (2000) Multiple levels of tobacco WIPK activation during the induction of cell death by fungal elicitins. *Plant J.*, **23**, 339–347.

Zurek, D.M., Mou, B., Beecher, B. & McClure, B. (1997) Exchanging sequence domains between S-RNases from *Nicotiana alata* disrupts pollen recognition. *Plant J.*, **11**, 797–808.

Index

ABC transporters 6, 9
ABP1 18
ABPHYLL 163
AGR1 8
Allium cepa 113, 129
ALTERED PHLOEM DEVELOPMENT 219
Antirrhinum 51, 168, 186, 189, 190, 244–5
APETELA 187–8, 189
Arabinogalactan protein (AGP) 90–94, 103
ARC1 264, 266
Asparagus offinalis 34
ASYMMETRIC LEAVES 159, 169
auxin
   ARFs 19, 208
   AUX1 4, 7, 8, 210
   AUX/IAA 19, 208
   AXR1 19
   and the cell wall 88, 89
   mechanism of action 18–21
   and phyllotaxis 15–16, 162–6
   response element 13, 205
   in the root 203, 205, 209, 210–11
   and sterol 208
   transport 2–12, 205, 208
   and vasculature formation 219–20

BELLRINGER 152
BIG protein 11
biophysics 98–101, 166
birch 149
BODENLOS 108
Brassicaceae 38, 39, 241, 259–60
brassinolide 30, 207–8
brefeldin A 6, 9, 10, 11, 206
BREVIPEDICELLUS 159–60
buckling 99–100
BY2 cells 11

calcium 124, 129, 250, 252–7, 265
calcium dependent protein kinase 248, 258
callose 130–31
calmodulin 265–6
calreticulin 130

CAPRICE 121, 212–14, 229, 233
carbon dioxide 95–6
carrot 91
Catheranthus roseus 94
CdiGRP 75
celery 184
cellulose 87
centrin 129
CENTRORADIALIS 186
CEPHALOPOD 208
Chara corallina 118
chemiosmotic model 4
chimeras 149–50
chitinase 93
CLAVATA 35–8, 42–3, 109, 151–9, 163, 170, 209–10, 264
CmPP16 122
CmPP36 133–4
competitive interaction 245, 247
connexin 118
CONSTANS 182–7
Cornus seicea 131
co-suppression 57–8, 61
cryptochrome 185
cucumber (*Cucumis sativis*) 52, 92
Cucurbit maxima 52, 53, 55, 122, 133
CUP SHAPED COTYLEDON 152
cuticle 87, 94–6, 259
cytokinin 161, 187, 219
cytoplasmic sleave 111, 113
cytoskeleton
   actin 10, 18, 98, 254–6
   plasmodesmata 119, 129
   tubulin 10, 118

DEFICIENS 51, 121, 189–92
desmotubule 111, 114–15, 117, 119, 129
DEVIL1 42
DICER 59–61
DIE NEUTRALIS 180
DNA methyltransferase 72
DORNROSCHEN 152
DR5 reporter gene 13, 207, 210

EIR1 8
embryogenesis 12–14, 91–3, 116–17, 200–207
endodermis 216
ENHANCER OF GLABRA 228
ENOD40 31–4, 40
epidermis 2, 32, 34, 38, 51, 94, 95, 97, 149, 189, 200–201, 204, 211–15, 225–31
ethylene 31, 44
expansin 100, 166
extensin 87

FACKEL 207–8
FASCIATA 152, 163
F-box protein 241, 244–5, 248
FCA 183
FIDDLEHEAD 95, 229
FLC 183–4
floral induction 179–88
FLOWERING LOCUS T 183
FLORICAULA 121, 189
FOREEVER YOUNG 152
FT 186–7
Fucus 97

GA1 183
gibberellin 92, 161, 183, 220, 233
GIGANTEA 184–5
GLABRA 211–14, 228–33
GLOBOSA 190
glycine rich protein 102
GNOM 12, 206, 211
G-protein coupled receptor 27
grafting 52–5, 62, 67, 71, 74, 76, 122, 150, 179, 180, 188, 247
guanine nucleotide exchange factors 10, 12, 206

HAIRY MERISTEM 153
HcPro 74
HD-ZIP 167–8, 219
heat shock protein 133
HIC 95
HOBBIT 209
Hordeum vulgarum (barley) 118–19, 129
HYDRA 208
hydrogen peroxide 41
hypersensitive response 41, 130, 256

Impatiens 193
INCREASED CHALCONE SYNTHASE EXPRESSION 229
INDETERMINATE 181

inflorescence 7, 178
inositide 252
ISE 119

jasmonate 30–31, 44, 220

KANADI 168, 219
kinase
  mitogen activated 27, 250, 254, 258
  serine/threonine 102
  tyrosine 27
  see also Wall associated kinase
kinase associated protein phosphatase 157, 263–4
KNOTTED1 and *KNOX* genes 50–51, 121, 133–4, 158–61, 170, 190
  and hormones 161–2

LACERATA 94
LATE FLOWERING 180
Latuca sativa 114
leaf formation 15–16, 162–9
LEAFY 51, 121, 131, 186–93
lipo-chitin 31
lipo-oligosaccharide 93
LRR-receptor kinase 35, 37, 42, 157, 170, 236

maize (*Zea mays*) 33, 117, 118, 129, 168, 181, 190
mechanical signalling 99–100, 163
meristem 17, 35–7, 50, 55, 121, 131
  shoot apical 147–66
  root apical 199–209
MGOUN 153
miRNA 61, 76, 168–70, 187
mitogen activated protein kinase 27, 250, 254, 258
MONOPTEROS 163, 208
MOUSE EARS 52, 122, 187
morphogen 21
movement protein 50, 121, 131–4
mRNA movement 49–53
myosin 118–19, 129, 133
MYB factor 228–9

nexin 264
Nicotiana alata 242
Nicotiana benthamiana 55, 62, 65–8, 73–4
Nicotiana clevelandii 128
Nicotiana tabacum (Tobacco) 6, 49, 75, 123, 129, 133
Nicotiana plumbaginifolia 246
Nitella 122

nitrate reductase 61–6, 69, 71, 73
nitric oxide 41
non-cell autonomous protein (NCAP) 121, 131–4
notch/delta 97
NPA binding protein 6, 9

Oligogalacturonic acid (OGA) 89

Papaveraceae 241, 249, 250, 256–62, 267
PCP 262
Pea 88, 89, 91, 180
Pectin 89, 119
Pectin Methyl Esterase (PME) 132
Peptide signals
   in the meristem 35–8, 154–8
   in the root nodule 31–3
   in self- incompatability 38–40, 261–3
   in the wound response 28–31
Perilla 179
Petunia 57, 242, 244–6
PHABULOSA 167–8
PHANTASTICA 163, 168–9
PHAVOLUTA 167–8
phloem 2, 7, 28, 30, 51–4, 65–8, 75, 116, 123, 131, 168, 180, 185–7, 218–19
phloem lectin2 55
phospholipase 93
PHOTOPERIOD 180
phragmoplastin 114
PhSDP1 248
phyllotaxis 15–16, 162–6
phytochrome 185
phytosulfokine (PSK) 34–5, 40, 42, 44
   binding protein 35
   and hormones 34, 44
PIN proteins 4, 7–8, 10–18, 22, 163–4, 206–7
PINHEAD 153
PINOID 163
plasmodesmata 49–50, 54, 55, 109–34, 225, 233
   branched 111
   components 117–20
   and flowering 187, 193
   primary 113, 191–2
   secondary 114, 115, 191–2
   simple 111
   size exclusion limit (SEL) 122–30
   structure 110–16
   trafficking 120–34
   and transcription factors 121–2, 190–91
   and viruses 121, 130

pollen 38–9
   SCR (SP11) 38–40, 261–5
   S-gene 244–6
   SLF 244–8
   SLG 259–62
   S-locus glycoprotein 38
   SLR 260
   SRK 38–40, 260–66
   S-RNase 241–4, 247–9
POLTEGEIST 157
polygalacturonic acid 87
polysaccharide signals 88–90
PP2 122
programmed cell death 250, 255–9
proteases 40
   aspartic protease (CDR1) 41
   caspase 256
   kex protease 40–41
   subtilisin-like 236
   serine protease 41
proteosome 245
post-transcriptional gene silencing (PTGS) 57–62, 72, 122
pyrophosphatase 253, 258

RAF kinase inhibitor 186
RALF 31, 40, 43
reaction diffusion model 232
receptor 27, 30, 35, 37, 42, 157, 170, 236, 260–66
   see also LRR-receptor kinase
REDUCED TRICHOME NUMBER 228
REVOLUTA 168
Rhizobium 32
Rho/Rac GTPase 38
ribonuclease 240–48, 257
rice 33–4, 42
RISC 59–60
RNA see mRNA, miRNA, rRNA, siRNA,
root
   apical meristem 199–209
   and auxin 13–17, 205–11
   and *CLAVATA* 209–10
   epidermal patterning 211–15
   ground tissue patterning 215–18
   hairs 32, 211–14, 228
   secondary 210–11
   vascular patterning 218–20
Rop 157
Rosaceae 240–41
rRNA 247
RNAi 57, 59, 60

salicylate 41
SBP 254
SCARECROW 192, 210, 215–18
SCR (SP11) 38–40, 261–5
Scrophulariaceae 240, 242
SDE 58, 60, 63, 68–74
Self-incompatability
    gametophytic 241–59
    sporophytic 38, 259–67
SELF PRUNING 186
Setcreasa purporea 123
SHEPHERD 157
SHORTROOT 51, 121–2, 192, 215–18
shoot apical meristem 35–7, 147–66
    mutants 152–3
SHOOT MERISTEMLESS 159
S-locus glycoproteins 260
silencing 56–77
    mechanism 56–61
    systemic 61–75
Sinapis alba 179, 188
siRNA 73, 76
SLF 244–8
SLG 259–62
S-locus glycoprotein 38
SLR 260
S-RNase 241–4, 247–9
SOC1 184, 188
Solanaeae 29, 38, 240–41
Solanum chacoense 243
Sorbus torminalis 17
SOS5 92
SPH 250–51
spinach 179
SRK 38–40, 260–66
statolith 18
STERILE NODES 180
sterols 167, 207–8
stigma 250, 257, 259–62
stomata 95, 225, 233–7
STOMATAL DENSITY AND DISTRIBUTION 236
STRUWWELPETER 153
Strychnos nuxvomica 110
sucrose 179, 180, 187
SULPHUR 75
suspension culture 11, 91
SUT1 52

SXD1 131
systemic acquired resistance 41
systemin 28–31, 40, 42, 44

tapetum 245, 262
tensegrity 98
TERMINAL FLOWER1 186
THL (thioredoxin-like) 263, 265
TIR1 19, 208
tomato (*Lycopersicon*) 28, 29, 33, 42, 186, 242, 246, 257
TOO MANY MOUTHS 236
tracheary element 218
transcription factor movement 50–53, 120–22, 189–93
TRANSPARENT TESTA GLABRA 215, 228–33
trichomes 225–32
TRIPTYCHON 215, 227–31
tropism 17
TTS 247

Ubiquitination 19, 117, 209, 244–5, 247, 264, 266–7
ULTRAPETELA 153

vascular patterning 218–20
Vicia faba 123
viroids 53–6
VirP1 55–6
Viruses 50, 52, 61, 70, 73, 75–6
vitronectin 101

wall associated kinase 92, 101–2
WEREWOLF 121, 211–14
WIGGUM 153
WOODEN LEG 219
WOX 158–9
WUSCHEL 37, 156–8, 209–10

xylogen 103
xyloglucan 87–8
XET/XTH 89
xylem 218

YABBY 168

Zinnia elegans 91–2